184
Advances in Polymer Science

Editorial Board:
A. Abe · A.-C. Albertsson · R. Duncan · K. Dušek · W. H. de Jeu
J. F. Joanny · H.-H. Kausch · S. Kobyashi · K.-S. Lee · L. Leibler
T. E. Long · I. Manners · M. Möller · O. Nuyken · E. M. Terentjev
B. Voit · G. Wegner

Advances in Polymer Science

Recently Published and Forthcoming Volumes

Crosslinking in Materials Science
Vol. 184, 2005

Phase Behavior of Polymer Blends
Volume Editor: Freed, K.
Vol. 183, 2005

Polymer Analysis/Polymer Theory
Vol. 182, 2005

Interphases and Mesophases in Polymer
Crystallization II
Volume Editor: Allegra, G.
Vol. 181, 2005

Interphases and Mesophases in Polymer
Crystallization I
Volume Editor: Allegra, G.
Vol. 180, 2005

Inorganic Polymeric Nanocomposites
and Membranes
Vol. 179, 2005

Polymeric and Inorganic Fibres
Vol. 178, 2005

Poly(arylene Ethynylenes)
From Synthesis to Application
Volume Editor: Weder C.
Vol. 177, 2005

Ring Opening Metathesis
Volume Editor: Buchmeister, M.
Vol. 176, 2005

Polymer Particles
Volume Editor: Okubo M.
Vol. 175, 2005

Neutron Spin Echo in Polymer Systems
Authors: Richter, D., Monkenbusch, M.,
Arbe, A., Colmenero, J.
Vol. 174, 2005

Advanced Computer Simulation
Approaches for Soft Matter Sciences I
Volume Editors: Holm, C. Kremer, K.
Vol. 173, 2005

Microlithography · Molecular Imprinting
Vol. 172, 2005

Polymer Synthesis
Vol. 171, 2004

NMR · Coordination Polimerization ·
Photopolymerization
Vol. 170, 2004

Long-Term Properties of Polyolefins
Volume Editor: Albertsson A.-C.
Vol. 169, 2004

Polymers and Light
Volume Editor: Lippert T. K.
Vol 168, 2004

New Synthetic Methods
Vol. 167, 2004

Polyelectrolytes with Defined
Molecular Architecture II
Volume Editor: Schmidt M.
Vol. 166, 2004

Polyelectrolytes with Defined
Molecular Architecture I
Volume Editor: Schmidt M.
Vol. 165, 2004

Filler-Reinforced Elastomers ·
Scanning Force Microscopy
Vol. 164, 2003

Liquid Chromatography ·
FTIR Microspectroscopy · Microwave
Assisted Synthesis
Vol. 163, 2003

Crosslinking in Materials Science

With contributions by
B. Ameduri · B. Boutevin · P. Czub · P. Penczek · J. Pielichowski
M. A. Rodríguez-Pérez · A. Taguet

This series presents critical reviews of the present and future trends in polymer and biopolymer science including chemistry, physical chemistry, physics and material science. It is adressed to all scientists at universities and in industry who wish to keep abreast of advances in the topics covered.

As a rule, contributions are specially commissioned. The editors and publishers will, however, always be pleased to receive suggestions and supplementary information. Papers are accepted for "Advances in Polymer Science" in English.

In references Advances in Polymer Science is abbeviated *Adv Polym Sci* and is cited as a journal.

The electronic content of *Adv Polym Sci* may be found at springerlink.com

Library of Congress Control Number: 2005924536

ISSN 0065-3195
ISBN 3-540-25831-0 **Springer Berlin Heidelberg New York**
ISBN 978-3-540-25831-5 **Springer Berlin Heidelberg New York**
DOI 10.1007/b136215

This work is subject to copyright. All rights are reserved, whether the whole or part of the material is concerned, specifically the rights of translation, reprinting, reuse of illustrations, recitation, broadcasting, reproduction on microfilm or in any other way, and storage in data banks. Duplication of this publication or parts thereof is permitted only under the provisions of the German Copyright Law of September 9, 1965, in its current version, and permission for use must always be obtained from Springer. Violations are liable for prosecution under the German Copyright Law.

Springer is a part of Springer Science+Business Media
springeronline.com
© Springer-Verlag Berlin Heidelberg 2005
Printed in Germany

The use of registered names, trademarks, etc. in this publication does not imply, even in the absence of a specific statement, that such names are exempt from the relevant protective laws and regulations and therefore free for general use.

Cover design: *Design & Production* GmbH, Heidelberg
Typesetting and Production: LE-T$_E$X Jelonek, Schmidt & Vöckler GbR, Leipzig

Printed on acid-free paper   02/3141 YL – 5 4 3 2 1 0

Editorial Board

Prof. Akihiro Abe
Department of Industrial Chemistry
Tokyo Institute of Polytechnics
1583 Iiyama, Atsugi-shi 243-02, Japan
aabe@chem.t-kougei.ac.jp

Prof. A.-C. Albertsson
Department of Polymer Technology
The Royal Institute of Technology
S-10044 Stockholm, Sweden
aila@polymer.kth.se

Prof. Ruth Duncan
Welsh School of Pharmacy
Cardiff University
Redwood Building
King Edward VII Avenue
Cardiff CF 10 3XF
United Kingdom
duncan@cf.ac.uk

Prof. Karel Dušek
Institute of Macromolecular Chemistry,
Czech
Academy of Sciences of the
Czech Republic
Heyrovský Sq. 2
16206 Prague 6, Czech Republic
dusek@imc.cas.cz

Prof. Dr. W. H. de Jeu
FOM-Institute AMOLF
Kruislaan 407
1098 SJ Amsterdam, The Netherlands
dejeu@amolf.nl
and
Dutch Polymer Institute
Eindhoven University of Technology
PO Box 513
5600 MB Eindhoven, The Netherlands

Prof. Jean-François Joanny
Physicochimie Curie
Institut Curie section recherche
26 rue d'Ulm
F-75248 Paris cedex 05, France
jean-francois.joanny@curie.fr

Prof. Hans-Henning Kausch
EPFL SB ISIC GGEC
J2 492 Bâtiment CH
Station 6
CH-1015 Lausanne, Switzerland
kausch.cully@bluewin.ch

Prof. S. Kobayashi
Department of Materials Chemistry
Graduate School of Engineering
Kyoto University
Kyoto 615-8510, Japan
kobayasi@mat.polym.kyoto-u.ac.jp

Prof. Kwang-Sup Lee
Department of Polymer Science &
Engineering
Hannam University
133 Ojung-Dong
Taejon 300-791,Korea
kslee@mail.hannam.ac.krr

Prof. L. Leibler
Matière Molle et Chimie
Ecole Supérieure de Physique
et Chimie Industrielles (ESPCI)
10 rue Vauquelin
75231 Paris Cedex 05, France
ludwik.leibler@espci.fr

Prof. Timothy E. Long
Department of Chemistry
and Research Institute
Virginia Tech
2110 Hahn Hall (0344)
Blacksburg,VA 24061, USA
telong@vt.edu

Prof. Ian Manners
Department of Chemistry
University of Toronto
80 St. George St.
M5S 3H6 Ontario, Canada
imanners@chem.utoronto.ca

Prof. Dr. Martin Möller
Deutsches Wollforschungsinstitut
an der RWTH Aachen e.V.
Pauwelsstraße 8
52056 Aachen, Germany
moeller@dwi.rwth-aachen.de

Prof. Oskar Nuyken
Lehrstuhl für Makromolekulare Stoffe
TU München
Lichtenbergstr. 4
85747 Garching, Germany
oskar.nuyken@ch.tum.de

Dr. E. M. Terentjev
Cavendish Laboratory
Madingley Road
Cambridge CB 3 OHE
United Kingdom
emt1000@cam.ac.uk

Prof. Brigitte Voit
Institut für Polymerforschung Dresden
Hohe Straße 6
01069 Dresden, Germany
voit@ipfdd.de

Prof. Gerhard Wegner
Max-Planck-Institut
für Polymerforschung
Ackermannweg 10
Postfach 3148
55128 Mainz, Germany
wegner@mpip-mainz.mpg.de

Advances in Polymer Science
Also Available Electronically

For all customers who have a standing order to Advances in Polymer Science, we offer the electronic version via SpringerLink free of charge. Please contact your librarian who can receive a password or free access to the full articles by registering at:

springerlink.com

If you do not have a subscription, you can still view the tables of contents of the volumes and the abstract of each article by going to the SpringerLink Homepage, clicking on "Browse by Online Libraries", then "Chemical Sciences", and finally choose Advances in Polymer Science.

You will find information about the

– Editorial Board
– Aims and Scope
– Instructions for Authors
– Sample Contribution

at springeronline.com using the search function.

Contents

Unsaturated Polyester Resins: Chemistry and Technology
P. Penczek · P. Czub · J. Pielichowski 1

**Crosslinked Polyolefin Foams:
Production, Structure, Properties, and Applications**
M. A. Rodríguez-Pérez 97

Crosslinking of Vinylidene Fluoride-Containing Fluoropolymers
A. Taguet · B. Ameduri · B. Boutevin 127

Author Index Volumes 101–184 213

Subject Index 235

Unsaturated Polyester Resins: Chemistry and Technology

Piotr Penczek[1] (✉) · Piotr Czub[2] · Jan Pielichowski[2]

[1]Industrial Chemistry Research Institute, ul. Rydygiera 8, 01-793 Warsaw, Poland
piotr.penczek@ichp.pl

[2]Cracow University of Technology, ul. Warszawska 24, 31-155 Cracow, Poland
pczub@usk.pk.edu.pl, pielich@usk.pk.edu.pl

1	Introduction	5
2	Major Raw Materials	5
2.1	Dicarboxylic Acids and Acid Anhydrides	6
2.2	Glycols and Other Polyhydric Alcohols	14
2.3	Crosslinking Monomers	19
2.4	Dicyclopentadiene	22
2.5	Manufacture of UPRs Using PET Scrap	28
3	Vinyl Ester Resins	30
4	Flexibilization	35
4.1	Incorporation of Segmental Modifiers into the UP Molecules	35
4.1.1	Terminal Poly(oxyethylene) Segments in the UPs	35
4.1.2	UPs with Incorporated Perfluoropolyether Segments	35
4.1.3	Copolymerizable Rubber Flexibilizers	36
4.2	Interpenetrating Polymer Networks	38
5	Decrease in Styrene Emission	41
5.1	With Low-Volatility Monomers	41
5.2	With Wax-Like Additives	42
6	Decrease in Flammability and Smoke Emission	43
7	Chemical and Alkali Resistance	46
8	Thermal Degradation	49
9	Chemical Analysis and Structural Investigation of Crosslinked Polyesters	50
9.1	Spectrometry	51
9.2	Other Methods	59
10	Curing of Unsaturated Polyesters and Vinyl Ether	62
10.1	Initiators: Peroxides	62
10.2	Promoters	63
10.2.1	Cobalt Accelerators	64
10.2.2	Aromatic Amines	64
10.3	Photopolymerization: Maleimides and Vinyl Ethers	66
10.4	Copolymerization Kinetics, Thermochemistry and Rheology	68

11	Low-Shrink and Zero-Shrink Systems	77
11.1	Compositions with Thermoplastic Additives	77
11.2	Mechanism of Decrease in Shrinkage	78
12	UPR-Based Nanocomposites	80
13	Major Application Processes	84
13.1	Composites with Glass Fibers and Natural Fibers	85
13.2	Composites Reinforced with Fillers	86
13.3	Molding Compounds	87
13.4	Powder Coatings from UPs	89
References		90

Abstract Results of investigations on novel formulations, structure-property relationships, curing, and compositions with fillers and reinforcing fibers (1997–2004) are reviewed with about 200 references to articles and several references to patents. The following topics are considered in particular: novel dibasic acids, glycols, crosslinking monomers, and curing systems, "vinyl ester" resins, fire retardant materials, IPNs and other systems comprising built-in thermoplastic polymers and oligomers with terminal functional groups. Information on unsaturated polyesters manufactured using PET scrap is given. Analytical (mainly spectrometric) methods for studying the chemical structure of crosslinked unsaturated polyester resins are presented. Approaches to the decrease in styrene emission on processing of unsaturated polyester resins are also discussed.

Keywords Unsaturated polyester resins · Reinforced polyesters · Poly(ethylene terephthalate) · Vinyl ester resins · Curing

Abbreviations

ΔH_r	heat released
ρ_v	depolarization ratio
AFM	atomic force microscopy
APP	ammonium polyphosphate
ATH	aluminum trihydroxide $Al(OH)_3$
BHET	bis(hydroxyethyl)terephthalate
Bis-GMA	2,2-bis[4-(3-methacryloyloxy-2-hydroxypropoxy)phenyl]
BMC	bulk molding compounds
BMI	bismaleimide
CDRE	convulsion difference resolution enhancement
CMDB	composite modified double-base
CoHx	cobalt hexanoate
CoNp	cobalt naphthenate
CP/MAS	magic angle spinning
CPD	cyclopentadiene
CTBN	carboxyl terminated poly(butadiene-co-acrylonitrile)
DAD	diode array detection
DCPD	dicyclopentadiene
DD	dipolar decoupling
DEF	diethyl fumarate
DEPT	distortionless enhancement by polarization transfer

DGEBA	diglycidyl ether of bisphenol A
DKGA	diketogulonic acid
DLS	dynamic light scattering
DMA	dynamic mechanical analysis
DMB	2,5-dimethyl-2,5-bis(2-ethylhexanoylperoxy)hexane
DMC	dough molding compounds
DSC	differential scanning calorimetry
DTA	differential thermal analysis
E'	storage modulus
E''	loss modulus
E_a	activation energy
EMTHPA	endomethylene tetrahydrophthalic anhydride
ESR	electron spin resonance spectroscopy
FID	free induction decay
FT-IR	Fourier transform infrared spectroscopy
FUPR	fluorine-modified unsaturated polyester resin
GC/FT-IR	gas chromatography-Fourier transform infrared spectroscopy
GC/MS	gas chromatography-mass spectrometry
G_{IC}	fracture energy
GPC	gel permeation chromatography
GPC-MALLS	multiangle laser light scattering detector
HDT	heat deflection temperature
HPLC	high-performance liquid chromatography
HPN	hybrid polymer network
IPN	interpenetrating polymer network
iTBN	maleimide terminated liquid butadiene-acrylonitrile rubber
k_0	frequency factor
LLCT	lyotropic liquid-crystalline thermoset
LOI	limiting oxygen index
LOM	laminated object manufacturing
LPA	low profile additives
LRP-NMR	low-resolution pulse ^{1}H-NMR
LSA	low shrink additives
LSE	low styrene emission
$\overline{M_n}$	number-average molecular weight
$\overline{M_w}$	weight-average molecular weight
MDI	methylene diphenyl diisocyanate
MEKP	methylethylketone peroxide
MI	maleimides
MMT	montmorillonite
MPD	2-methyl-1,3-propane diol
MTDSC	modulated temperature differential scanning calorimetry
MTGA	modulated thermogravimetric analysis
NBR	acrylonitrile-butadiene rubber
NPI	Novolac-type polyisocyanate
PB	pentabromoethylbenzene
PCL	poly(ε-caprolactone)
PD	1,2-propanediol
PDO	*tert*-butylperoxy-2-ethyl hexanoate
PET	poly(ethylene terephthalate)

PFPE	perfluoropolyether
phr	weight parts of additive per 100 parts by weight of resin
PPF	poly(propylene fumarate)s
PTSA	*p*-toluenesulfonic acid monohydrate
PU	polyurethane
PVAc	poly(vinyl acetate)
Py-GC	pyrolysis-gas chromatography
RDA	rheometric dynamic analyzer
RO	rapeseed oil
RTM	resin transfer molding
S	antimony trioxide, Sb_2O_3
SBR	styrene-butadiene rubber
SCRIMP	Seemann composites resin infusion molding process
SIN	semi-interpenetrating polymer network
SLS	static light scattering
SMC	sheet molding compounds
TTT	time-temperature-transformation
SPE	solid-phase extraction
T_1^H	proton spin-lattice relaxation time
$T_{1\rho}^H$	spin-lattice relaxation time in the rotating frame
T_2	spin-spin relaxation time
T_g	glass transition temperature
tan δ	loss tangent
TCTFE	1,1,2-trichloro-1,2,2-trifluoroethane
TDI	toluene diisocyanate
TEM	transmission electron microscopy
TFA	trifluoroacetic acid
TGA	thermogravimetric analysis
TMS	tetramethylsilane
TSR	thermal scanning rheometry
TX/PCL	fluorinated macromers/poly(ε-caprolactone)
TXCL	poly(ε-caprolactone)-perfluoropolyether-poly(ε-caprolactone) *block* copolymer
UP	unsaturated polyesters
UPR	unsaturated polyester resin
VARTM	vacuum-assisted resin transfer molding
VE	vinyl ether
VEUH	vinyl-ester-urethane hybrid
VER	vinyl ester resin
VTBN	vinyl terminated poly(butadiene-*co*-acrylonitrile)
x	order of reaction
ZSA	zero shrink additive

1
Introduction

Unsaturated polyester resins (UPRs) have been known for many years. The production of UPRs started in the 1930s. Recently, their manufacture has reached a peak level. UPRs are, along with polyurethanes, the most important crosslinkable polymeric materials. The importance of UPRs is due to their important fields of application, mainly in glass fiber reinforced plastics. The rapid increase in the share of UPRs in the plastics market, comprising also highly filled materials, coatings, and cast objects etc., is due to their simple processing.

The chemistry of UPRs involves the synthesis of unsaturated polyesters (UPs) by polyesterification or step-by step ionic copolymerization. The thus synthesized UP is then dissolved in an unsaturated monomer and crosslinked applying the radical polymerization approach. Thus, the chemistry of UPRs involves the polycondensation or ionic polymerization methods and crosslinking by peroxide or photochemically initiated radical polymerization.

Thanks to the various types of chemical reactions being applied in the manufacture and processing of UPRs and to the versatility of industrial applications, the progress of research and development in UPRs is very fast. The industrial progress in UPRs is accompanied by intense research, design and processing activities making the UPR industry an important component of polymeric materials science and technology.

2
Major Raw Materials

A classification of methods for the synthesis of unsaturated polyesters on the basis of conceptions of condensation and polycondensation as well as addition and polyaddition has been proposed [1]. The presented methods were characterized taking into account a regularity of the distribution of unsaturated bonds and the appearance of side reactions. A model to estimate the average number of chain branches and of chain ends of UP prepolymers has also been proposed [2]. Fundamental molecular parameters, i.e. hydroxyl and carboxyl values, Ordelt saturation (reaction of hydroxyl groups with double bonds) extent, mass polydispersity index, short- and long-chain branch distribution, and composition of starting reactants were included in the proposed model. The real molar mass, especially the molecular mass of the linear backbone chain, as well as the carboxyl and hydroxyl functionalities of UP prepolymers [3] could be estimated using the described model. The obtained results should be very useful for developing UP sheet-molding compounds (SMC) thickening technology.

2.1
Dicarboxylic Acids and Acid Anhydrides

The introduction of dicarboxylic acids or acid anhydrides with cycloalkene configuration into the polyester chain results in an increase in impact strength, chemical resistance and resistance against UV light as well as a decrease in refractive index and surface tackiness. UP resins were prepared (Scheme 1) from *cis*-4-cyclohexene dicarboxylic anhydride (tetrahydrophthalic anhydride), diethylene glycol, propylene glycol and 2,2-di(4-hydropropoxyphenyl)propane [4]. An improvement of mechanical properties, shortening of drying time of the casting surface, lowering of refractive index, more than twofold decrease in water absorption as well as a considerable increase in the Martens temperature of cured UPRs were observed when phthalic anhydride was replaced with tetrahydrophthalic anhydride. Next, partial substitution of maleic anhydride (Table 1) with an eutectic mixture of anhydrides of cyclic non-aromatic dicarboxylic acids (hexahydrophthalic anhydrides and three isomeric tetrahydrophthalic anhydrides) was studied [5]. Crosslinked UPRs prepared from the mixture of acid anhydrides were characterized by improved mechanical properties (Table 2) and considerable resistance to sunlight, particularly in regard to the impact strength and heat resistance. Epoxyfumarates formed by the addition of acrylic or methacrylic acid or acid esters of maleic or fumaric acid to epoxy resins (Scheme 2) are an important group of chemically resistant resins, sharing advantages of UPRs and epoxy resins [6]. The synthesis consists of the following stages:

- addition of an alcohol to maleic anhydride to form an alkyl hydrogen maleate/fumarate;

- addition of the thus obtained acid maleate (hydrogen maleate) to the liquid epoxy resin;

- catalytic *cis-trans* isomerization of the thus obtained addition product (the maleate) to form the corresponding fumarate;

- dissolution in styrene of the thus obtained fumarate followed by peroxide-initiated radical copolymerization (crosslinking).

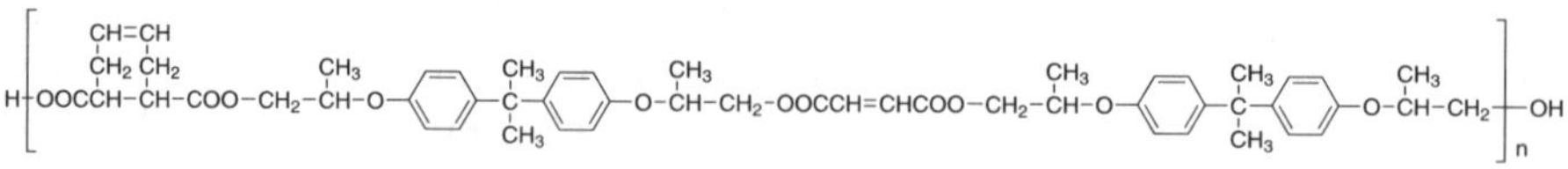

Scheme 1

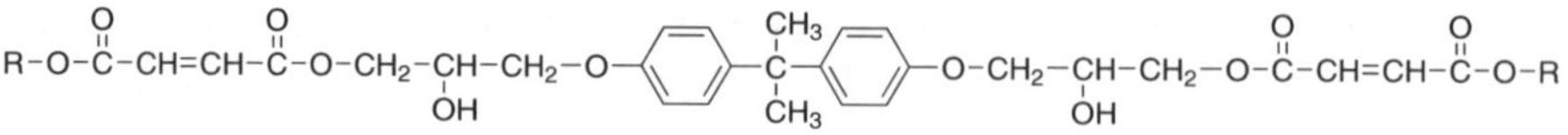

Scheme 2

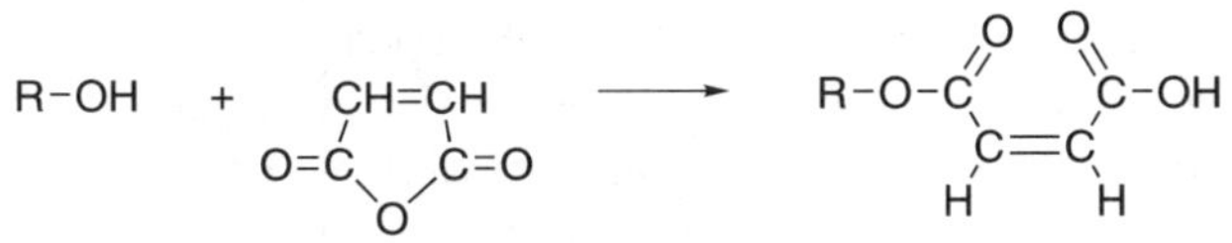

Scheme 3

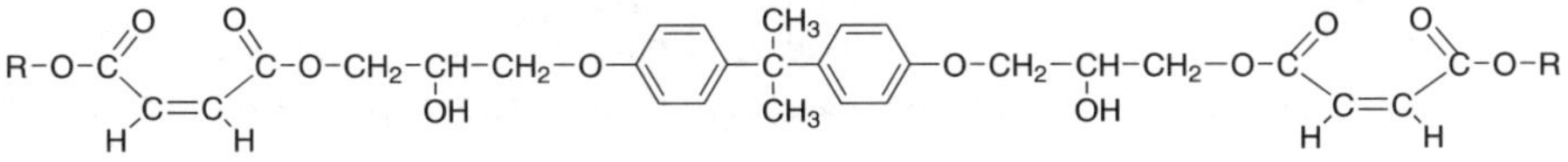

Scheme 4

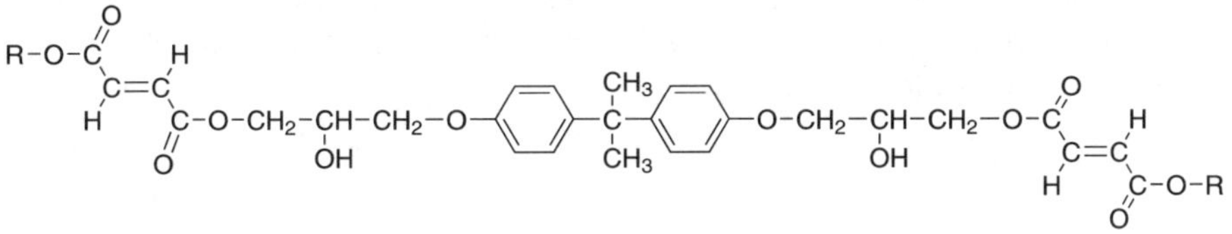

Scheme 5

Table 1 Composition of the studied UPRs. Reprinted from (1995) Polimery 40:669 [5] with permission

Resin	Mixture of anhydrides	Maleic anhydride	Diethylene glycol	Propylene glycol	Glycerol	Xylene
1	1	1	2.1	–	–	0.072
2	1	1	–	2.1	–	–
3	1	1	–	2.0	0.07	–

Components [mol]

Table 2 Properties of cured UPRs prepared from a mixture of acid anhydrides (according to Table 1) and commercial reference resin. Reprinted from (1995) Polimery 40:669 [5] with permission

	Resin			
	1	2	3	Polimal 103
Flexural strength [MPa]	67	69	72	60
Compression strength [MPa]	190	82	102	102
Static stress at break [MPa]	32	35	26	20
Impact strength [kJ/m^2]	9.6	3.4	4.0	2.3
Heat deformation temperature (Martens) [°C]	58	75	76	55
Water absorption [%]	0.34	0.2	0.1	0.3
Hardness [MPa]	85	147	160	103

Similar to the vinyl ester resins, the cured epoxyfumarate resins are distinguished by enhanced chemical resistance (e.g. in aqueous 20% NaOH at 60 °C), heat deflection temperature and flexibility. The chemical composition of the R group (methyl, ethyl, n-butyl, benzyl, cyclohexyl) influences the properties of the crosslinked epoxyfumarate resins [6]. If the R group contains bromine (e.g. tribromoneopentyl or 2,3-dibromopropyl), the cured resins are fire retardant [7]. Moreover, the brominated resins are distinguished by increased Martens heat deformation temperature and low water absorption.

To further increase the crosslinking density and thus the Martens heat deformation temperature, an allyl group was built into the molecule of epoxyfumarate resin (Scheme 6) [8]. A Martens heat deformation temperature exceeding 100 °C could be reached.

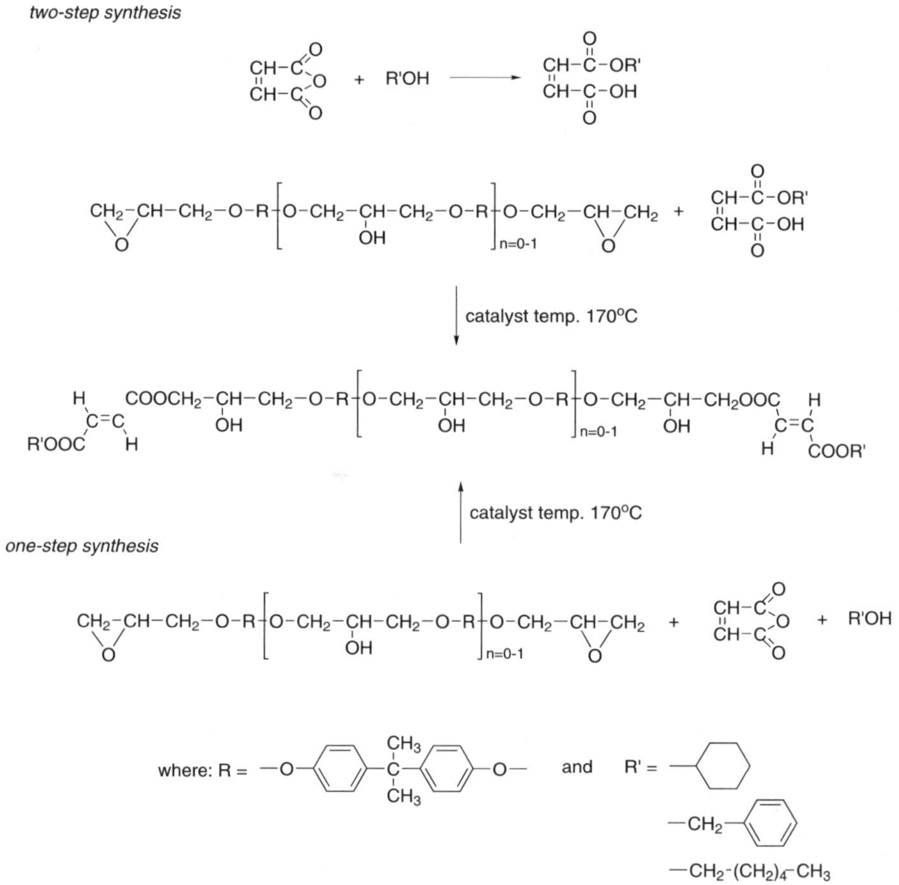

Fig. 1 One- and two-step syntheses of the studied epoxyfumarate resins. Reprinted from (2000) J Appl Polym Sci 77:3077 [11] with permission

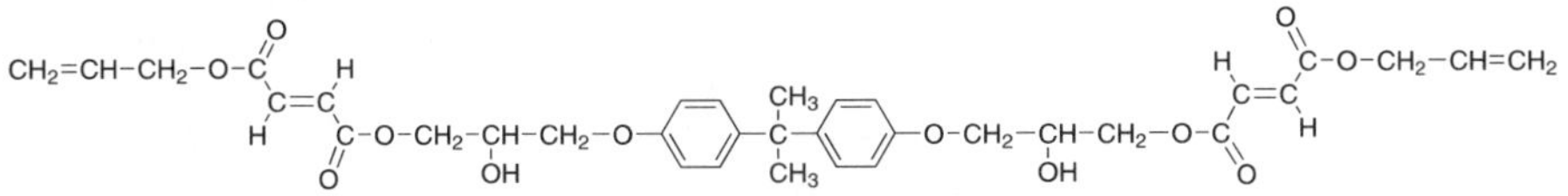

Scheme 6

The synthesis of epoxyfumarate resins by adding an acid ester of maleic acid to the commercial low-molecular weight epoxy resin (Epidian 5) with simultaneous isomerization of maleate groups to fumarate ones was presented [9]. The resins were synthesized in the two-step procedure using the acid ester of maleic acid prepared separately with epoxy resin or in the one-step synthesis (Fig. 1). The acid ester of maleic acid was obtained in the reaction of maleic anhydride and cyclohexanol, although use of benzyl alcohol [10] and *n*-hexanol was also reported [11].

2,4,6-Tris(dimethylaminomethyl)phenol was used as a catalyst for the reaction of carboxyl and epoxy groups. The studied resins were crosslinked using styrene. Next, piperidine was added as the isomerization catalyst of maleate groups to fumarate ones (Table 3).

The structure of the synthesized resins was ascertained by ^{1}H-NMR spectroscopy and the content of the *trans* isomer was determined from the NMR spectra according to the method of Curtis using the areas of the signals due to fumarate at about 6.9 ppm and maleate at about 6.4 ppm. It was found that the isomerization of maleate bonds did not occur without piperidine [10]. Moreover, unsaturated epoxymaleate resins were significantly less reactive than the epoxyfumarate resins. The properties before curing and of the crosslinked resins obtained in one- and two-step procedures are similar (Table 4), although the resins prepared in one-step synthesis were character-

Table 3 Composition of the synthesized epoxyfumarate resins. Reprinted from (1998) J Appl Polym Sci 68:1423 [9] with permission

Substrates [g]	Resin Number			
	1	2	3	4
Epoxy resin (Epidian 5)	29.10	26.64	29.72	27.29
Cyclohexanol	–	–	14.63	13.40
Maleic anhydride	–	–	15.20	13.90
Acid cyclohexanol maleate	30.46	27.95	–	–
Styrene	40.00	45.00	40.00	45.00
2,4,6-Tris(dimethylamino-methyl)phenol	0.149	0.136	0.149	0.136
Piperidine	0.298	0.273	0.298	0.273
Hydroquinone	0.0075	0.0075	0.0075	0.0075

Table 4 Thermomechanical properties of the crosslinked epoxyfumarate resins. Reprinted from (1998) J Appl Polym Sci 68:1423 [9] with permission

Properties	Resin number (according to Table 3)			
	1	2	3	4
Initial decomposition temperature [°C]	180	180	160	160
Heat deformation temperature (Martens) [°C]	82.5	80.5	79	80
Impact strength (Charpy) [kJ/m^2]	3.6	4.5	2.7	2.8
Ball indentation hardness [MPa]	119.5	107.8	111.1	117.2
Tensile strength [MPa]	44.8	31.5	38.1	39.1
Flexural strength [MPa]	70.4	72.6	84.2	75.8

ized by higher flexural strength. On the other hand, the resins obtained in the two-step procedure exhibited particularly high tensile strength.

The studied epoxyfumarate resins absorb organic solvents and become more flexible. Generally, epoxyfumarate resins are characterized by higher reactivity and higher tensile and flexural strength as well as better chemical resistance than the corresponding epoxymaleates. It was also shown that the chemical structure of alcohol used for the synthesis of the acid maleate ester affected the *cis-trans* isomerization temperature. Nitrogen-containing unsaturated epoxyfumarate resins (Fig. 2), prepared from monobutyl maleate and tetraglycidyl methylenedianiline in the one- and two-step procedure, were also studied [12].

No significant differences between the properties of resins synthesized according to either procedure were found. The resin obtained in the single-step method without the use of the inhibitor (hydroquinone) was, however, more easily curable than the other resins. The resulting N-containing epoxyfumarate resins were characterized by a good chemical resistance to inorganic liquids.

Various anhydrides and acids, i.e. phthalic anhydride, maleic anhydride, succinic acid, adipic acid and sebacic acid, with various glycols, i.e. ethylene glycol, diethylene glycol, triethylene glycol, were used for the synthesis of a series of UPRs for casting large-sized objects [13]. Different types of aromatic and aliphatic dibasic acids with various combinations of polyhydric alcohols were applied to obtain resins with a lower heat of crosslinking with styrene and suitable mechanical properties. It was found that the maximum curing temperature (T_{max}) was related to the molecular weight of the glycol used for the synthesis. It decreased with increasing molecular weight of glycol, al-

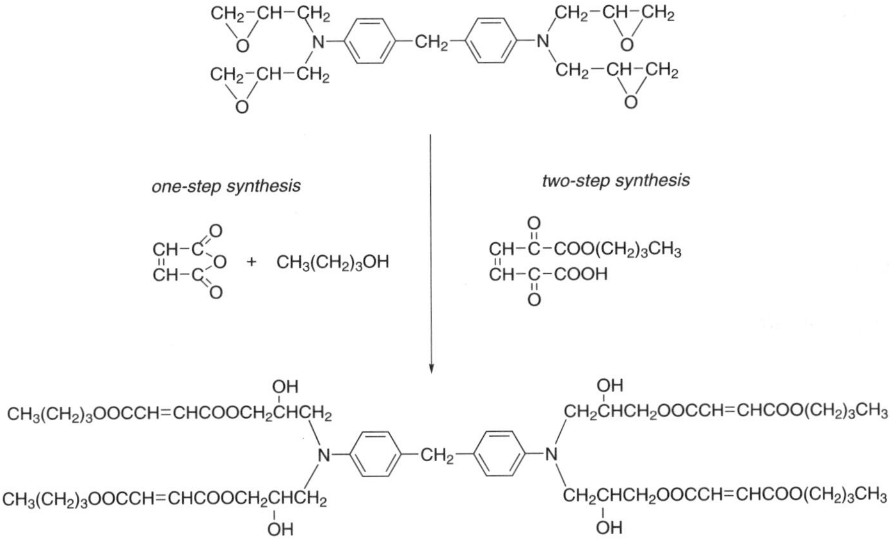

Fig. 2 One- and two-step syntheses of the epoxyfumarate resin

though the time to peak temperature increased. On the other hand, the type of the saturated acid as well as the molecular weight of the glycol affected T_{max}, Young's modulus and compressive strength. An equation representing the variation of compressive strength as a function of equivalent polymerizable double bonds was proposed and could be used to predict the compressive strength of any particular UP formula.

The novel UP resins were synthesized by a three-step polyesterification process from mixed glycol (based on ethylene glycol and neopentyl glycol), an aliphatic dibasic acid, isophthalic acid, and maleic anhydride [14]. Succinic acid, suberic acid and sebacic acid were used as dibasic acids. It was found that the increase in the chain length of dibasic acid resulted in an increase in gel time, elongation, nitroglycerine absorption (from double-base propellants) and thermal conductivity (Table 5). On the other hand, it also leads to a decrease in exotherm peak temperature, tensile strength, bond strength with composite modified double-base (CMDB) propellants, hardness and heat resistance.

The results obtained by the investigation enabled authors to select the best UPRs for inhibition of CMDB propellants.

Maleopimaric acid, diethylene glycol and maleic anhydride in the molar proportions 1 : 4 : 3 were used in the synthesis of unsaturated polyester prepolymer [15]. Levopimaric acid was obtained by the acid isomerization of natural rosin or abietic acid and used for the synthesis of maleopimaric acid (Fig. 3). The reaction of maleopimaric acid with a suitable excess of diethylene glycol, carried out at a temperature reasonably lower (220 °C) than the

Table 5 Properties of the synthesized novel UPRs based on dibasic acids. Reprinted from (1997) React Funct Polym 34:145 [14] with permission

Properties	Resin based on		
	Succinic acid	Suberic acid	Sebacic acid
Gel time [min]	51.0	71.0	74.0
Exotherm temperature [°C]	64.1	58.2	56.0
Bond strength* [kg/cm^2]	7.6	4.5	3.1
Elongation [%]	8.33	48.0	49.2
Tensile strength [kg/cm^2]	118.0	11.0	10.0
Shore hardness (D scale)	66.0	55.0	48.0

*- with CMDB propellants

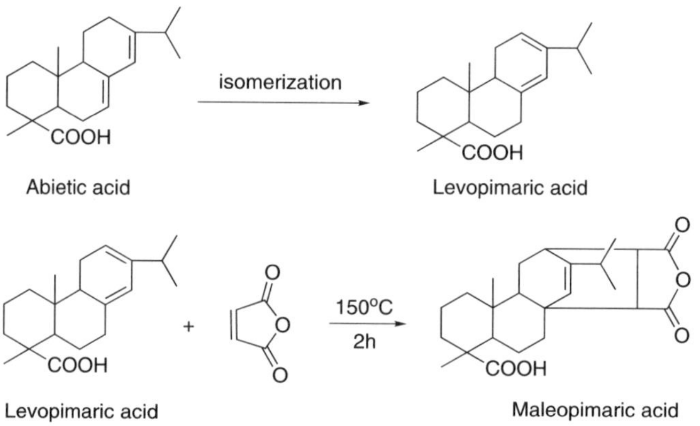

Fig. 3 Synthesis of maleopimaric acid

melting point of maleopimaric acid (229.5 °C) to prevent the decomposition of the acid anhydride, was performed as the first step of polycondensation. In the second step, the esterification with maleic anhydride was carried out at 170 °C in the presence of dibutyltin oxide as the catalyst. The lower temperature of esterification prevented maleic anhydride from sublimating from the reaction medium. The used catalyst reduced the amount of unreacted maleopimaric acid.

The synthesized UP prepolymer ($\overline{M_w}$ = 2250 g/mol, $\overline{M_n}$ = 879 g/mol, I_p = 2.56) was crosslinked with styrene. Similar mechanical properties (Table 6) were obtained for the chosen commercial product and the tested resin. The resin based on maleopimaric acid was characterized, however, by the higher Vicat softening temperature.

Polyalicyclic N-(2-hydroxyethyl)maleimidopimaric acid (Scheme 7) was obtained from maleopimaric acid and used jointly with maleic anhydride for the synthesis of the unsaturated polyesterimide (Scheme 8) [16]. The thus

Table 6 Thermomechanical properties of tested UPRs. Reprinted from (1993) Eur Polym J 29:491 [15] with permission

	Commercial UPR (SPRA 1004)	The Synthesized UP + 30% styrene	The Synthesized UP + 40% styrene
Flexural strength σ [MPa]	96	85	90
E Modulus [GPa]	3.9	3.5	3.4
Vicat softening temperature [°C]	77	109	109

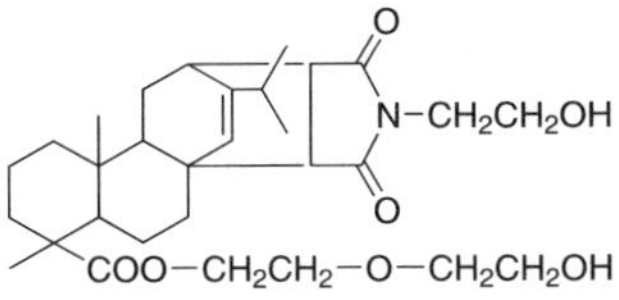

Scheme 7

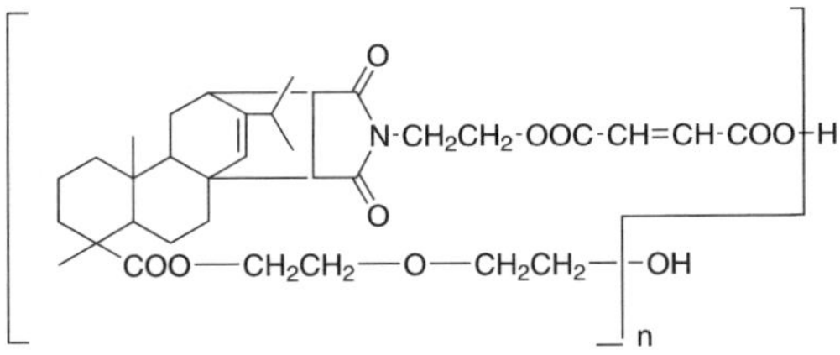

Scheme 8

obtained resin was converted into crosslinked products by radical copolymerization with styrene initiated with organic peroxides or hydroperoxides. The resulted products were characterized by an elevated alkali resistance and a relatively high heat deformation temperature. Moreover, decomposition of the studied polyesterimide resin begins at a slightly higher temperature in comparison with the UPRs based on propoxylated bisphenol A.

Maleic anhydride adducts of anthracene and cyclopentadiene (Fig. 4) were used in the synthesis of UPRs and applied directly during the polycondensation reaction of maleic anhydride with ethylene glycol or diethylene glycol (Table 7) [17].

The optimum molar ratio of substrates for UPR synthesis was found to be 1.10 : 0.75 : 0.30 (glycol : maleic anhydride : anthracene/cyclopentadiene). Polycondensation reactions were carried out with the addition of xylene, in the temperature range of 140–150 °C. The polyesters were crosslinked with styrene (45 wt %) in the presence of cumyl hydroperoxide and cobalt naphthenate. The results of determination of mechanical and electrical properties

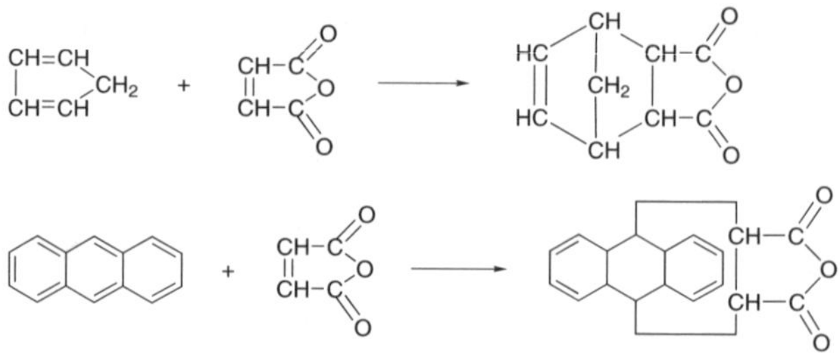

Fig. 4 Synthesis of maleic anhydride adducts of cyclopentadiene and anthracene. Reprinted from (1995) Polimery 40:624 [17] with permission

Table 7 Compositions of UPRs based on maleic anhydride adducts. Reprinted from (1995) Polimery 40:624 [17] with permission

Substrate	Amount [mol]					
	1	2	3	4	5	6
Glycol	1.10	1.10	1.10	1.10	1.10	1.10
Maleic anhydride	0.55	0.60	0.75	0.55	0.55	0.55
Phthalic anhydride	0.30	0.25	–	0.10	0.10	–
Anthracene	0.20	–	0.30	–	0.40	–
Cyclopentadiene	–	0.20	–	0.40	–	0.50

of modified UPRs (Table 8) showed that the amount of introduced anthracene and cyclopentadiene adducts did not affect the final properties of cured resins.

Generally, UPRs prepared using maleic anhydride adducts were characterized with higher thermal resistance and better dielectric properties than the reference resin, although the mechanical properties were similar.

2.2
Glycols and Other Polyhydric Alcohols

Among the glycols used for the synthesis of UPs, 2-methyl-1,3-propanediol (MPD) is preferred both due to its high reactivity in polyesterification and the improved properties of cured UPRs using MPD. The increased polyesterification rate of MPD with maleic anhydride, phthalic anhydride and isophthalic acid was ascertained [18]. Kinetic studies were carried out.

UPRs with enhanced chemical, thermal and mechanical resistance were synthesized from neopentyl glycol and isophthalic acid [19]. Phosphorus-containing catalysts, especially a composition of *ortho*-phosphoric acid with molybdenum trioxide, were found to be the most effective in decreasing polyesterification time and giving light-colored products. Next, the properties

Table 8 Synthesis of maleic anhydride adducts of cyclopentadiene and anthracene. Reprinted from (1995) Polimery 40:624 [17] with permission

Substrate	Resin (according to Table 7)						Ref. resin
	1	2	3	4	5	6	
Molecular weight $\overline{M_n}$ [g/mol]	940	1175	–	1000	1020	970	–
Acid value [mg KOH/g]	33–38	35–41	39–42	42–46	36–41	33–35	–
Viscosity at 80 °C [cP]	25000	28700	24100	26200	25200	23100	–
Flexural strength [MPa]	50–70	50–65	60–66	55–65	52–63	60–75	70–100
Compression strength [MPa]	125	128	139	125	143	135	40–90
Brinell hardness [MPa]	152	172	169	141	160	150	80–120
Vicat softening temperature [°C]	170	165	175	168	176	165	85–120
Volume resistivity [10^{16} Ωm]	1.60	1.80	–	1.11	1.65	1.20	50
Dielectric loss [10^6 Hz]	0.008	0.0067	0.0058	0.007	0.0063	0.0075	0.02–0.03
Permittivity at 20 ± 2 °C [10^6 Hz]	3.6	3.5	3.6	3.4	3.8	3.7	4.4–5.2
Breakdown resistance at 20 ± 2 °C [mV/m]	25–26	23–25	25–26	23–26	25–26	24–26	13–19
Water absorption after 24 h [%]	0.10	0.15	0.20	–	0.08	0.10	0.05–0.06

of isophthalic-neopentyl resins and the commercial *ortho*-phthalic resin were compared (Table 9).

The resin based on neopentyl glycol was characterized by higher impact and flexural strength as well as higher chemical and thermal resistance.

Soluble unsaturated polyesters were synthesized from maleic anhydride and glycidol in dimethoxyethane. At 80 °C, an acid anhydride ring opening occurred and monomaleate ester was formed. In the second step, at 120 °C, the epoxy ring was opened by the COOH group (Fig. 5), formed in the first step [20].

Table 9 Properties of isophthalic-neopentyl resins and a commercial UPR. Reprinted from (1990) Polimery 35:24 [19] with permission

Properties	Isophthalic-neopentyl resin	*ortho*-Phthalic resin (Polimal 109)
Molecular weight $\overline{M_n}$ [g/mol]	2500	1800
Acid value [mgKOH/g]	21	35
Styrene content [%]	38	35
Viscosity at 25 °C [mPa·s]	850	820
Impact strength [kJ/m^2]	7.5	2.0
Flexural strength [MPa]	98.5	70.8
Heat deformation temperature (Martens) [°C]	74	59
Heat deflection temperature (HDT) [°C]	85	68

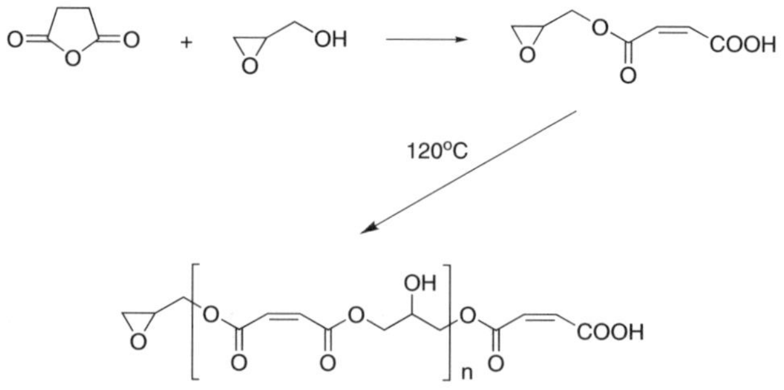

Fig. 5 Synthesis of monomaleate ester. Reprinted from (2003) J Polym Sci Polym Chem 41:2549 [20] with permission

The synthesized UPs exhibited a molecular weight in the range of 6000–18 000 and could be crosslinked with vinyl monomers (styrene in particular).

Unsaturated diols prepared by the transesterification of diethyl adipate and diols: cis-2-butene-1,4-diol and 2-butyne-1,4-diol (Scheme 9) were used for the synthesis of UPRs. Products with the molecular weight of $\overline{M_w}$ = 500–2000 g/mol and a low polydispersity were obtained [21]. The chemical structure of the synthesized hydroxypolyesters was proved by elemental analysis, IR and ^{1}H-NMR spectroscopy. The resins were also characterized using thermogravimetry (TGA) and gel permeation chromatography (GPC). The advantages of the transesterification method of the UPRs synthesis comparatively to the direct polycondensation method were pointed out: lower diol consumption, decreased reaction temperature and time, the absence of decarboxylation reactions, higher molecular weight, lower polydispersity, easy removal of alcohol, and an easy process control. However, the use of a catalyst and the higher costs of the diethyl adipate which should be prepared a priori from the corresponding diacid and ethyl alcohol are the disadvantages of this procedure.

The acetylated derivatives of pentaerythritol (Scheme 10) were obtained and used as the glycol component in the synthesis of UPRs [17]:

A molar ratio of pentaerythritol to acetic anhydride from 1 : 1.37 to 1 : 3.5 and a temperature in the range of 100–150 °C were found to be the best reaction conditions for the synthesis of pentaerythritol diacetate. UP resins based on acetylated derivatives of pentaerythritol diacetate were obtained in the presence of a small amount of o-xylene at 140–150 °C with p-toluenesulfonic acid as the catalyst, under dry conditions. It was found that pentaerythritol diacetate also contained free pentaerythritol and pentaerythritol monoacetate, which caused a premature crosslinking of the synthesized resin. Thus, ethylene glycol and diethylene glycol were used in the synthesis of the UPR together with pentaerythritol diacetate.

$$HO-CH_2-CH_2-CH_2-CH_2-[O-\overset{O}{\overset{\|}{C}}-(CH_2)_4-\overset{O}{\overset{\|}{C}}-O-CH_2-CH_2-CH_2-CH_2]_n-OH$$

$$HO-CH_2-CH=CH-CH_2-[O-\overset{O}{\overset{\|}{C}}-(CH_2)_4-\overset{O}{\overset{\|}{C}}-O-CH_2-CH=CH-CH_2]_n-OH$$

$$HO-CH_2-C\equiv C-CH_2-[O-\overset{O}{\overset{\|}{C}}-(CH_2)_4-\overset{O}{\overset{\|}{C}}-O-CH_2-C\equiv C-CH_2]_n-OH$$

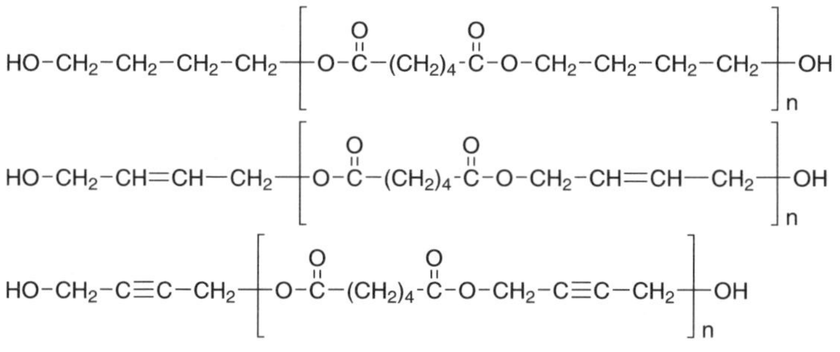

Scheme 9

$$n(CH_3CO)_2O + C(CH_2OH)_4 \longrightarrow (CH_3COOCH_2)_n\text{-}C(CH_2OH)_{4-n} + nCH_3COOH$$

Scheme 10

The mechanical properties of the UPR prepared using pentaerythritol diacetate were comparable with the properties of a commercial UP resin (Table 10). Novel unsaturated *para*-linked aromatic polyesters and copolyesters (Scheme 11) containing fumaroyl units were synthesized by the interfacial

Table 10 Mechanical properties of UPR prepared using pentaerythritol diacetate. Reprinted from (1995) Polimery 40:624 [17] with permission

Property	UPR synthesized using pentaerythritol diacetate	Reference UPR
Impact strength [kJ/m^2]	5.8–7.5	6–12
Flexural strength [MPa]	55–85	70–100
Tensile strength [MPa]	24.5–36.0	40–65
Compressive strength [MPa]	85–115	90–140
Ultimate elongation [%]	3.8–5.5	5–8
Brinell hardness [MPa]	73–105	80–120
Vicat softening temperature [°C]	80–110	85–120

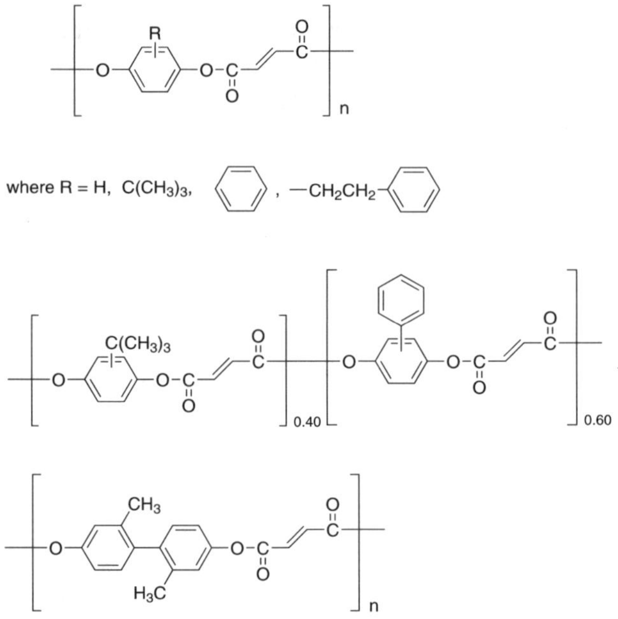

where R = H, C(CH$_3$)$_3$, —C$_6$H$_5$, —CH$_2$CH$_2$—C$_6$H$_5$

Scheme 11

and melt polycondensation from fumaroyl chloride and *tert*-butyl-, phenyl-, and phenethylhydroquinone or 2,2′-dimethylbiphenyl-4,4′-diol [22].

Next, the obtained polyesters and copolyesters were dissolved in styrene or vinyl acetate forming lyotropic liquid-crystalline thermoset (LLCT) systems, capable of thermal crosslinking by the addition of a free-radical initiator. The thermal and liquid crystal properties of the novel UPs were investigated. It was also reported that UPRs prepared from propylene glycol, maleic anhydride and isophthalic acid were used as a matrix for nematic liquid-crystal droplets [23].

2.3
Crosslinking Monomers

The influence of styrene concentration in UPR on the miscibility with unsaturated polyester and the mechanical properties was investigated [24]. A commercial UPR containing 38 wt % styrene was used to prepare samples with 6–58 wt % styrene by partial evaporation of styrene in vacuo or dilution with fresh styrene. Sets of curves presenting the effect of styrene content on the glass transition temperature and the mechanical properties of the cured UPR vs. temperature were presented. Phase separation in the crosslinked UPR with increase in styrene concentration was observed.

The viscosity of UPRs, being solutions of UPs in styrene, presents an important property, which influences the processing. It is obvious that the viscosity of UPR depends mainly on the unsaturated polyester-styrene ratio. Moreover, it depends on the temperature and on the molecular weight of the unsaturated polyester. The viscosity vs. temperature and logarithmic viscosity vs. reciprocal of temperature dependencies were determined [25]. The shear stress vs. shear rate of UPRs dependence follows the Arrhenius-type correlation. The UPRs viscosity depends on the compatibility of the unsaturated polyester with styrene. The difference was exemplified by viscosity determinations of styrene solutions of UPs made of maleic anhydride and 1,2-propylene glycol or maleic anhydride, 1,2-propylene glycol and isophthalic acid or maleic anhydride, diethylene glycol and neopentyl glycol. The number average molecular weight of the investigated polyesters was in a range between 1490 and 4370, whereas the weight average molecular weight was between 5420 and 35 050.

Styrene is predominantly used as the crosslinking monomer in UPRs. Some other polyunsaturated monomers, e.g. triallyl isocyanurate, have been applied jointly with styrene in order to increase the glass transition temperature of crosslinked UPRs. Recently, bismaleimide (Scheme 12) joined the monomers being used in UPRs exhibiting enhanced heat resistance [26, 27]. The presence of bismaleimide results in an increase in crosslinking density and in the overall rigidity of the network. Moreover, the yielding and fracture behavior are improved. During the investigation of the fracture behavior

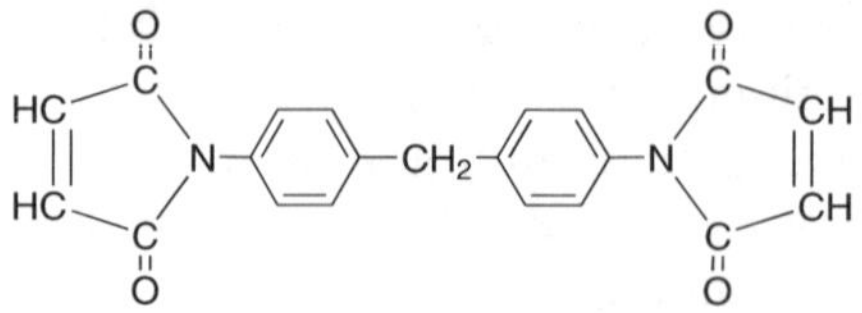

Scheme 12

Table 11 Some thermal properties of cured UPRs with partial replacement of styrene with the bismaleimide (Scheme 12) [27]

Concentration of bismaleimide Scheme 12 [%]	Ball indentation hardness [MPa]	Initial decomposition temperature [°C]	Glass transition temperature [°C]
0	144.4	200	78.4
5	150.3	220	179.9
10	156.5	240	181.9
15	159.3	280	184.1

Table 12 Properties of used crosslinking monomers. Reprinted from (1995) Polimery 40:636 [28] with permission

Properties	Styrene	Vinyltoluene	2-Hydroxyethyl methacrylate	Glycerol α-allyl ether
Ignition temperature [°C]	32	53	102	98
Boiling temperature [°C]	145	168	67*	245
Vapor pressure at room temperature [hPa]	9	5	–	–
Emission [%]	> 15	> 15	–	–
Maximum permissible concentration [ppm]	20–50	50–100	2000	–
LD_{50} [mg/kg]	5000	> 4000	5888	–

* - under the pressure of 0.47 kPa (3.5 mm Hg)

of the crosslinked polymers comprising an unsaturated polyester, styrene and the bismaleimide, three distinct types of crack propagation mode were observed [26]: stable, brittle propagation; unstable, brittle propagation; and stable, ductile propagation. Incorporation of bismaleimide into a UPR affects

Table 13 Results of DSC study of the curing process of unsaturated polyesterimide resins. Reprinted from (1995) Polimery 40:636 [28] with permission

Monomer	Resin *	T_{onset} [°C]	T_{max} [°C]	T_{end} [°C]	ΔH [J/g]	E_a [kJ/mol]	n	$\ln(K_0)$ [s^{-1}]
Styrene	A	130.0	162.7	178.3	94.8	176.6	1.02	44.4
	B	91.3	128.4	154.9	48.2	165.1	1.38	44.0
Vinyltoluene	A	132.0	153.1	176.6	106.9	–	1.31	86.9
	B**	98.5	128.7	144.7	25.3	153.2	0.88	41.1
		145.5	163.9	180.9	9.6	165.4	0.90	40.7
2-hydroxyethyl methacrylate	A	95.8	124.6	151.7	183.9	206.6	1.58	58.2
	B	88.0	110.9	142.6	179.2	209.9	1.65	61.6
Glycerol α-allyl ether	A	70.0	112.6	150.4	–	78.5	1.05	19.2
	B	81.3	119.0	146.2	107.7	92.2	0.96	23.0

* – A – the resins synthesized in two-step procedure,
  B – the resins obtained in one-step synthesis
** – Two exotherm signals were recorded

some important thermal properties of model UPRs (Table 11). A commercial standard *ortho*-phthalic UPR was used.

Vinyltoluene, 2-hydroxyethyl methacrylate and glycerol monoallyl ether (Table 12) were tested as the crosslinking monomers for the curing of unsaturated polyesterimide resins [28]. The monomers were chosen taking into account their lower volatility and toxicity in comparison to styrene.

The enthalpy (ΔH) and activation energy (E_a), pre-exponential coefficient [$\ln(K_0)$], and order of reaction (n) for each monomer were determined using the DSC method (Table 13).

It could be assumed on the basis of the results of the DSC study on kinetics of the free-radical copolymerization of unsaturated polyesterimide resins that the reactivity of the tested monomers occurs in the sequence: glycerol α-allyl ether > vinyltoluene $\geq$ styrene > 2-hydroxyethyl methacrylate. On the other hand, the curing rate depends on the type of polyesterimide resin and varies for the products of one- and two-step synthesis.

2.4
Dicyclopentadiene

Dicyclopentadiene (DCPD) is a very interesting raw material, readily available from the products of crude oil refining. Traditionally, it has been used in the production of hydrocarbon resins, pesticides, elastomers and fire-retardant additives. However, due to easy availability from cracking processes, DCPD became an important raw material for synthesis of UPRs, mainly in the USA. During the UP synthesis, DCPD undergoes different reactions depending on the temperature. DCPD is a stable compound up to 150 °C, but at a temperature above 150 °C (especially above 170 °C), the de-dimerization reaction (retrodiene reaction) proceeds and cyclopentadiene (CPD) is produced (Fig. 6). In the presence of maleic anhydride, CPD undergoes the Diels–Alder reaction and endomethylene tetrahydrophthalic anhydride (EMTHPA) is formed. Further reaction of EMTHPA with glycols is possible (Fig. 7). At a temperature below 120–140 °C and in an acidic environment, DCPD undergoes nucleophilic addition to carboxylic acids, glycols or water (Fig. 8). The reaction proceeds via the reactive double bond in the norbornene ring due to rearrangement followed by formation of a non-classic stable carbocation. Nu-

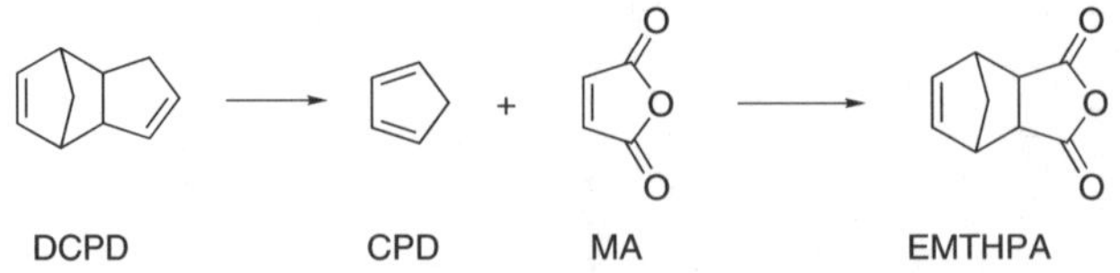

Fig. 6 Formation of endomethylene tetrahydrophthalic anhydride (EMTHPA). Reprinted from (1999) Polimery 44:745 [29] with permission

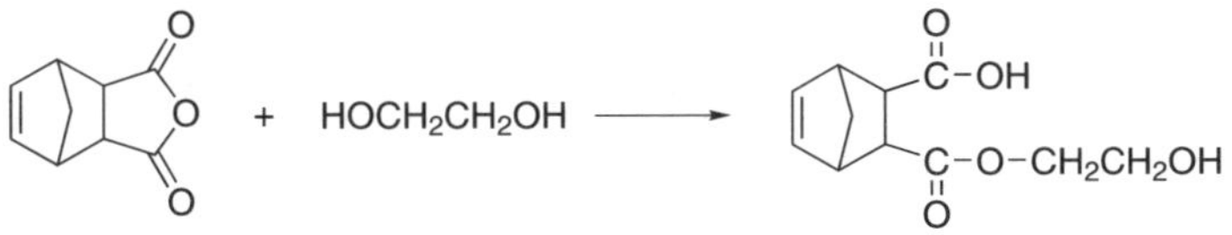

Fig. 7 Reaction of endomethylene tetrahydrophthalic anhydride with ethylene glycol. Reprinted from (1999) Polimery 44:745 [29] with permission

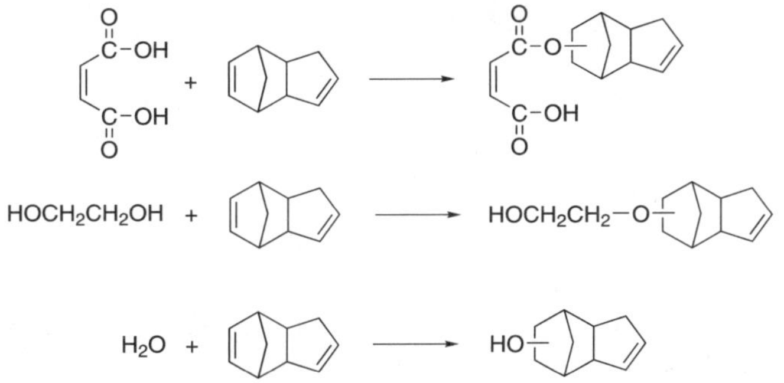

Fig. 8 Nucleophilic addition of dicyclopentadiene to carboxylic acid, glycol and water. Reprinted from (1999) Polimery 44:745 [29] with permission

cleophilic addition to the cyclopentene ring could occur under more drastic conditions (Fig. 9). Four basic methods of synthesis of UPRs modified with DCPD are known. In the "beginning" method, the reaction of maleic anhydride and phthalic anhydride, glycols and DCPD is carried out for 2–3 h, in a temperature range of 120–140 °C under nitrogen [29]. The acid catalyzed DCPD addition to carboxyl and hydroxyl groups occurs. Next, the reaction is continued for 5–6 h at the temperature of 200 °C and the condensation water is removed. In the "Halfester" method, maleic anhydride, phthalic anhydride and glycols are mixed at the beginning and heated under nitrogen for 1–2 h, at a temperature of 120–140 °C [30]. The acid anhydrides react with glycols at this stage (Fig. 10). Next, DCPD is added slowly and the reaction is continued for 1 h. The addition of carboxyl and hydroxyl groups to DCPD occurs.

In the "anhydride" method, a mixture of maleic anhydride and DCPD is heated under nitrogen at a temperature of 180 °C [31]. In the retrodiene reaction of DCPD, cyclopentadiene is formed. Then it reacts with maleic an-

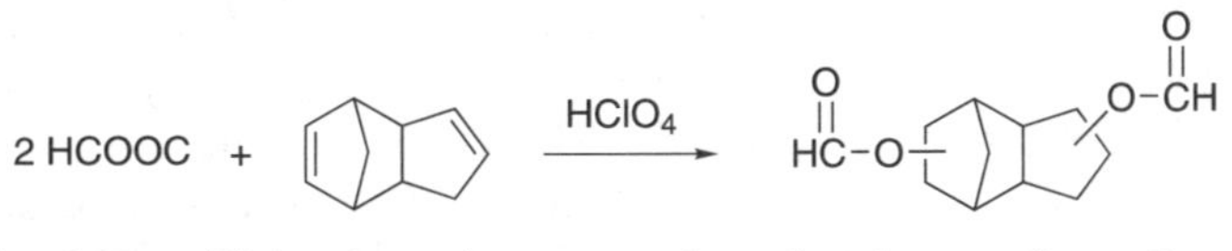

Fig. 9 Nucleophilic addition to cyclopentene ring of cyclopentadiene. Reprinted from (1999) Polimery 44:745 [29] with permission

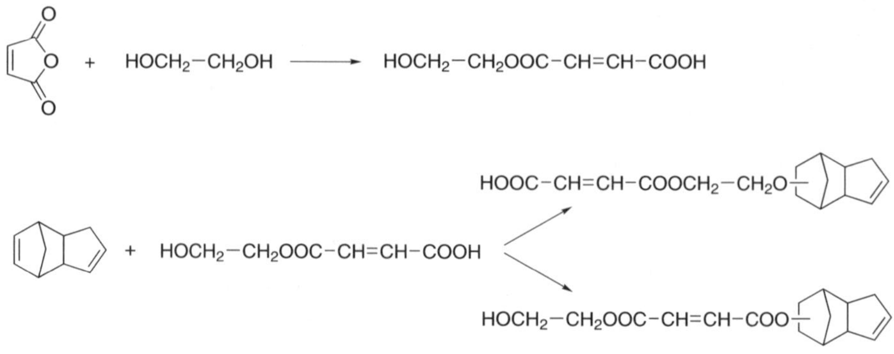

Fig. 10 Reactions occurring in the "Halfester Method"

hydride giving endomethylene tetrahydrophthalic anhydride (Fig. 6). Next, phthalic anhydride and glycol are added and the reaction mixture is heated up to 200 °C, while water is removed. In the "end" method, the polyesterification reaction of maleic and phthalic anhydrides with glycols is carried out at 200 °C under nitrogen [32]. When the acid value of 35 mg KOH/g is achieved, the reaction temperature is decreased down to 160–180 °C. The methods based on the Diels–Alder reaction are not as popular because the products are dark brown and the smell of the DCPD remains. The synthetic procedures are complicated and the reactivity of the synthesized UPRs is low. The beginning method is economical (saving 15% of the raw materials cost) and more popular due to easy availability and easy processing (even at low temperatures) of the raw materials. The products prepared using these methods are characterized by increased chemical resistance, high UV stability, good thermal properties, excellent dielectric properties and adhesion to glass fibers and steel, and good compatibility with other resins (e.g. urea-formaldehyde and maleic resins).

Three different UPRs were prepared from maleic and phthalic anhydrides and ethylene and diethylene glycols [29]. Two of them were modified with DCPD used in the amounts of 6 and 12 wt % and one was mixed with resins that contain 6 wt % DCPD. Then, the resins containing 2, 4, 6, 8, and 10 wt % DCPD were prepared and examined at -10 °C and at room temperature to establish miscibility with styrene, gel time, maximum copolymerization temperature and time elapsed to attain the maximum temperature (Table 14).

It was found that DCPD enhanced solubility of the resins in styrene almost to an unlimited level, even at -10 °C. The presence of DCPD resulted in an increased gel time and time to achieve T_{max} and significantly decreased maximum copolymerization temperature. Next, ball indentation hardness was determined for cured UPRs (Table 15). Generally, the final hardness and the hardness after 2 days were higher for resins with higher DCPD contents.

Table 14 Gel time and maximum copolymerization temperature in relation to the amount of DCPD built into UP. Reprinted from (1999) Polimery 44:745 [29] with permission

DCPD content [wt.%]	Styrene content [wt.%]	Gel time [min]	Maximum copolymerization temperature t [min]	T_{max} [°C]
0	35	14.5	30.5	106.4
6	35	15.0	40.0	86.6
6	40	12.0	43.0	83.2
6	45	10.5	45.0	79.1
6	50	10.0	61.5	61.2
12	35	16.5	42.0	79.1

Table 15 Ball indentation hardness of cured UPRs. Reprinted from (1999) Polimery 44:745 [29] with permission

DCPD content [wt.%]	Styrene content [wt.%]	6 h	1 day	2 days	6 days	10 days
0	35	45.6	33.9	59.0	76.2	91.2
2*	35	21.5	49.2	67.2	76.6	95.7
4*	35	32.1	59.0	67.0	95.3	119.6
6*	35	35.1	54.1	69.1	92.1	119.4
8*	35	21.8	56.6	78.1	98.2	123.5
10*	35	32.1	59.3	77.5	95.7	123.3
12	35	37.2	54.5	77.8	103.2	136.7
6	35	14.7	49.5	67.8	83.2	92.2
6**	40	13.2	66.3	74.2	83.9	93.4
6**	45	22.8	52.4	70.9	89.0	99.5
6**	50	9.5	54.5	73.6	95.7	95.7

* – Resins obtained from mixed UPRs containing 0 wt.% and 12 wt.% DCPD
** – Resins prepared by dilution with the resin containing 6 wt.% DCPD and the resin from direct synthesis with styrene

DCPD built into UPR enhanced the degree of drying of the coating surface and their Persoz pendulum hardness (Table 16). Comparable properties could be achieved for UPRs based on propylene glycol, but the use of dicyclopentadiene is more effective from the economical point of view.

DCPD/vegetable oil derived UPRs were synthesized in the standard way from maleic anhydride, phthalic anhydride, ethylene glycol, and 1,2-propylene glycol. The UP thus obtained was dissolved in styrene. To flexibilize the resin, 5–20% rapeseed oil was incorporated into the polyester [33]. More-

Table 16 The degree of drying of the coating surface and Persoz pendulum hardness of the surface of UPR in relation to the amount of DCPD built into the UPR. Reprinted from (1999) Polimery 44:745 [29] with permission

DCPD content [wt.%]	Styrene content [wt.%]	Degree of drying of the coating surface				Persoz pendulum hardness of the surface after 10 days
		4 days	7 days	10 days	Postcuring (2 h, 80 °C)	
0	35	0	0	0	0	–
2*	35	0	0	0	0	–
4*	35	0	0	0	0	–
6*	35	0	0	1	1	0.17
8*	35	0	1	2	2	0.28
10*	35	1	2	3	3	0.38
12	35	1	2	3	3	0.45
6	35	0	0	1	1	0.12
6**	40	0	1	2	2	0.21
6**	45	0	1	2	2	0.20
6**	50	1	2	3	3	0.28

* – Resins obtained from mixed UPRs containing 0 wt.% and 12 wt.% DCPD
** – Resins prepared by dilution with styrene of the resin containing 6 wt.% DCPD and the resin from direct synthesis

over, 10% dicyclopentadiene was built-in. The formulation and the properties of the uncured resins are given in Table 17.

In Table 18, mechanical properties of the cured resins are presented.

The incorporation of rapeseed oil improved the mechanical properties which depend on flexibilization: tensile, flexural and impact strength and elongation at break as well. Dicyclopentadiene increased the flexural and compression strength, hardness and heat deflection temperature (HDT) and the chemical resistance, which is, in general, better in water and in aqueous solutions of acids than in alkali (10% KOH).

Low viscous UP was designed using an appropriate glycol and adipic acid [34]. Moreover, two different kinds of copolymerizable double bonds were incorporated: maleic/fumaric and a terminal cyclopentadiene adduct, in direct neighborhood to the maleic/fumaric unsaturation. Unfortunately, detailed formulations have not been given in the article.

Acrylic derivatives of 5,6-dihydrodicyclopentadiene could be applied as the reactive diluents for coating and adhesive compositions to replace high volatile aliphatic acrylates. They could be synthesized from acrylic acid and DCPD in the presence of acidic catalysts and inhibitors (e.g. hydroquinone) (Fig. 11) [35]: or in acrylic acid esterification with the reaction product of DCPD with water or glycol in the presence of acidic catalyst (e.g. BF_3 or p-toluenesulfonic acid) (Fig. 12) [36, 37]:

Table 17 Formulations and properties of uncured resins with built-in rapeseed oil and DCPD. Reprinted from (2004) Fatipec Congr 2:617 [33] with permission

Component [wt. %]	Resin sample No							
	1	2	3	4	5	6	7	8
Maleic anhydride	25.0	25.0	25.0	25.0	25.0	25.0	25.0	25.0
Phthalic anhydride	34.5	31.3	28.1	21.8	28.1	27.2	23.8	17.3
Ethylene glycol	18.2	17.4	16.6	14.9	16.6	14.9	14.1	12.4
1,2-Propylene glycol	22.3	21.3	20.3	18.3	20.3	18.2	15.2	20.3
Rapeseed oil	–	5.0	10.0	20.0	–	5.0	10.0	20.0
DCPD	–	–	–	–	10.0	10.0	10.0	10.0
Unsaturated polyester								
Acid value [mg KOH/g]	11.9	19.3	23.6	24.8	12.3	29.8	25.7	20.6
Hydroxyl value [mg KOH/g]	78.3	80.5	75.8	59.6	66.4	56.2	43.2	39.8
Softening temperature [°C]	54	49	47	45	59	60	57	40
Unsaturated polyester resins [% styrene]								
Viscosity at 23 °C [mPa]	393	432	466	695	303	570	800	897
Gel time at 25 °C [min]	45	48	59	61	30	43	45	47

Table 18 Mechanical properties of the cured UPRs with built-in rapeseed oil (RO) and dicyclopentadiene (DCPD). Reprinted from (2004) Fatipec Congr 2:617 [33] with permission

Resin sample No (Table 17)	1	2	3	4	5	6	7	8
RO [wt.%]	–	5	10	20	–	5	10	20
DCPD [wt.%]	–	–	–	–	10	10	10	10
Mechanical properties								
Tensile strength [MPa]	27.5	37.0	40.3	42.7	24.6	29.4	33.3	34.8
Elongation at break [%]	1.6	2.3	3.4	4.8	1.7	1.5	1.8	3.1
Flexural strength [MPa]	68.9	71.4	88.0	90.2	75.4	82.7	91.2	92.8
Impact strength [kJ/m^2]	6.0	6.1	6.4	6.8	4.2	4.6	5.2	5.6
Compression strength [MPa]	120.0	115.4	105.0	98.5	125.4	129.5	112.0	108.3
Hardness [MPa]	143.1	141.9	115.8	104.4	148.5	152.3	130.1	126.1
HDT [°C]	58.9	54.5	53.5	50.2	60.7	60.0	57.3	52.0

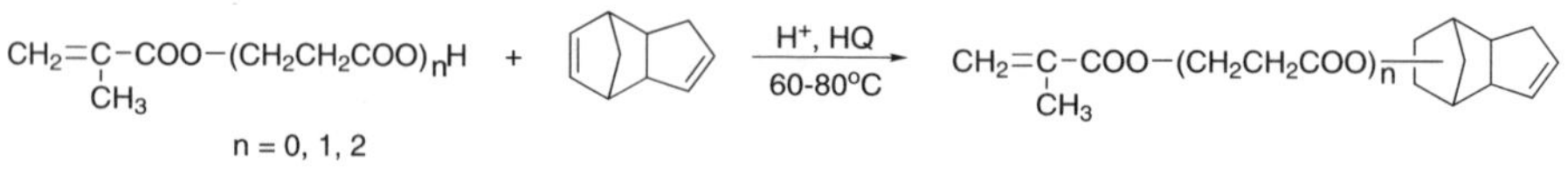

Fig. 11 Synthesis of acrylic derivatives of 5,6-dihydrodicyclopentadiene from acrylic acid and DCPD

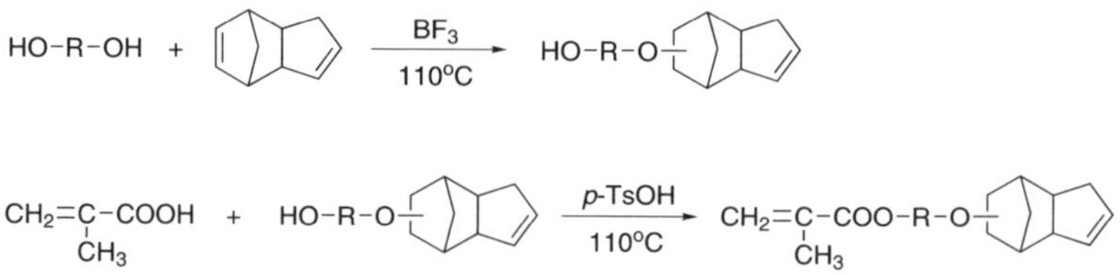

Fig. 12 Acrylic acid esterification of the reaction product of DCPD with water or glycol

2.5
Manufacture of UPRs Using PET Scrap

PET scrap was utilized for the manufacturing of UPRs [38]. The manufacturing process, which consisted of the reaction of glycolyzed PET with acid anhydrides (maleic anhydride in particular) and a glycol, was first patented in Poland [39] and then described by Ostrysz in the articles [40, 41]. A vast patent and literature review on PET bottle scrap can be found in the book [42], where numerous patents and papers on PET scrap utilization were collected. Most of the information contained in that review concerns the manufacture of UPRs from PET waste. The review is based on 159 citations. The methods of chemical recycling of waste PET were described in detail [43]. UPRs based on glycolyzed PET are widely applied as for example a superficial layer increasing chemical and thermal stability of reinforced plastics (gel coat resins), wax-free varnish, binders for glass fiber mat, pre-pregs and pultrusion products; molding compound [44]; polymer concrete and polymer mortar [45]; materials for sewerage systems [46], artificial marble, putties for car body repair [47] and adhesive cartridges for mining [48]. The reaction kinetics [49, 50] and conditions of glycolysis of PET wastes [43, 51] as well as the glycolysis products were studied in detail [52–54].

After many years from the first papers on UPRs from PET scrap having appeared in the early 1970s in Poland, the polycondensation of PET glycolyzate with maleic anhydride followed by the dissolution in styrene of the UP thus obtained was described [55]. The glycolysis of PET waste was performed with ethylene glycol, diethylene glycol or propylene glycol in the presence of zinc acetate as the catalyst. It was found that the extent of the glycolysis depends on the moles of the glycol present when glycolyzed on a weight basis and was the highest for ethylene glycol [56]. On the other hand, the products of PET waste glycolysis with propylene glycol and diethylene glycol exhibited a higher number average molecular weight and a broader molecular weight distribution. In another article, the glycolysis of PET with 1,4-butanediol and triethylene glycol was described [57]. To synthesize UPRs using PET recyclate, PET was first glycolyzed with 1,2-propylene glycol, diethylene glycol or their mixtures [58]. The propylene glycol/diethylene glycol ratio affects the curing characteristics and the mechanical properties of UPRs from glycolyzed PET. Diethylene glycol decreases the tensile modulus and increases the toughness of the cured UPR. It was also shown that the extent of depolymerization decreased and the gelation time was delayed with an increasing amount of diethylene glycol [54]. Moreover, the incorporation of diethylene glycol units greatly increases the tensile toughness of UPR. With the use of different glycol compositions, it is possible to control the extent of depolymerization, gelation time and the brittleness of UPRs, these being very important features for the application of UPRs as matrices in fiber-reinforced composite systems.

Different metal acetates, i.e. cobalt acetate, $Co(CH_3COO)_2 \cdot 4H_2O$; manganese acetate, $Mn(CH_3COO)_2 \cdot 4H_2O$; cupric acetate, $Cu(CH_3COO)_2 \cdot H_2O$; sodium acetate, $Na(CH_3COO)_2 \cdot 3H_2O$, and zinc acetate, $Zn(CH_3COO)_2 \cdot 2H_2O$ were examined as the PET glycolysis catalysts [53]. Zinc acetate was found to be the most effective catalyst. It might facilitate the bond scission of polymer chains and subsequently increase the extent of depolymerization, giving lower oligomers as the glycolysis products. The polycondensation reactions of depolymerized PET waste was found to follow third-order kinetics [49]. It was demonstrated that the reactivity of the terephthalic-acid-based systems increased with a decrease in the amount of terephthalic acid moieties. Moreover, the reaction rate increased with temperature and decreased with the extent of polymerization. The kinetics of glycolysis of PET melts with ethylene glycol carried out in a pressurized reactor at the temperature above 245 °C was studied [50]. A first order (in both ethylene glycol and ethylene diester concentration) kinetic model was proposed. It was found that ethylene glycol did not play a significant role as an internal catalyst in the glycolysis of PET. Moreover, the zinc salts used did not seem to influence the glycolysis rate at a temperature above 245 °C, they catalyzed however the melt hydrolysis in an earlier study and appeared to have a catalytic effect on glycolysis below 245 °C. The authors suggest that the difference in catalytic behavior may be explained by the differences in the solubility of PET and its oligomers in water and in ethylene glycol.

The curing reactions of UPRs based on glycolyzed PET and maleic anhydride were studied by differential scanning calorimetry and various kinetic parameters were obtained from dynamic data using the Kissinger expression [59]. It was demonstrated that the polymerization heat, associated with styrene and polyester double bonds, can be calculated by extrapolating the heat of reaction obtained from different styrene contents.

Water-extended UPRs were prepared using depolymerized PET scrap. Two different UPRs were mixed together: the first one with built-in maleic anhydride and another one with incorporated sebacic acid. The thus prepared UPs were cured with styrene monomer and benzoyl peroxide as the initiator and applied for making decorative art objects [60].

Recently, the thermal characteristics of cured UPRs based on recycled PET was given [61]. The results of tensile tests performed at various temperatures were discussed.

3
Vinyl Ester Resins

"Vinyl ester" resins are related to UPRs. They consist of a styrene monomer and the addition products of epoxy resins with methacrylic acid. The vinyl esters are usually cured using a peroxide initiator at elevated temperature [62] or at ambient temperature using a peroxide initiator with a cobalt promoter. Vinyl ester resin can also be photopolymerized [63]. Photocurable vinyl ester systems with methacrylic acid/phenyl glycidyl ether addition product as a diluent were used with ($\pm$)-camphorquinone photoinitiator and $N,N,3,5$-tetramethylaniline photoreducer [64]. The effect of the monomethacrylate diluent and other components of the resin on the photopolymerization rate and the postcuring in the dark was investigated.

The structure of vinyl ester resins is based on the low-molecular-weight epoxy resin (DGEBA) esterified with methacrylic acid. VERs with a different structure are based on triglycidyl maleopimarate [65]. Thanks to the particularly voluminous structure of maleopimaric acid, more styrene can be used in that resin than usual. At a styrene monomer content as high as 70 wt %, satisfying thermomechanical properties (i.e. T_g and deformation in the high elastic area) and chemical resistance were achieved. In general, the maleopimaric VERs with methacrylic groups exhibit better thermomechanical properties and higher chemical resistance with the corresponding acrylates.

Aliphatic and cycloaliphatic polyesters exhibit much higher UV resistance than the aromatic ones [66]. A cycloaliphatic dimethacrylate was prepared by the addition of methacrylic acid to the cycloaliphatic diepoxide, 3,4-epoxycyclohexylmethyl, 3,4-epoxycyclohexanecarboxylate (ERL-4221, Scheme 13) in the presence of triphenylphosphine catalyst.

The diepoxide was used in excess in relation to methacrylic acid. Then the remaining epoxy groups were reacted with glutaric acid. That chain extension yielded the oligomeric cycloaliphatic vinyl ester resin (Scheme 14).

Radical copolymerization of the cycloaliphatic vinyl ester resin with methyl methacrylate monomer led to a crosslinked polymer characterized by high UV resistance.

Styrene solutions of cycloaliphatic epoxy-based divinyl ester resin (45, 50 and 55 wt % styrene) were prepared and modified with TDI at room temperature [67] (Table 19).

Modification with TDI led to an increased viscosity of the resins, shorter gelation time in comparison with the parent resin, and to reduced exotherm peak temperature. The modified resins as well as the resins with a higher styrene concentration were more stable at elevated temperatures. An increase in TDI concentration also caused a significantly higher T_g and thermal resistance (by about 40 °C; Table 20). However, the addition of TDI did not affect the hardness. The decrease in flexural strength of the prepared resins was

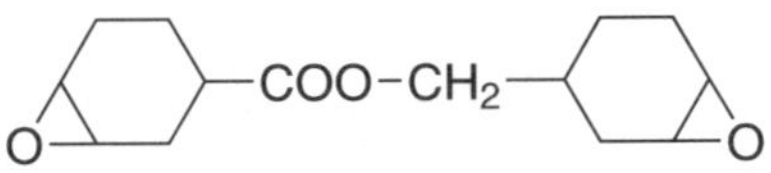

Scheme 13

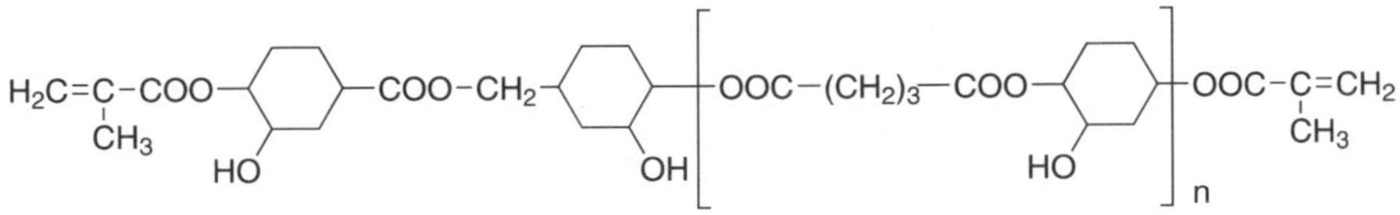

Scheme 14

Table 19 Properties of the modified divinyl ester resins in the non-cured state. Reprinted from (2001) J Appl Polym Sci 81:2062 [67] with permission

Resin No	Styrene concn. [%]	TDI concn. [%]	Density [g/cm³]	Viscosity [mPa s]	Acid value [mg KOH/g]	Gelation time [min]	Peak exotherm temp. [°C]	Stability at 60 °C [days]
1	45.0	–	1.04	38.0	5.3	48	197	1.0
2	45.0	0.1	1.05	51.2	9.1	17	170	1.5
3	45.0	0.2	1.04	52.5	10.4	15	168	1.5
4	45.0	0.5	1.04	53.1	9.3	13	163	1.5
5	45.0	1.0	1.04	56.4	9.9	13	162	1.5
6	50.0	0.1	1.03	45.6	8.1	23	156	8.5
7	50.0	0.2	1.03	46.8	7.1	20	155	8.5
8	50.0	0.5	1.03	47.6	7.4	18	152	8.5
9	50.0	1.0	1.03	48.2	7.0	17	149	8.5
10	55.0	0.1	1.02	36.4	4.6	32	141	9.0
11	55.0	0.2	1.02	37.6	5.2	26	138	9.0
12	55.0	0.5	1.02	38.6	4.9	21	133	9.0
13	55.0	1.0	1.02	39.8	4.9	20	133	9.0

Table 20 Properties of the cured resins. Reprinted from (2001) J Appl Polym Sci 81:2062 [67] with permission

Resin No.	Thermal resistance (Martens) [°C]	Glass transition temperature T_g [°C]	Impact strength (Charpy) [kJ/m^2]	Ball indentation hardness [MPa]	Tensile strength [MPa]	Flexural strength [MPa]
1	–	101.3	–	128.8	–	113.00
2	94.0	138.2	2.46	140.0	82.01	138.12
3	96.0	139.0	2.50	145.3	85.27	122.24
4	99.0	140.9	2.63	149.4	88.96	109.88
5	102.5	142.5	2.85	150.8	92.11	108.33
6	96.5	–	3.83	151.6	81.23	130.10
7	99.0	–	4.41	153.3	82.45	127.52
8	101.0	–	7.39	152.4	84.45	121.25
9	104.0	–	8.4	153.2	92.03	120.00
10	96.0	–	3.61	145.5	67.48	118.12
11	97.0	–	4.13	147.8	72.34	113.43
12	99.0	–	4.84	149.8	79.00	104.88
13	102.0	–	5.39	147.7	83.91	103.35

negligible. The crosslinked polymers were characterized by a high impact resistance and, unexpectedly, by a high tensile strength value. The results of the investigation proved that the epoxy-based divinyl ester resins can be easily modified with diisocyanates and this type of modification could improve the properties.

Two liquid rubbers, i.e. carboxyl-terminated poly(butadiene-co-acrylonitrile) (CTBN) and vinyl terminated poly(butadiene-co-acrylonitrile) (VTBN) were used for the modification of VER cured with styrene [68]. CTBN rubber is almost unreactive with VER and styrene, and the phase separation, similar to that occurring in UPRs modified with a low profile additive, was observed during the crosslinking process. Some toughening was achieved for the system containing 5 wt % CTBN. The morphology of small rubber particles included in the crosslinked VER/styrene matrix was found for VTBN.

Further toughening of VERs modified with a CTBN elastomer by introducing an epoxy-terminated butadiene-acrylonitrile copolymer diluted with a styrene monomer (Hycar ETBN 1300x40 from BF Goodrich) was presented [69]. The improvement of toughness achieved by the incorporation of the CTBN elastomer (Table 21) extends the range of VER applications from corrosion resistant equipment to automotive, marine and infrastructure markets.

An increase in the fracture energy (G_{IC}) of a standard bisphenol-A VER and a significant increase in the mechanical properties of the CTBN-

Table 21 Mechanical properties of VERs. Reprinted from (1996) Reinf Plastics 10:50 [69] with permission

	Bisphenol-A vinyl ester resin	Flexible vinyl ester resin	CTBN elastomer modified vinyl ester resin
Tensile strength [MPa]	76–86	28	62–76
Tensile elongation [%]	4.9–5.2	22	8–10
Tensile modulus [MPa]	3170–3273	2005	2480–3170
G_{IC} [J/m^2]	100	659	409
T_g [°C]	120	56	105

Table 22 Improvement of the toughness of VERs. Reprinted from (1996) Reinf Plastics 10:50 [69] with permission

	Bisphenol-A VER	Bisphenol-A VER + ETBN 1300x40	CTBN elastomer modified VER	CTBN elastomer modified VER + ETBN 1300x40
Total rubber parts	0	12.5	8	12.5
Tensile strength [MPa]	76–86	34	62–76	41–42
Tensile elongation [%]	4.9–5.2	3.5	8–10	10–12
Tensile modulus [MPa]	3170–3273	1654	2480–3170	1929–2203
G_{IC} [J/m^2]	120	910	410–550	1770–2180
T_g [°C]	120	121	95–125	114–120

elastomer-modified VER were observed after the addition of ETBN 1300x40 (Table 22).

The best morphology for toughening purposes was obtained in the case of the VER modified with ETBN 1300x40 and was due to small rubbery particles varying from a nanometer to a few microns size.

4
Flexibilization

4.1
Incorporation of Segmental Modifiers into the UP Molecules

Several possibilities have been proposed for adding oligomeric segments at the chain ends (i.e. in the terminal position) or as the pendant groups along the chain of a UP. Particular properties are obtained, if a polyether segment is located in the terminal position. Perfluorinated polyether segments impart unique properties to UPs, if randomly incorporated into the UP chain. Incorporation of elastomeric segments may enhance the properties which are connected with flexibility. However, it does not adversely affect the glass transition temperature.

4.1.1
Terminal Poly(oxyethylene) Segments in the UPs

A carboxyl-terminated unsaturated polyester was esterified with poly(oxyethylene) diols [70]. Using poly(oxyethylene) diol with a number average molecular weight of 2000, partly crystalline block copolymers were obtained. The unsaturated block copolyetherester contained predominantly one terminal polyether group per molecule. Block unsaturated copolyetheresters with shorter poly(oxyethylene) terminal groups of $\overline{M_n}$ = 350 and 550 were also obtained. The effect of the composition of those UPRs on the mechanical and thermal properties was investigated [70]. The block copolyetherester with $\overline{M_n}$ = 2000 can be dissolved in styrene monomer, thus forming a UPR with acceptable viscosity at as little as 20% styrene.

4.1.2
UPs with Incorporated Perfluoropolyether Segments

The incorporation of saturated segments into molecules of unsaturated polyesters is not limited to poly(oxyethylene) blocks. Application of hydroxyl-terminated segments of perfluoropolyethers represents another approach [71]. In contradistinction to the poly(oxyethylene) segments (Scheme 15): the perfluoropolyethers (Scheme 16): contain two reactive terminal functional groups and thus are randomly incorporated along the chain of unsaturated

$$HO-[CH_2CH_2O]_n-CH_3$$

Scheme 15

$$\text{HO}\{\text{CH}_2\text{CH}_2\text{O}\}_n\text{CH}_2\text{CF}_2\text{O}\{\text{CF}_2\text{CF}_2\text{O}\}_p\{\text{CF}_2\text{O}\}_q\text{CF}_2\text{CH}_2\text{O}\{\text{CH}_2\text{CH}_2\text{O}\}_n\text{H}$$

Scheme 16

polyester and not exclusively at the ends of the molecule. The block fluorinated polyetherester macromonomers are copolymerized with styrene.

The materials exhibit some unusual properties. The structure of the fluorinated polyether segments affect important properties of the UPRs. Phase separation was observed. At a low molecular weight (1100) of the fluorinated macromonomer, a very high adhesion of the fluorine-rich phase with the matrix was found.

The morphology and its effect on resilience, on the behavior in flexural and impact tests were investigated. In general, an increase in toughness with respect to standard UPRs was observed.

Fluorine-modified UPRs (FUPRs) were prepared from conventional UPRs and poly(ε-caprolactone)-perfluoropolyether-poly(ε-caprolactone) block copolymers (TXCL) [72]. The presence of poly(ε-caprolactone) (PCL) blocks leads to an enhancement of compatibility with respect to pure perfluoropolyether macromers. It was found that the morphology of crosslinked FUPRs was strongly affected by the curing rate and by a critical balancing of the fluorinated macromer/poly(ε-caprolactone) (TX/PCL) ratio and the molecular weight of the TXCL copolymers. A slight plasticization effect was observed by increasing the molecular weight and the PCL segment length of the TXCL copolymers. On the other hand, the best toughening effect of the TXCL copolymers was found for an intermediate TX/PCL ratio. Moreover, a significant reduction of the water diffusion coefficient value for FUPRs was shown.

4.1.3
Copolymerizable Rubber Flexibilizers

To improve the fracture properties, a copolymerizable liquid rubber flexibilizer was used. The well-known commercial amino-terminated butadiene-acrylonitrile copolymer (ATBN) reacted with maleic anhydride to form maleimide-terminated liquid butadiene-acrylonitrile rubber [73]. A standard two-stage procedure was applied. In the first stage, maleamic acid was formed by the opening of the anhydride ring by the amino group. Then, in the second stage, maleimide rings were closed by the dehydration of maleamic acid amide groups with sodium acetate dissolved in acetic anhydride (Fig. 13). Thus, a maleimide-terminated liquid butadiene-acrylonitrile rubber (iTBN) was obtained. An unsaturated polyester, styrene and iTBN composition was radically copolymerized. The morphology of the crosslinked copolymers was investigated using TEM and SEM.

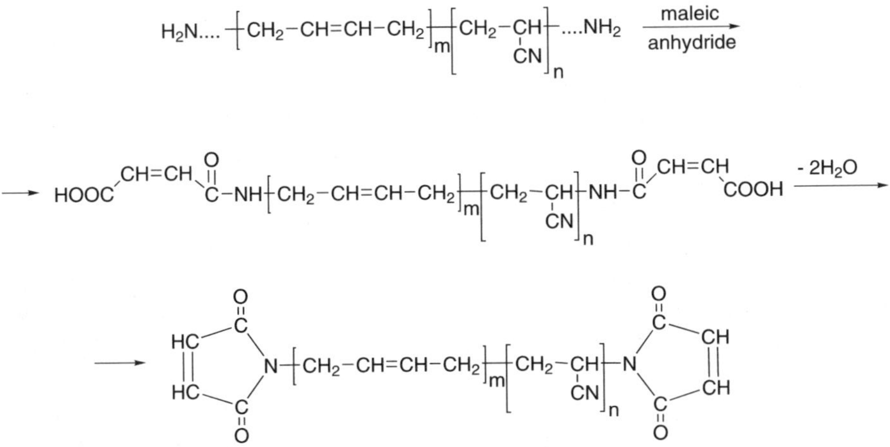

Fig. 13 Synthesis of maleimide terminated liquid butadiene-acrylonitrile rubber

The addition of the rubber component resulted in a decrease in the modulus and yield stress. However, toughening and an improvement in the impact behavior were observed. The phase separation and formation of the rubbery domains played an important role.

In a previous contribution [74], a hydroxyl-terminated polybutadiene rubber was transformed into an isocyanate-terminated polybutadiene. That modification changed the morphology of the thus-modified UPR and resulted in an enhancement of toughness.

Blending of UPRs with functionalized rubber was applied to improve the toughness, the impact and the tensile strength [75]. The most advantageous complex of mechanical properties, in particular those relating to flexibility and the impact characteristics, was achieved using nitrile rubber grafted with maleic anhydride. UPR with incorporated hydroxy-terminated polybutadiene or natural rubber, or epoxidized natural rubber exhibited inferior mechanical properties. Another approach to the flexibilization of UPRs consisted in the blending of the UPR with the reaction product of polycaprolactone diol with maleic anhydride. The thus flexibilized UPRs were applied as binders in SMC (sheet molding compounds). The flexibilization resulted in an increase in the flexural and impact strength. From the range of the flexibilizer added to the UPR of between 3 and 50% by weight, the optimum level of 20% was chosen. At that level, the mechanical properties of the cured molding compounds, the thickening behavior, the processing characteristics, and the fracture behavior were determined. The phase separation on curing of the blends was observed.

Cured vinyl ester resins are fragile after curing like standard UPRs and thus need flexibilization. The flexibilization of vinyl ester resins is made using the known reactive rubber: vinyl-terminated liquid elastomeric butadiene-acrylonitrile copolymer (VTBN) [76]. The macrostructure of the VTBN flexibilized vinyl ester resins is mainly affected by the composition and the concentration of the VTBN component, by the molecular weight, and by the curing temperature. These factors influence the morphology of the crosslinked vinyl ester-styrene-VTBN copolymers, in particular the ratio and the shape of the resin-rich and elastomer-rich domains, which are formed as the result of phase separation.

Still another way to increase the toughness of vinyl esters consists of formation of the semi-interpenetrating polymer networks (SIN) comprising DGEBA dimethacrylate, styrene, a diisocyanate, and an epoxy resin, the epoxy component containing, for example, cyclohexylene units [77]. The boat/chair conformation of the cyclohexylene units influenced the toughness characterized by fracture toughness and fracture energy.

Apart from the modification of UPRs through incorporation of various oligomeric segments at the end or along the chain of unsaturated polyester, the effect of structural changes in standard unsaturated polyesters on the properties of UPRs was investigated. Thus, the influence of molecular weight of unsaturated polyesters on the copolymerization with styrene and the properties of the cured UPRs was studied. The standard polyester composed of maleic anhydride, isophthalic acid, 1,2-propylene glycol, and diethylene glycol in the equimolar ratio was the object of that study [78]. Increasing the polycondensation time resulted in an increase in the number average molecular weight in the interval of 480–1700 and in the maleate-fumarate isomerization degree from 85 up to 95%. The molecular weight was shown to affect the compatibility with styrene. Moreover, it affects the glass transition temperature as determined by DSC and the water absorption. It seems that the effects of molecular weight itself of the unsaturated polyester are superimposed by the influence of different *cis-trans* isomerization degree and by the content of the OH and COOH end groups as well. That approach resembles the older publications on the influence of the position of unsaturated bonds at the ends ("endenes") or in the center ("centrenes") of the polyester molecule on the properties after crosslinking with styrene [79–83].

4.2
Interpenetrating Polymer Networks

UPRs can form interpenetrating polymer networks (IPNs). High mechanical strength was achieved, if one of the components was a polyurethane [84]. Also hybrid networks consisting of UPRs and polyurethanes were investigated by many authors. In any case, an improvement of properties by incorporating polyurethanes into crosslinked UPR [24, 85, 86] was found.

Polyurethane and UPR-hybrid IPN networks were studied by several authors [86–88]. The thermomechanical properties of the networks were investigated. Influence of hard domains on mechanical properties was the main factor studied [86]. DSC measurements were applied to evaluate the miscibility and the morphology of the polyurethane-UPR IPNs [87]. The study was supported by DMTA. The same methods were applied for the detailed investigation of these IPNs comprising the gelation, supermolecular structure, segmental and domain structure, static mechanical properties, thermal stability and swelling in organic solvents [88]. Interpenetrating polymer networks (IPNs) were obtained from UPRs and the acrylate-modified polyurethane [89]. The morphology and multi-phase structure were characterized. A multicomponent IPN system was created using an epoxy resin, a bismaleimide, a cyanate ester, and components which form a UPR [90]. As the cyanate ester, 1,1-bis(3-methyl-4-cyanatophenyl)cyclohexane was applied, whereas N,N'-bismaleimido-4,4′-diaminodiphenylmethane was used as the bismaleimide component. Materials with enhanced thermal degradation resistance and high heat deflection temperature were obtained.

A composite network was synthesized from a hydroxy-terminated unsaturated polyester derived from maleic anhydride, phthalic anhydride and excess 1,2-propylene glycol, styrene and an urethane prepolymer derived from poly(oxypropylene) diol of a polyester diol based on adipic acid and excess diethylene glycol, the oligomeric diols being reacted with excess toluene diisocyanate (TDI). Those three components react with each other to form a network that is crosslinked with polystyrene bridges and urethane linkages. The authors call such tridimensional polymers "hybrid polymer networks" (HPN), to be distinguished from IPNs, wherein no linkages appear between the networks. It was shown by TEM microphotography that the polymers were microheterogeneous and rubber microdomains were separated from the glassy UPR matrix. The obtained structure exhibited improved mechanical strength in tensile and impact tests. Optimum impact strength, tensile strength and flexural strength appear at a certain weight ratio of the components.

A different approach consisted of the formation of a HPN by the reaction of an unsaturated polyester as described above with a diisocyanate hard extender [86]. The hard extender was created using a stiff glycol molecule (ethylene glycol or 1,6-hexanediol) and MDI (Scheme 17):

Interpenetrating polymer networks (IPNs) consisting of a UPR, ethylene glycol diacrylate, polyurethane from toluene diisocyanate and castor oil and other components were investigated [91]. The effect of composition of the IPNs on their mechanical properties and the morphology was studied. Sorption and diffusion of organic solvents through IPNs based on UPR and castor-oil based polyurethane were tested for various weight compositions of PU and UPR, different crosslinking density (NCO/OH ratio) and various hydroxyl values of the used polyols [92]. It was found that all factors affect the sorption behavior, i.e. diffusion coefficient increases with both an increase in NCO/OH ratio and in UPR contents. The sorption coefficient showed a reverse trend. Generally, all the IPNs studied showed high crosslinking densities and low diffusion and sorption values.

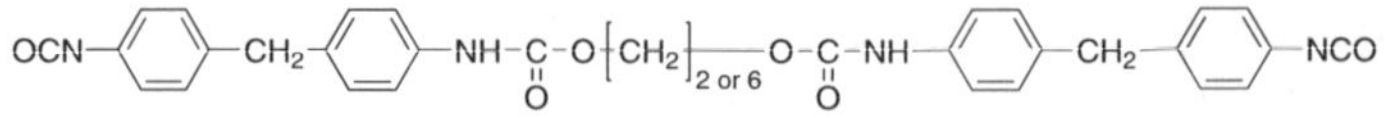

Scheme 17

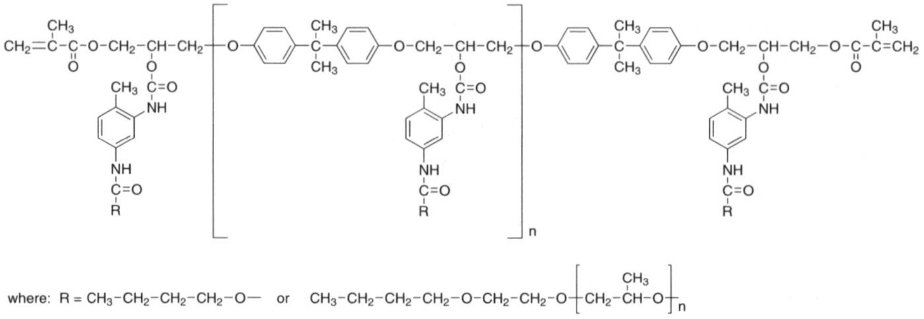

Scheme 18

where: R = CH$_3$-CH$_2$-CH$_2$-CH$_2$-O— or CH$_3$-CH$_2$-CH$_2$-CH$_2$-O-CH$_2$-CH$_2$-O-[CH$_2$-CH-O]$_n$ (with CH$_3$ branch)

Graft VERs (Scheme 18) consisting of different side chains formed from TDI and butanol (BO-g-VER) or poly(oxypropylene) with different molecular weight ($\overline{M_w}$ = 200 and 390, called 200-g-VER and 390-g-VER) were synthesized and used with polyurethane (formed with TDI) to obtain SINs, in which there were no chemical bonds between the two networks [93]. The morphology of the prepared SINs was studied using DSC and the SEM observation.

It was found that the interpenetration and/or compatibility between the two networks depended on the morphology of graft VERs and the best interpenetration was obtained for 200-g-VER. Moreover, the compatibility of SINs would be improved due to the higher content of urethane groups existing in grafts which should mix well with groups in the PU networks (BO-g-VER). The presence of BO-g-VER and 200-g-VER resulted in an increase in the tensile strength and the elastic modulus of the investigated SINs (Table 23).

Table 23 T_g value (from DSC measurements) and mechanical properties of PU/VER SINs. Reprinted from (2000) Eur Polym J 36:735 [93] with permission

Type of SIN	Composition [wt/wt]	T_{g1} [°C]	T_{g2} [°C]	Tensile strength [MPa]	Elongation at break [%]	Elastic modulus [MPa]
PU/BO-g-VER	100/0	232	–	9.4	489	12
	70/30	249	303	16.2	239	73
	50/50	248	302	20.0	68	262
PU/200-g-VER	70/30	–	350	15.5	261	55
	50/50	–	312	18.5	78	245
PU/390-g-VER	70/30	272	319	8.3	282	46
	50/50	–	303	9.7	97	97

Although the elongation at break of SINs was seriously decreased, the BO-g-VER and 200-g-VER phases reinforced the PU phase.

The inferior results obtained for the PU/390-VER systems were caused by the microphase separation of the 390-g-VER network.

A novel approach to the modification of UPRs with thermoplastic oligomers was presented [94]. Poly(ethylene terephthalate) (PET) oligomer was used for the modification of properties of a commercial UPR. The PET oligomer was extracted from the polycondensation product of bis(hydroxyethyl)terephthalate (BHET). It was then dissolved in hot UPR after fine disintegration to increase the dissolution rate. After the copolymerization of the UP-styrene system, containing dissolved PET oligomer, a semi-interpenetrating network was formed. Addition of the PET oligomer resulted in an enhancement of flexibility, thus presenting an increase in the elongation at break of the hardened UPR.

5
Decrease in Styrene Emission

5.1
With Low-Volatility Monomers

Styrene is well known as a volatile and toxic component of UPRs. Thus, regulations in most countries limit the concentration of styrene in air. Nevertheless, styrene monomer evaporates during the processing of UPRs in open mold. The evaporation of styrene takes place mainly on spraying and impregnating of the reinforcement during the hand-lay-up laminating. Strong ventilation contributes to styrene evaporation to the surrounding. The existing chromatographic methods for determining styrene in the industrial-area air were elaborated by Galina and Potoczek [95]. The new selective and sensitive methods using the surface acoustic wave chemical sensors or based on measuring of absorption of ultrasonic-modulated laser beams were also presented.

There are two methods to decrease styrene concentration in air. First, styrene replacement with less volatile and non-toxic monomers is possible. More than 20 years ago, an attempt was made to replace styrene in UPRs with a system of monomers consisting of glycol dimethacrylates and so-called VeoVa, i.e., vinyl ester of a branched monocarboxylic acid [96]. A second method is presented below.

5.2
With Wax-Like Additives

The systems with styrene evaporation suppressants work well only if the surface of the processed UPR is left undisturbed, allowing the waxy barrier film to be formed and stay intact. As soon as laminators start working on the item, the anti-emission film is broken and becomes ineffective.

In a general approach the styrene evaporation suppressants are dissolved in the UPRs. It is soluble in the resin, i.e., in the solution of unsaturated polyester in styrene and in styrene itself as well. It is, however, insoluble in the polyester. In the course of processing, styrene monomer evaporates from the surface and the suppressant forms a thin film, which exhibits barrier properties in relation to styrene. This mechanism is well known as far as solution of paraffin wax in UPRs is concerned. Nevertheless, paraffin wax cannot be applied as a styrene evaporation suppressant because of its antiadhesive properties and the decrease in interlaminar adhesion in glass fiber reinforced UPRs. Thus, a suitable styrene evaporation suppressant should contain an adhesion promotor with a waxy consistence.

The general properties of a commercial styrene evaporation suppressant are presented in an article [97]. The evaporation efficiency of that suppressant was determined. The laboratory determination of the evaporation is carried out by weighing the UPR before and after the weight loss in an air stream. The effect of temperature and air stream rate on the evaporation was investigated [98].

The interlaminar strength of the additivated UPRs was evaluated using qualitative, semiquantitative and quantitative methods [99]. Both static and dynamic (impact) methods were applied. Following quantitative static measurement methods were used: shear strength, peel strength and split resistance, whereas the determination of flexural strength delivered semiquantitative results. The dynamic quantitative methods comprised shear strength and separation resistance. The semiquantitative dynamic method was based on the impact strength test. Moreover, a peel strength evaluation method has been recommended.

Since commercialization in the mid 1980s by Byk-Chemie of the styrene evaporation suppressant BYK S 740 which was elaborated at the Industrial Chemistry Research Institute in Warsaw, Poland, some new information about low styrene-emission (LSE) UPRs was launched, cf., e.g., refs. [100, 101].

Styrene has been classified as a possible carcinogen. Thus, an operatively simple method was needed to determine the styrene concentration in air which the workers busy with laminating are occupationally exposed to. The elaborated method of analysis consists of a diffusive sampler, which can be analyzed using thermal desorption-gas chromatography, and is applied for styrene absorption from the ambient environment [102].

6
Decrease in Flammability and Smoke Emission

Many ways have been developed to decrease the inflammability of cured UPRs. The most important method consists of incorporating brominated raw materials or adding brominated fire retardants. The use of brominated compounds, in particular polybrominated diphenyl ether, has been criticized for toxicological reasons [103]. As the source of bromine in UPRs, 2,3-dibromo-2-butene-1,4-diol (Scheme 19):

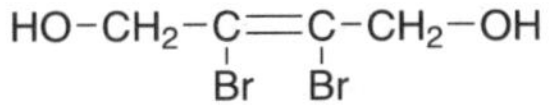

$$HO-CH_2-\underset{Br}{C}=\underset{Br}{C}-CH_2-OH$$

Scheme 19

and dibromoneopentyl glycol diallyl ether (Scheme 20):

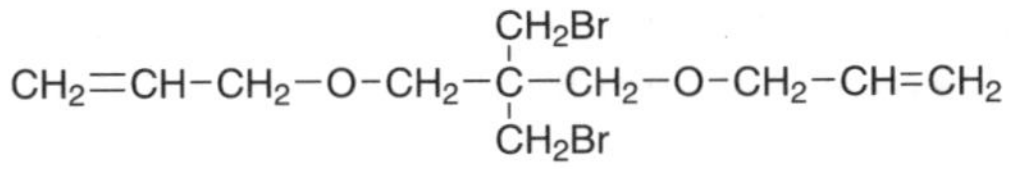

$$CH_2{=}CH-CH_2-O-CH_2-\underset{CH_2Br}{\overset{CH_2Br}{C}}-CH_2-O-CH_2-CH{=}CH_2$$

Scheme 20

have been proposed [104].

Phosphorus compounds, e.g. ammonium polyphosphate or encapsulated red phosphorus particles were also mentioned as efficient fire retardants [105, 106]. As an additive, which combines the fire retardant function of nitrogen and phosphorus and exhibits synergism of nitrogen and phosphorus, melamine diphosphate was described [107]. Bromine compounds are synergistic fire retardants, if used jointly with antimony oxide [108].

Another group of fire retardants presents additives, which evolve water on heating, namely aluminum hydroxide [109] and magnesium hydroxide [110]. A disadvantage of such additives consists of the fact that they need to be used in relatively large amounts, thus increasing the viscosity of UPR and making the laminating difficult. However, they show an important advantage: they sharply decrease the smoke emission on fire. Many years ago, the strongly acidic zinc borate that contains water of crystallization was commercialized as a fire retardant for UPRs [111]. Only recently, "plate-shaped" carbon black particles were shown to be applicable as efficient fire retardants for IPNs composed of styrenated UPRs and an amine cured epoxy resin [112]. In a study of the thermal degradation of carbon black containing IPN, the conventional or the modulated thermogravimetric analysis was applied. TGA apparatus was used. The activation energy of the decomposition process was determined. It was found that the activation energy increased when carbon black was added.

In addition to carbon black, expandable intercalated graphite flake was shown to be an effective intumescent fire retardant additive for UPR [113].

It decreases the flammability of the cured halogen-free UPR when added at the level of 10 phr. Expandable graphite is manufactured using natural crystalline graphite. Because no covalent bonding exists between the layers, other molecules can be inserted between them. In the commercial process, sulphuric acid is inserted into the graphite. Thereafter, the flakes are washed and dried. The intercalant is trapped inside the graphite lattice, so the final product is a dry, pourable, non-toxic material with negligible acidity. When the intercalated graphite is exposed to heat or flame, the inserted molecules decompose to generate gas. The gas forces apart the carbon layers and the graphite expands. The expanded graphite is a low density, non-burnable, thermal insulation. Commercial expandable graphite is named GRAFGUARD 220-80N. The decomposition temperature amounts to 220 °C.

An effect of expandable graphite and the expandable graphite combined with other fire retardants on the flammability of a hardened standard halogen-free unsaturated polyester resin is presented in Table 24.

The expandable graphite was particularly effective, if added in conjunction with ammonium polyphosphate as a synergist. The cast profiles made of rigid UPR with expandable graphite added were observed to eject burning sparks. However, the generation of sparks was eliminated, if expandable graphite (7.5-10.0 phr) was used together with the synergists: ammonium polyphosphate (5-10 phr) or pentabromoethylbenzene (5 phr) and antimony trioxide (2.5-3.0 phr). As far as fire retardation is concerned, the following behavior was observed: self-extinguishing; after-flame time 0 sec; non-burnt length 80 mm.

Among inorganic fire retardants, zinc hydroxystannate $ZnSn(OH)_6$ should be mentioned. $ZnSn(OH)_6$ is a commercial product (Flamtard H) manufactured by Alcan Chemicals Europe. Some other tin compounds: tin(IV) oxide SnO_2 and zinc stannate $ZnSnO_3$ can also be used [114]. Mixtures of zinc hydroxystannate $ZnSn(OH)_6$, zinc stannate $ZnSnO_3$ and zinc borate or zinc molybdate were used in UPRs as fire retardants and smoke suppressants [115]. The heat release rate, the time required to ignite the glass fiber reinforced UPR, the weight loss rate, and the surface extinction area were determined using the cone calorimeter. The same fire retardants and smoke suppressants in combination with $Al(OH)_3$, $Mg(OH)_2$ and Sb_2O_3 were investigated in the chlorine and bromine containing UPRs and reinforced polyesters [116]. Standard testing methods were applied. In the same series of publications, a study of the effect of $ZnSn(OH)_6$ on the kinetics of pyrolysis of a halogenated UPR was presented [117]. It was found that tin compounds were effective fire retardants in halogenated UPR systems [114]. As the halogenated components of the UPR systems with tin fire retardants, HET acid, dibromoneopentyl glycol and tetrabromophthalic anhydride were used. Formation of tin halides from the thermal degradation of Sn/Cl and Sn/Br systems was evidenced [114, 118].

Table 24 Flammability of non-halogenated unsaturated polyester resin with expandable graphite and other fire retardants. Reprinted from (2000) Flame Retardants 2000, Interscience Commun, London 2000, p 105 [113] with permission

Sample No.	Expandable graphite [phr]	Auxiliary fire retardant [phr] I	II	After-flame time [s]	Non-burnt length [mm]	Limiting Oxygen Index (LOI) [%]	Self-extinguishing	Sparkling
I-1	5.0	–	–	365	0	21.6	No	Yes
I-2	7.5	–	–	7	78	–	Yes	Yes
I-3	10.0	–	–	10	80	23.3	Yes	No
I-4	15.0	–	–	0	80	23.3	Yes	No
II-1	5.0	ATH, 15.0	–	394	24	–	No	Yes
II-2	10.0	ATH, 15.0	–	0	80	23.0	Yes	No
II-3	15.0	ATH, 15.0	–	3	79	23.4	Yes	No
III-1	5.0	APP, 5.0	–	13	73	–	No	Yes
III-2	5.0	APP, 10.0	–	10	75	–	Yes	Yes
III-3	10.0	APP, 5.0	–	0	80	–	Yes	No
III-4	5.0	APP, 15.0	–	0	80	–	Yes	No
IV-1	5.0	PB, 3.0	S, 2.0	0	74	–	Yes	No
IV-2	5.0	PB, 10.0	S, 3.0	7	80	–	Yes	No

ATH – Aluminum trihydroxide $Al(OH)_3$
APP – Ammonium polyphosphate
PB – Pentabromoethylbenzene
S – Antimony trioxide Sb_2O_3

The application of tin compounds as fire retardants in UPRs has been criticized due to the need of using halogenated components. Halogen-free systems have been preferred. Nevertheless, tin compounds became the object of investigation as fire retardants and smoke suppressants in UPRs [114, 117, 119]. Among others, the cone calorimeter was applied [120]. Within those investigations, an attempt was made to consider the flammability of cured UPRs on a background of the thermal degradation processes and kinetic calculations [117].

A similar approach was previously applied to the study of the thermal degradation of UPRs based on dihydrodicyclopentadienyl-terminated unsaturated polyesters [121]. The thermal degradation products were identified using coupled pyrolysis gas chromatography/mass spectrometry analysis. It was found that end groups capping with dicyclopentadiene enhanced the thermal stability [122]. Pyrolysis of standard UPRs was studied using thermogravimetry, differential thermal analysis, differential scanning calorimetry, and pyrolysis coupled with gas chromatography [123]. The effect of additives: glass and carbon fiber, silica, carbon black, metal oxide fire retardants, smoking suppressants, and lubricants was investigated.

Smoke emission (optical density of smoke) during combustion of cured UPRs can be decreased by using additives [124]. MoO_3, zinc borate or water-evolving additives [$Al(OH)_3$, $Mg(OH)_2$ and SnO_2] are suitable. The smoke decreasing efficiency of the additives in brominated UPRs depends on the combustion temperature. MoO_3 is efficient at high temperature (700 °C), whereas zinc borate and SnO_2 are efficient at a low temperature of pyrolysis (300–400 °C). The efficiency of $Al(OH)_3$ is high over a broad range of the pyrolysis temperature, it should be used, however, in large amounts.

Thermal degradation of UPRs used as a component of a binder in propellants was investigated [125].

7
Chemical and Alkali Resistance

As cheap and easy to process materials, unsaturated polyester resins are widely used as coatings as well as reinforced composites for the manufacture of for example water pipes, hulls of boats, yachts and large ships, solvent reservoirs, tanks, and other uses. It is well known that UPRs contain ester groups (especially maleate/fumarate and phthalate) in the chain and thus exhibit a relatively high sensitivity to hydrolysis. The hydrolysis can initiate cracking, which leads to macroscopic cracks. Vinyl esters containing methacrylic ester groups exhibit much better hydrolytic stability.

It was found that the flexibilization of chemically resistant UPRs resulted in an increase in the chemical resistance [126]. The comparison between standard alkali resistant UPRs based on propoxylated bisphenol A, which are brittle and tend to cracking, and their flexibilized analogues has confirmed the advantageous effect of flexibilization on the chemical resistance. As the flexibilizing components of the unsaturated polyester from propoxylated bisphenol A and maleic anhydride, dimerized fatty acids, poly(1,2-oxypropy-lene) diol and poly(1,4-oxybutylene) diol were applied. It is assumed that the flexibilization counteracts the formation of microcracks. The microcracks open up a path to the inside of the cured UPR, thus resulting in the chemical degradation of the crosslinked polymer.

Polyester prepolymers based on a binary, ternary and quaternary combination of maleic anhydride, isophthalic acid and various glycols (ethylene, propylene, neopentyl, diethylene) were synthesized and used as model compounds to predict the hydrolysis behavior of the corresponding styrene crosslinked networks [127]. Thus, samples of the prepared materials were immersed in water at 100 °C and the hydrolysis rate was determined by titration of acids at chain ends. The obtained data on structure-stability relationships and the proposed kinetic equations of the network hydrolysis gave a good order of magnitude for the rate constant, there was, however, a systematic overestimation. The rate constants were of second order and were independent of hydrophilicity. It was also found that fumarates are 5–20 times more reactive than phthalates as far as hydrolysis is concerned. No significant differences in reactivity of used glycols except for the more reactive ethylene glycol were observed. The hydrolytic stability of the four unsaturated polyester networks with chain ends modified by isocyanate or dicyclopentadiene were checked for comparison [128]. The results of gravimetric and infrared spectrometry analyses demonstrate similar stability of modified and standard polyester but for a given hydrolysis rate, the modification of chain ends with an isocyanate decreases the rate of weight loss very clearly. Although the use of dicyclopentadiene for the control of chain ends does not reduce the weight loss rate, it helps to decrease the hydrolysis rate.

Another combination of experimental and theoretical methods was used for the estimation of the solubility parameters of a bulk molding compounds based on unsaturated polyester with poly(vinyl acetate) as the low profile additive and glass fibers, calcium carbonate and various additives [129]. The BMC samples were dried in vacuo at 50 °C (constant weight) and were placed in an organic solvent saturated atmosphere for 500 hours in a thermostated chamber at 30 °C. The authors decided not to immerse the samples in solutions of selected chemicals because they assumed that the extraction was limited; the difference between the solvent concentration at the surface of the composite is insignificant and the equilibrium volume is the same. Moreover, the process is slower and could be observed by regular weighing. The Hildebrand solubility parameter calculated from the molar additive laws and determined from unrelaxed elastic constants as well as partial solubility parameters from a sorption test using bidimensional solubility maps showed that inorganic fillers played no significant role, and the composite characteristics are intermediate between polyester and poly(vinyl acetate). These parameters determined with maximum discrepancy between the applied methods of less than 10%, could be used to predict the behavior of the material in the presence of solvents of known characteristics, e.g. for a rational choice among a range of paint solvents.

The effects of solvent exposure on the viscoelastic properties of several vinyl ester resins (phenolic-novolac epoxy, propoxylated bisphenol-A fumarate, urethane and bisphenol-A epoxy based) and various unsaturated polyester resins (terephthalic or isophthalic acid with a standard glycol based) containing 10 wt % glass fiber were studied [130]. The results of dynamic mechanical analysis showed that the influence of exposure time to the solvent as well as the influence of temperature depended on the styrene content and chemical composition of the studied resins, while the amount of cobalt octoate used for the synthesis as the accelerator had no influence on the viscoelastic properties of the prepared materials after solvent exposure. It was also found that not fully cured urethane vinyl ester and the terephthalic acid-based unsaturated polyester resins showed excellent resistance to sulfuric acid exposure. However, interactions between the tested resins and petroleum could possibly occur through intermolecular bonding between the non-polar chains of the cured resins and the solvent.

A highly water sensitive permeameter associated with a new method of determination of the transport properties in polymer films was used for the study of diffusion and permeation properties of liquid water through an isophthalic acid-maleic anhydride-propanediol-based polyester resin [131]. The water migration was discussed on the basis of the change in the infrared spectra and chemical structure of the resin during the curing process. The concentration-dependent parameters related to the water transport in a UPR were determined and would be used for prediction of the steady-state or the transient flux of water through a polyester layer in the real application cases. It was also found that the water sorption by the resin decreased the glass transition temperature of the water saturated material by 20 °C and led to a diffusivity enhancement by the plasticization effect. Moreover, the two methods: the differential permeation and microgravimetry techniques were used for the study of the same penetrant-UPR system and gave valuable information on the behavior of water vapor in the sorption and diffusion in UPRs [132].

The microstructure of *iso*-phthalate polyester films before and after exposure at room temperature to an alkaline solution was studied using tapping mode atomic force microscopy (AFM) [133]. The results of structural characterization as well as chemical analyses using attenuated total reflection FT-IR, total carbon analysis and liquid chromatography-mass spectroscopy showed that the base-catalyzed hydrolysis of polyester was a heterogeneous process. The formation of pits occurs as a result of hydrolysis and the number and size of pits increases with exposure time.

The enzyme-catalyzed hydrolysis of phthalic units containing polyesters could be used as a potential tool for the block length sequence analysis [134]. Thus, copolyesters undergo two kinds of degradation process in the presence of lipase from *Chromobacterium viscosum*: a fast and selective enzymatic hydrolysis of the fumaric ester functions and a slow but non-selective uncatalyzed hydrolysis of fumaric and phthalic ester functions.

8
Thermal Degradation

Thermal degradation of UPRs prepared by the reaction of straight-chain as well as branched-chain glycols with maleic anhydride and either phthalic anhydride, isophthalic acid or terephthalic acid was studied using various methods, i.e. thermogravimetry (TG), differential thermal analysis (DTA), pyrolysis-gas chromatography (Py-GC) and differential scanning calorimetry (DSC) [135]. The effects of composition, the cure conditions and incorporating fillers on the degradation processes of UPRs were analyzed. It was found that degradation of the resins started at about 200 °C and proceeded with major weight loss at about 400 °C. The addition of fillers or glass fiber reinforcement decreases the thermal stability of UPRs and the temperature of char oxidation, these changes are, however, negligible. Similarly, the curing method and regime affect the stability of the resins and the composition of the degradation products only to a small extent. It is also well known from the previous research [136] that the structure of the initiator affects the thermal properties of UPRs cured with MEKP being more thermally stable than those cured with benzoyl peroxide.

TG, DTA, isothermal TG, IR and hot stage polarizing microscopy studies [137] of the thermal degradation of diethylene-glycol-based UPRs synthesized by one-step, two-step and three-step polyesterification processes showed that the mode of thermal degradation of the prepared UPRs was different in air and in nitrogen. The degradation of resins starts with the breaking of ester linkages (decarboxylation process with CO_2 forming) and occurs in a two-step first order reaction in air and in a single-step reaction in nitrogen. The thermostability, both in air and in nitrogen, and the fire behavior of UPRs synthesized from prepolymers: the maleate/*iso*-phthalate/propylene glycol type, the maleate/*iso*-phthalate/propylene glycol/neopentyl glycol type and the maleate/*ortho*-phthalate/propylene glycol/neopentyl glycol/ethylene glycol/diethylene glycol type with the chain end capped with dicyclopentadiene (Scheme 21) were studied [138]. The critical temperature of the thermal degradation was determined and the products released during pyrolysis were identified using a coupled PY/GC/MC technique to study the thermal and thermooxidative mechanisms. Water, carbon monoxide and carbon dioxide were found to be

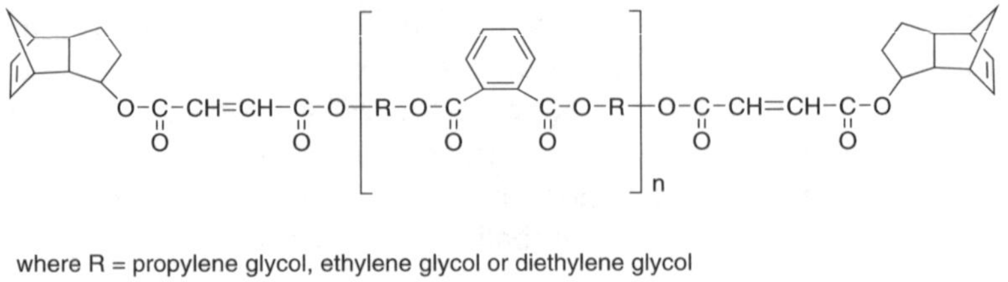

$$\text{[structure: DCPD-O-C(=O)-CH=CH-C(=O)-O-[R-O-C(=O)-C}_6\text{H}_4\text{-C(=O)-O-R-O-C(=O)-CH=CH-C(=O)-O-]}_n\text{-DCPD]}$$

where R = propylene glycol, ethylene glycol or diethylene glycol

Scheme 21

the most important degradation products, although styrene and its derivatives as well as the products based on DCPD structure and on polycyclic molecules were also detected. The ester linkage breaking was proposed to be the major degradation step, and the activation energy was calculated for different reaction rates. Moreover, it was proved that the modification of UPRs with DCPD and the post-curing process increased the oxygen index, i.e. the material behavior in fire was better than that of the DCPD-free UPRs.

Conventional and modulated thermogravimetric analysis (TGA, MTGA) was used for the study of thermal degradation behavior of carbon-black containing IPNs based on unsaturated polyester/epoxy resins [139]. The decomposition process of the IPNs consists of two non-interfering degradation processes of epoxy resin and UPRs and the polyester seemed to decompose first, but the epoxy component degraded at a higher conversion. The addition of a carbon black flame retardant, chosen because of its ability to expand at elevated temperature, increases the activation energy and changes the mechanism of IPN decomposition. Carbon black would inhibit the diffusion of oxygen and heat into the IPNs and their porous structure is able to absorb the volatile gaseous and liquid products of pyrolysis.

The thermal stability of UPRs prepared with styrene as the monomer reflects the extension of phase segregation of the mixture and therefore could be controlled by the styrene content [24]. The thermogravimetric curves of standard UPRs cured with different amounts of styrene show a single thermal degradation process in argon and the temperature corresponding to the maximum degradation rate increases with the styrene content up to 38 wt %. It is well known that crosslinked resins undergo spontaneous decomposition near to 300 °C, even in the absence of oxygen. All vinyl copolymers are degraded into monomeric units at high temperatures.

9
Chemical Analysis and Structural Investigation of Crosslinked Polyesters

Infrared and nuclear magnetic resonance spectroscopy are commonly used methods for the structural investigation of the by-products and the finally crosslinked UPRs. Ultraviolet spectroscopy is applied for the purity control of the starting materials. The new techniques were developed, or adapted to the UPRs analysis: the low-resolution NMR, ^{1}H-^{13}C correlation spectra, high resolution ^{13}C-NMR in the solid state and different polarization techniques with high power dipolar decoupling giving more information about the structure and enabling a detailed study of the morphology of crosslinked UPRs and also of UPRs modified with other polymers, as well as of different types of IPNs with UPRs. The results obtained from the complex spectroscopy studies are very often compared and/or completed by the rheological measurements, thermal (DSC, TG) and dynamic mechanical thermal analysis (DMTA) [140]. Spectroscopic techniques and other different analytical methods (e.g. X-ray diffraction; SEM, TEM and microscopic observations, static and dynamic light scattering) are presented in this chapter.

9.1
Spectrometry

Polymerization kinetics of polyurethane and vinyl ester resin SINs were studied [141] by using FT-IR spectroscopy in a temperature range from 60 up to 110 °C. A decrease in the characteristic absorption peaks at 2274 cm^{-1} corresponding to the isocyanate group, styrene at 910 cm^{-1} and vinyl ester (VER) at 812 cm^{-1} was observed (Fig. 14) as the reaction of PU/VER proceeded.

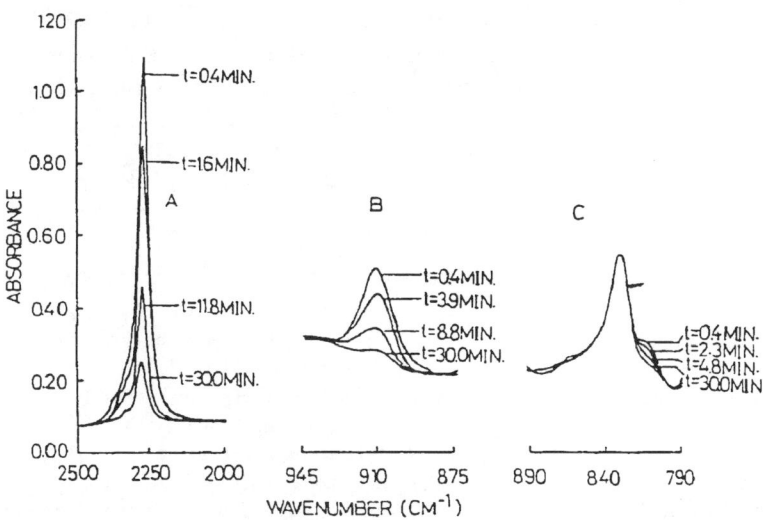

Fig. 14 Decrease in characteristic absorption peaks as the reaction proceeds of PU/VERA (the resin with pendant hydroxyl group capped with acetyl groups) (50/50) SIN at 100 °C: **a** isocyanate group at 2274 cm^{-1}; **b** styrene at 910 cm^{-1}; **c** vinyl ester at 812 cm^{-1}. Reprinted from (1996) J Appl Polym Sci 59:1417 [141] with permission

The C-H peak at 2924 cm^{-1} and the phenyl peak of styrene at 700 cm^{-1} were chosen as the internal standards for PU and styrene and the absorbance area determined by the tangent base-line method was used to calculate the reaction conversion from the change of the normalized absorbance. The pendant hydroxyl groups of vinyl ester resin (VER) were capped with acetyl groups to minimize the possibility of chemical bond formation between both networks and therefore infrared spectroscopy was assisted by ^{1}H-NMR analysis. Irrespective of the kinetic study of PU/VER SINs, FT-IR could be also applied to optimize the operational parameters involved in the production of such materials by RIM technology. The olefinic absorption at 989 cm^{-1} (bond stretching vibration) [142] and 1650 cm^{-1} (bond deformation vibration) with the inner *cis* (^{1}H-NMR: chemical shift 6.24 ppm) and *trans* (^{1}H-NMR: chemical shift 6.82 ppm) stereoisomeric forms of the ester unit could be used for the confirmation of the unsaturated nature of the polyester [143]. The two small NMR signals for outer fumarate (6.78 ppm) and maleate (6.20 ppm) protons are useful for the determination of the molecular weight of UPs. The main absorptions in the IR spectra of polyesters are listed in Table 25.

^{1}H-NMR (200.13 MHz with a sweep width of 3600 Hz) and ^{13}C-NMR (50.33 MHz with a sweep width of 13.5 kHz) spectroscopy were used to study the microstructure of fumarate-based polyesters for use in bioresorbable bone cement composites and the impact of their microstructure on the physical properties as well as hydrolysis of the cements [145]. Poly(propylene fumarate)s (PPF) were synthesized by the transesterification polycondensation of diethyl fumarate (DEF) with different diols: ($\pm$)-1,2-propanediol (PD), (S)-(+)-1,2-propanediol, 2-methyl-1,3-propanediol, and 2,2-dimethyl-1,3-propanediol in the presence of *p*-toluenesulfonic acid monohydrate (PTSA) or metal-containing catalysts: zinc chloride, aluminum trichloride, ti-

Table 25 The main absorptions in the IR spectra of UPs [143, 144]

Absorption [cm^{-1}]	Assignment
3560 (small, broad)	Stretching OH terminal hydroxyl
3450 (small, broad)	Stretching OH terminal carboxyl
3100 (small, broad)	Stretching $=$ C – H olefinic
2870 (strong, sharp)	Stretching *sym* methylene
2800 (medium, sharp)	Stretching *asym* methylene
1700–1740 (strong, sharp)	Stretching C $=$ O ester
1650 (medium, sharp)	Stretching – C $=$ C – olefinic
1580 (small, sharp)	Stretching – C $=$ C – aromatic and olefinic
1480 (medium, sharp)	Bend methylene
1150–1380 (medium, sharp)	Stretching – C – O – C –
1185 (strong, sharp)	Stretching – C – O – (C $=$ O) –
670, 750, 880 (medium, sharp)	*cis*-olefinic and aromatic residues

tanium tetrabutoxide and titanium tetrachloride. Applying high-resolution NMR analysis with the distortionless enhancement by the polarization transfer technique (DEPT, evaluation delay of 3.704 ms) and ^{1}H-^{13}C correlation spectra showed that the extent of formation of branched structures associated with the addition of hydroxyl end groups to the unsaturated bonds in the polyester depended on the acidity of the catalyst. Polycondensation in the presence of a metal-containing catalyst, which can coordinate with the hydroxyl groups of 1,2-propanediol, gives a polymer with a more regular structure and with a higher content of double bonds than the polymerization with PTSA as the catalyst. The different regiospecificity in the case of using PTSA is a consequence of the different modes of addition of 1,2-propanediol. A decreasing number of double bonds results from the side reactions.

The structure of thermotropic block copolyesters containing aliphatic unsaturated units was studied using ^{13}C and ^{1}H NMR with selective homo- and heteronuclear irradiations and off-resonance ^{13}C{^{1}H} experiments [146]. Copolyesters were prepared by interfacial polycondensation between fumaroyl chloride, 2-methyl-1,4-dihydroxybenzene and a variable amount of the α,ω-dichlorocarbonyl oligomer corresponding to the flexible UP block, and the aliphatic UP with unsaturations in the main chain, poly(2-methyl-1,2-propanediyl fumarate) (Scheme 22):

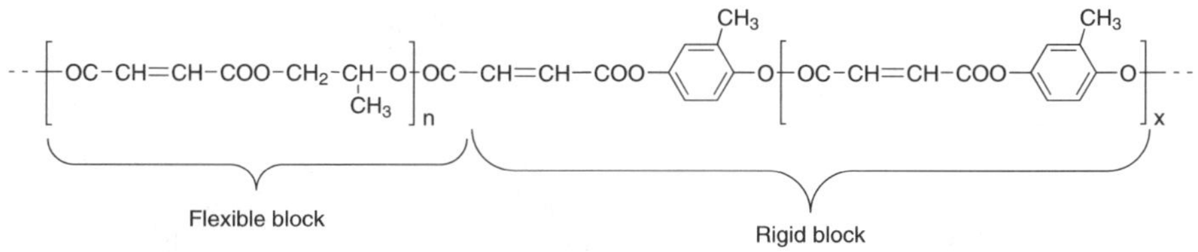

Scheme 22

and with unsaturations in side groups, poly(oct-7-ene-1,2-diyl adipate) (Scheme 23):

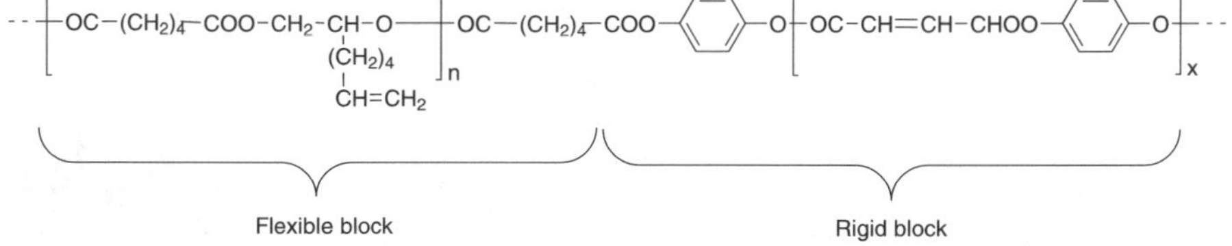

Scheme 23

were obtained. The carboxylic nature of the end groups in α,ω-dicarboxy polyester (precursor of the α,ω-dichlorocarbonyl oligomer) is shown by the absence of CH₂OH and CHOH signals (60–65 ppm for ^{13}C and 3.5–4.0 ppm for ^{1}H), as well as the presence of a COOH resonance in the ^{13}C-NMR spectrum (Fig. 15), additionally confirmed by the end groups chemical titration. ^{1}H-NMR spectroscopy also provides information about the extent of the side reaction of hydroxyl groups on the double bonds, be-

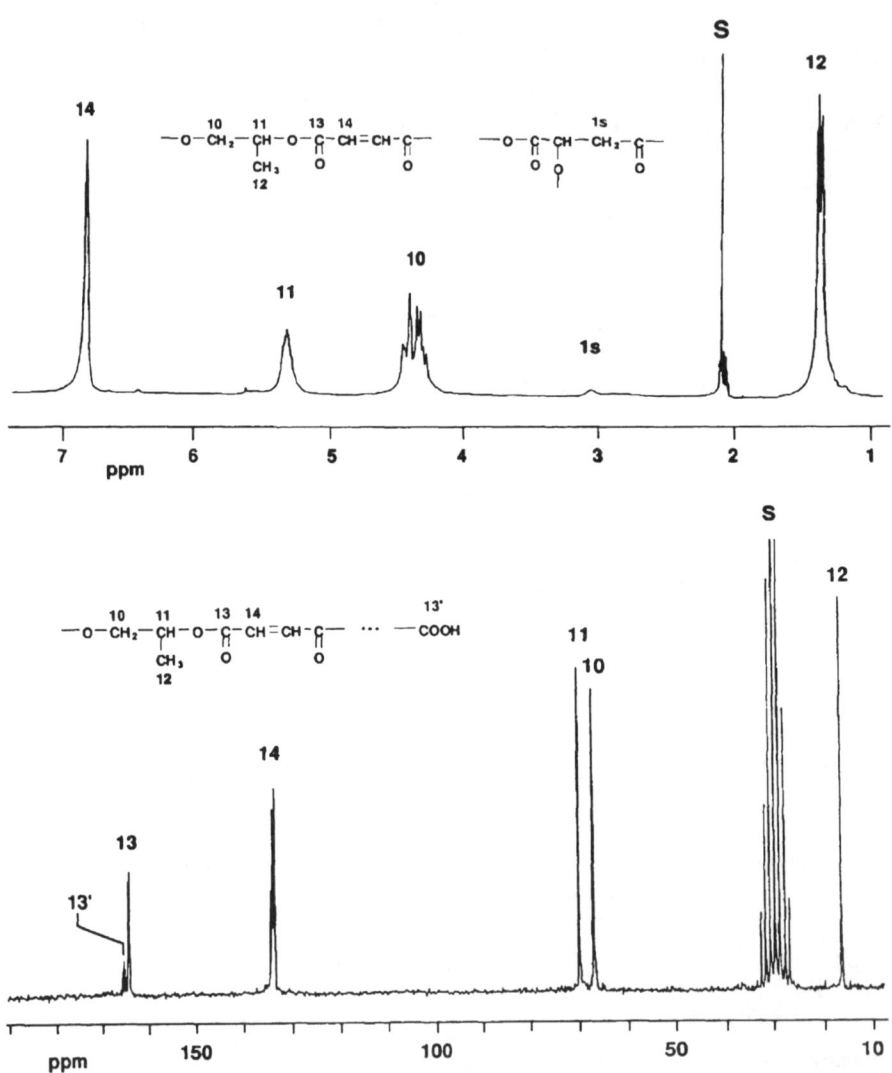

Fig. 15 ^{1}H-NMR (250 MHz) and ^{13}C-NMR (62.9 MHz) spectra (CD₃COCD₃, ref. TMS, S – solvent) of α,ω-dicarboxy polyester. Reprinted from (1995) Eur Polym J 31:733 [146] with permission

low 6% of the total number of double bonds (H^{1S} signal at 3.05 ppm, Fig. 15). The ^{13}C-NMR spectra (e.g., Fig. 16) of the final block copolyesters, recorded from trifluoroacetic/CDCl$_3$ solutions fit with the chain structure of (Schemes 22, 23) and present signals (starred small peaks observed in the 110–155 ppm range) corresponding to 4-hydroxy-3-methylphenyl fumarate and 4-hydroxy-2-methyl fumarate end groups. The glass transition temperature of copolyesters and the melting processes were investigated by DSC and the thermal crosslinking in the nematic state was studied using IR spectroscopy. Anisotropic melts with nematic textures for some branched polymers were observed by polarizing microscopy and the nematic mesophase was confirmed by X-ray diffractometry.

The ^{1}H-NMR (300 MHz) analyses performed by using a CDCl$_3$/1,1,2-trichloro-1,2,2-trifluoroethane (TCTFE) mixture as the solvent were applied for the monitoring of the degree of synthesis and structural investigation of the UPRs modified with perfluoropolyethers [71]. The number-average molecular weight ($\overline{M_n}$) of the used telechelic-fluorinated macromers based on perfluoropolyethers (HO–R$_H$–PFPE–R$_H$–OH) was determined by ^{19}F-NMR.

Some low-molecular-weight model molecules were prepared by the reaction between the diols (1,2-propanediol and telechelic macromers) and maleic and phthalic anhydrides. Model compounds were used to assign the

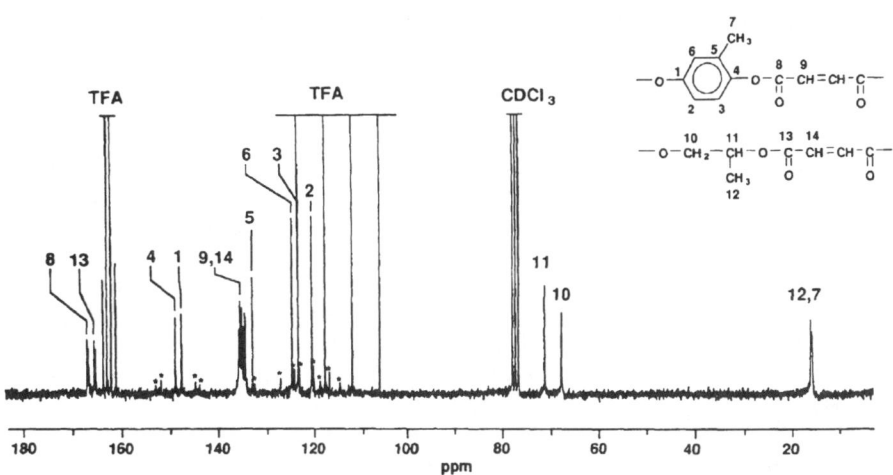

Fig. 16 ^{13}C-NMR (62.9 MHz, CDCl$_3$/TFA, ref. TMS, TFA–trifluoroacetic acid) spectrum of block copolyester (Scheme 22) contains 75 wt % of rigid block. Reprinted from (1995) Eur Polym J 31:733 [146] with permission

peaks in the ^{1}H-NMR spectra of the final resins, as described below:

$- CF_2 - CH_2 - O - (CH_2 - CH_2 - O)_{0.5} - CH_2 - C\underline{H}_2 - O - (C = O) -$
at 4.46 ppm (broad singlet)

$- CF_2 - C\underline{H}_2 - O - (C\underline{H}_2 - C\underline{H}_2 - O)_{0.5} - C\underline{H}_2 - CH_2 - O - (C = O) -$
at 3.5–4.0 ppm (broad multiplet)

$- O - (C = O) - C\underline{H} = C\underline{H} - (C = O) - O -$
at 6.25 ppm (*cis*, broad singlet)
at 6.85 ppm (*trans*, broad singlet)

and in Table 26, where the typical ^{1}H-NMR signals in spectra of polyesters are presented. The maleate (*cis*)-fumarate (*trans*) isomerism in dicarboxy diesters and UPs from maleic anhydride and various diols (propylene glycol, propylene oxide and propoxylated bisphenol A) was studied by ^{1}H (300 MHz) and ^{13}C (75 MHz) nuclear magnetic spectroscopy with tetramethylsilane (TMS) as an internal standard [147]. The *cis-trans* isomerism and the presence of fumarate unsaturation sites play a key role in determining the physical and chemical properties of the crosslinked UPRs, owing to a higher reactivity of the *trans* isomer in copolymerization with vinyl monomers. The assignments of the resonances were confirmed by using DEPT (Distortionless Enhancement by Polarization Transfer) experiments. The spectral resolution (consequently the accuracy of integration) and peak position determination were improved by using the CDRE (Convulsion Difference Resolution Enhancement) method. Thanks to the sensitivity of the carbon atoms in diols to the maleate/fumarate isomerization, it was found that the extent of isomerization at a given temperature appeared to be intimately associated with the chemical structure of the diol units, and was governed by the distance be-

Table 26 The main chemical shifts in the ^{1}H-NMR spectra of UPRs [71, 143, 144, 148]

Resonance signal (δ) [ppm]	Assignment
7.2–7.6 doublets	$- C\underline{H} = C\underline{H} -$ aromatic
6.82–6.85 singlet	$- C\underline{H} = C\underline{H} -$ fumarate
6.24 singlet	$- C\underline{H} = C\underline{H} -$ maleate
4.35 singlet	$- (C = O) - O - CH(CH_3) - C\underline{H}_2 - O - (C = O) -$
4.22 triplet	$- O - C\underline{H}_2 - (CH_2)_4 - C\underline{H}_2 - O -$
3.5–5.5 multiplet	$- O - C\underline{H}_2 - C\underline{H}_2 - O -$ glycol
2.5–3.0 singlet	$- (C = O) - (C\underline{H}_2 - C\underline{H}_2) - (C = O) -$
1.4–1.5 triplet	$C\underline{H}_3 -$ group
1.1–1.4 multiplet	$- (C = O) - O - CH(C\underline{H}_3) - CH_2 - O - (C = O) -$
1.32 multiplet	$- O - CH_2 - (C\underline{H}_2)_4 - CH_2 - O -$

tween the hydroxyl groups, their structure (primary or secondary) as well as the molecular mobility.

The structural changes upon crosslinking of UPRs derived from maleic anhydride and propoxylated bisphenol A were followed by measuring the proton spin-lattice relaxation time, T_1^H and spin-lattice relaxation time in the rotating frame, $T_{1\rho}^H$ vs. crosslinking temperature and styrene content, using high resolution ^{13}C-NMR in the solid state with cross polarization along with high power Dipolar Decoupling (DD) and Magic Angle Spinning (CP/MAS) [149]. It was found that the crosslinking effect was governed by the number of residual unsaturation sites, the length of styrene subchains as well as the concentration of styrene and polyester chain ends. The extent of the crosslinking effect can be determined by subtracting the calculated values of relaxation times related exclusively to the composition (so called copolymer effect) from the experimental values resulting from the crosslinking and combined copolymer effects. The ^{13}C-NMR results of the further study of styrene crosslinked mixed polyesters were in good agreement with T_g values obtained from DSC analyses as well as with some mechanical properties and showed that the extent of the crosslinking effect under fixed curing conditions (i.e., for a given polystyrene segment sequence length and curing temperature) might be associated also with the changes in crosslinking density, controlled by the variations in the UP composition [150]. The ^{13}C-NMR (50.3 MHz) spectroscopy in solid state with CP/MAS was applied for the determination of residual unsaturation in styrene-cured UPRs containing fumarate, *iso*-phthalate and propylene glycol structural units [151]. Problems appear with overlapping peaks in ^{13}C-NMR spectra of solid styrene-cured UPRs, but the used interrupted decoupling pulse sequence reduced the level of interference by suppressing the signals from the CH and CH_2 groups. The signals at recorded spectra were assigned as described below:

129 ppm – CH groups in the phenyl rings of styrene units and the aromatic rings of isophthalic units;

131 ppm – non-protonated aromatic carbon in *iso*-phthalate units;

144 ppm – aromatic rings in structural units derived from styrene;

165 ppm – carboxylic carbon in unreacted fumarate and isophthalic units;

172 ppm – carboxylic carbon in the saturated reaction products derived from fumarate structural units.

It was found that decreasing the styrene content below 47% by weight resulted in residual polyester unsaturation, whose degree was estimated from the relative area of peaks at 172 ppm and 165 ppm. Both ^{13}C (50.28 MHz) and ^{1}H-NMR (200 MHz) spectroscopies (in $CDCl_3$ solution with TMS as a standard) were used jointly for analysis of the copolyesterification in bulk between phthalic anhydride, oleic acid and neopentyl glycol [152]. Application of the high resolution NMR techniques enabled the identification and quantitative determination of the different polymeric sequences and monomeric structures formed during the process of copolyesterification.

Low-resolution pulse ^{1}H-NMR nuclear magnetic resonance (LRP-NMR) was used to study the curing behavior of five UPRs with different molar ratios of styrene to the double bonds in the polyester chain [153]. The method is based on the monitoring and characterizing of the ^{1}H-NMR signal—Free Induction Decay (FID) by the constant T_2 (spin-spin relaxation time) and, because the decay of the FID signal is sensitive to the motion of the molecules, three components could be seen in the crosslinking reaction: styrene (very mobile, $T_2 \approx 3$ s), free polyester molecules (less mobile, $T_2 \approx 150$ ms), and cured resin (immobile, $T_2 \approx 0.04$ ms). At the gel point, the third component could be detected and the proton population of the cured resin started to increase, while the proton mobilities of styrene and free polyester were rapidly decreasing. The crosslinking of UPRs was characterized by the gel time, the curing time and the maximum temperature. The results obtained from the standard (ISO 584-82) method based on changes in electrical conductivity with temperature and low resolution ^{1}H-NMR spectroscopy were in good agreement with each other. They showed that the gel and the curing time decreased when the molar ratio of styrene to double bonds in UP decreased, while the maximum temperature of the crosslinking decreased, respectively. Subsequent investigation, correlated with the rheological measurements of viscoelastic properties, led to a conclusion that the time value taken from the LRP-NMR method was not actually the gel time, but the time at which the resins transform from the microgel formation stage to the transition stage [154].

The XPS method was applied for the investigation of the morphology of crosslinked fluorine-modified UPR (FUPR) systems, prepared from conventional UPRs and poly(ε-caprolactone)-perfluoropolyether-poly(ε-caprolactone) block copolymers (TXCL) [72]. A very strong surface enrichment in fluorinated segments, which increases by increasing the TX/PCL ratio, was demonstrated by XPS analysis.

Apart from the structure investigation of crosslinked polyesters, different methods were used to check the purity of raw materials for making polyesters. Solid-Phase Extraction (SPE) and analytical techniques such as High-Performance Liquid Chromatography (HPLC) coupled with Diode Array Detection (DAD), Gas Chromatography-Mass Spectrometry (GC/MS) and Gas Chromatography-Fourier Transform Infrared Spectroscopy (GC/FT-IR) were jointly applied for the identification of impurities affecting commercial ethylene glycol UV transmittance [155]. The elaborated method allows one to determine the concentration of impurities which were estimated to be less than 2 µg/ml. That method could be used as a practical technique for the identification of impurities in commercial ethylene glycol. Conventional Gel Permeation Chromatography (GPC) using a single column packed by a mixed gel packing with polystyrene calibration and GPC connected with the Multi-angle Laser Light Scattering Detector (GPC-MALLS) provides good molecular characterization of UPRs, as well as even the trace amounts of their high-molecular weight by-products [156]. A slight tendency to overestimate $\overline{M_n}$ and to underestimate $\overline{M_w}$ values of samples containing high-molecular weight fractions was indicated.

9.2
Other Methods

Different methods were developed and applied for the structure analysis and characterization of the polyesters properties. The morphology of the unsaturated polyester and vinyl ester networks was studied using the ArF excimer laser (λ = 193 nm, pulse duration time 16 ns, in air under atmospheric pressure) surface treatment, followed by SEM observations [157]. Samples were analyzed, after laser treatment, by profilometry to determine the ablation threshold. The differences between the thresholds at which ablation of various consistent phases of the materials occurs were used. That method enabled the determination of the two-phase structure of the vinyl ester matrices and the organized structure network of UPs. SEM analysis was also used for investigation of the compatibility of uncured fluorinated UPR systems. The results were in good agreement with the macroscopic results obtained by visual inspection [72].

Structure-property relationships in crosslinked polyester-clay nanocomposites, prepared by dispersing methyl tallow bis-2-hydroxyethyl quaternary ammonium chloride-modified montmorillonite in prepromoted polyester resin and subsequently crosslinking using the methylethylketone peroxide initiator at room temperature and at several clay concentrations, were analyzed by X-ray diffraction combined with Transmission Electron Microscopy (TEM), thermal (TGA) and dynamic mechanical analyses as well as by the determination of mechanical and optical properties [158]. In all cases the formation of a nanocomposite and the morphology of a dispersion of intercalated/exfoliated aggregates of clay sheets in the resin matrix were confirmed (Fig. 17). In the absence of reflection irrespective of clay concentration in the scattering curves for all the polyester-clay nanocomposites was ascertained.

Dynamic mechanical analysis is the most widely used technique for the investigation of mechanical properties and the structure-property relationships in polymeric materials. The dynamic mechanical results expressed as storage modulus (E'), loss modulus (E'') and loss tangent ($\tan \delta$) in the function of temperature demonstrate for example the phase composition, phase transition with glass transition temperature and the structural relaxation processes. The phase segregation in the cured UPRs with an increase in styrene concentration and the dependence of glass transition temperature of UPRs

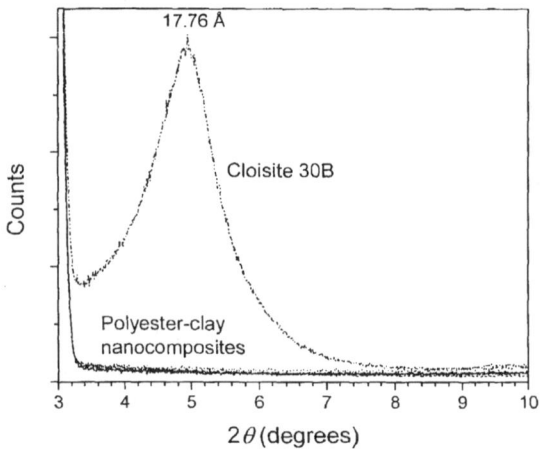

Fig. 17 X-ray scattering profiles for organically modified montmorillonite clay (Closite®30B) and crosslinked polyester nanocomposites containing 1, 2, 5, and 10 wt % clay. Reprinted from (2002) Polymer 43:3699 [158] with permission

on styrene content were studied using dynamic mechanical tests [24]. It was found that the phase segregation was governed by the crosslink density and by the immiscibility of UPR and polystyrene.

Transmission Electron Microscopy (TEM) and Atomic Force Microscopy (AFM) followed by dynamic mechanical parameter tests were used for the investigation of the morphology of unsaturated polyester/diol-polyurethane hybrid polymer networks (UP/PU HPNs). The effect of phase separation on mechanical properties was described [142].

The results of DC-electrical conductivity, confirmed by DSC measurements, were used for the monitoring of radiation and thermally initiated crosslinking of UPRs [159]. Interpretation using conductivity data itself instead of the commonly used logarithmic data form (Fig. 18), as proposed by the authors, showed the influence of the upper liquid-liquid transition on the rate of radiation induced reaction, as well as two reaction rate maxima of thermally initiated crosslinking with the rate increase caused by heat release in the first part of the process.

The relaxation processes of styrene-butadiene rubber (SBR) and acrylonitrile-butadiene rubber (NBR) modified with 5 phr UPR were studied using the ultrasonic technique at 2 and 5 MHz in a range between 180 and 346 K [148]. Measurements of the ultrasonic velocity and attenuation, made at higher frequencies because of the extremely high ultrasonic absorption, showed a main relaxation peak and a secondary one due to the small-scale segmental motion of butadiene. It was also found that the blending of rubber with UPR increased the glass transition temperature and the apparent activation energy of the main relaxation process in SBR. Moreover, it increased the activation energy in NBR, while it has no effect on the T_g of NBR. DSC,

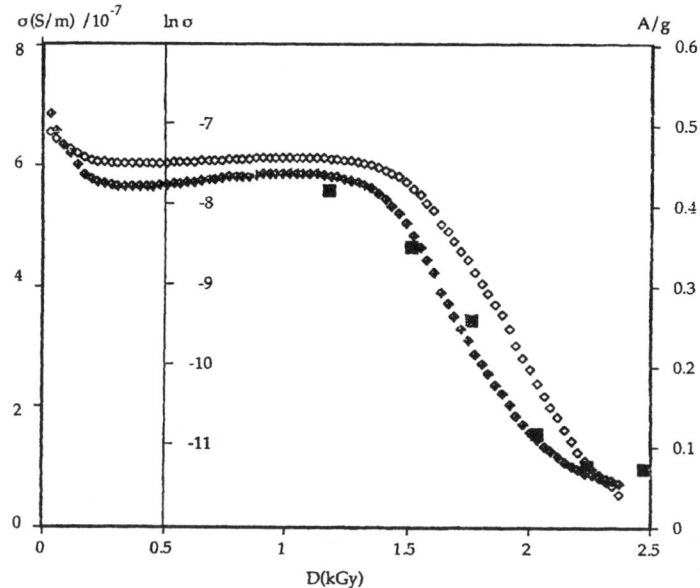

Fig. 18 Change of electrical conductivity of UPR during radiation induced crosslinking at the dose rate B = 0.345 kGy/h at 290 K in non-logarithmic (*solid symbols*) and logarithmic scale (*open symbols*). The logarithm of conductivity is calculated from the data presented in non-logarithmic form. The changes of free styrene content after extraction are also shown (*large solid squares*). Reprinted from (1999) Radiat Phys Chem 54:95 [159] with permission

mechanical and swelling data confirmed the observed improvement of rubber characteristics by UPR addition.

Static (SLS) and dynamic light scattering (DLS) methods were jointly applied to study the microgelation phenomenon in the early stage of curing of a UP/styrene system [160]. The change of shape and anisotropy of the UP molecules in the styrene monomer and the compatibility of UP with styrene before gelation were analyzed by the measurement of molecular weight $\overline{M_w}$, depolarization ratio ρ_v, second virial coefficient A_2 (SLS) and the particle-size distribution (DLS). The formation of microgel due to the intramolecular crosslinking reactions inside the UP coils in the early stage of curing and a strong decrease in compatibility between the partially cured UP and the styrene monomer as the intermolecular crosslinking process among the microgel particles were demonstrated. Moreover, two modes of microgel particle-size distribution during curing were analyzed and discussed.

Vibration damping tests were carried out using polymer concrete made from an isophthalic UPR to prove that the filled polyester resin was an appropriate material for the manufacturing of machine tool beds with respect to damping [161]. Vibration analyses of polyester samples with different ratios of quartz filler and cast iron samples showed that the critical damping ratio of polymer concrete was approximately four to seven times higher than that of the cast metal.

10
Curing of Unsaturated Polyesters and Vinyl Ether

Standard linear unsaturated polyesters contain approximately ten unsaturations per chain and are usually transformed into a crosslinked polymer network by free radical copolymerization with a vinyl monomer in the presence of a reinforcement, fillers and other additives. The crosslinking process is very complex because of possible simultaneous reactions, i.e. styrene-UP copolymerization, styrene homopolymerization and UP homopolymerization, which occur depending on the curing conditions. Styrene is the commonly used monomer, although other commercially important crosslinkers including vinyl esters, dimethacrylates and divinylbenzenes may also be used. Moreover, new initiators, promoters and comonomers are still searched. The curing process, its kinetic and physical changes in a resin system are studied analytically and described with mathematical models.

10.1
Initiators: Peroxides

Novel functionalized peroxides which may be used as UPR curing agents as well as initiators for polymerization reactions and as monomers for polymerizations to form peroxy-containing polymers were elaborated [162]. Initiators may be prepared by reacting hydroxy-containing tertiary hydroperoxides with diacid halides, dichloroformates, phosgene, diisocyanates, acid anhydrides and lactones to form the functionalized peroxides. These reaction products may be further reacted, if desired, with dialcohols, diamines, aminoalcohols, epoxides, epoxy alcohols, epoxy amines, diacid halides, dichloroformates and diisocyanates to form additional functionalized peroxides. The use of monoperoxyoxalates of the structure (Scheme 24) as initiators

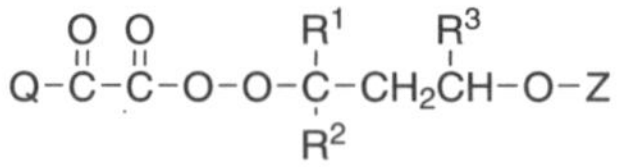

$$Q-\overset{O}{\overset{\|}{C}}-\overset{O}{\overset{\|}{C}}-O-O-\overset{R^1}{\underset{R^2}{C}}-CH_2\overset{R^3}{\underset{}{C}}H-O-Z$$

where: R^1, R^2, R^3 - alkyl groups of 1 to 4 carbons
Q - a group comprising Cl, Br or R-O

Scheme 24

for the curing of UPRs was proposed [163]. Dibenzoyl peroxide was described as an initiator for road paint compositions containing UPRs [164].

Different organic peroxides, i.e. 3,3-bis(*tert*-butylperoxy)ethyl butyrate, 1,1-bis(*tert*-butylperoxy)-3,3,5-trimethylcyclohexane, *tert*-butyl perbenzoate, hexamethyltetraoxacyclo-nonanone, 1,3-bis(*tert*-butylperoxy-*iso*propyl)benzene, dicumyl peroxide, and 2,5-dimethyl-2,5-di(*tert*-butylperoxy)hexane, with a minimum curing temperature of between 140–190 °C were proposed as crosslinking initiators for powder coatings which are applicable to temperature sensitive and metallic substrates [165].

New additives for UPR processing were developed. Peroxide initiators were synthesized by Gawdzik et al. from an aqueous solution of hydrogen peroxide, methylethylketone and ethylene glycol or 1,4-butanediol [166].

10.2
Promoters

Unsaturated polyester resins are usually crosslinked at an elevated temperature or at room temperature in combination with suitable initiating systems. An organic peroxide initiator and a cobalt salt or aromatic amine as accelerators are predominantly used for curing at room temperature. Some typical aromatic amine accelerators, e.g. aniline and its derivatives, are classified as toxic substances, therefore the accelerating effects of other compounds were tested. Different organic peroxides with ferrocene and its derivatives were used to initiate the crosslinking process of UPs [167] and a mechanism of the formation of active radicals was proposed (Fig. 19).

It was found that the promoting activity of the tested ferrocene compounds occurred in the sequence: acetylferrocene > ferrocene > benzoylferrocene > 1,1'-diethylferrocene > 1,1'-dibenzoylferrocene > 1,1'-diacetylferrocene > 1,2-diferrocenylethane > *N,N*-dimethylamino-methylferrocene > chloroferrocene. Ferrocene and its derivatives were able to accelerate the decomposition of benzoyl peroxide, but were inactive to methylethylketone peroxide, butylhydroperoxide, 7-cumylhydroperoxide and *tert*-butylhydroperoxide.

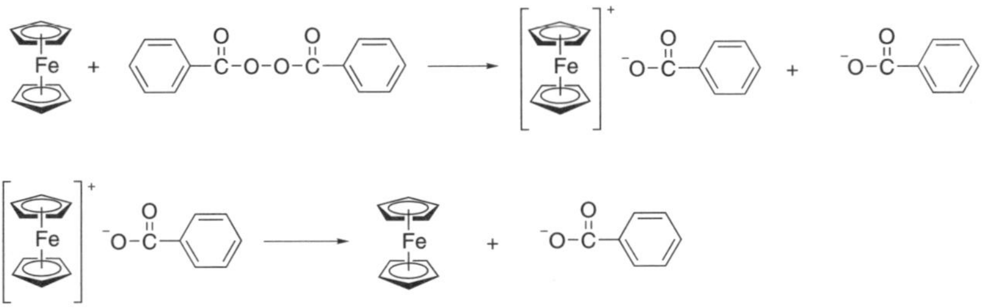

Fig. 19 Decomposition of dibenzoyl peroxide due to ferrocene effects. Reprinted from (1995) Eur Polym J 31:1099 [167] with permission

10.2.1
Cobalt Accelerators

The curing of UPR initiated with MEK peroxide and Co octoate as the promoter was studied by DSC [168]. The kinetic analysis of an asymmetrical DSC peak was performed. It was assumed that the asymmetrical shape of the DSC curves resulted from two independent reactions. When two independent reactions were considered, the fit of the experimental data was better.

2,4-Pentanedione was applied as either a co-promoter in the copolymerization of a UP or a vinyl ester resin with styrene [169]. The activity of 2,4-pentanedione consists of the chelating of metal compounds, e.g. cobalt carboxylates. It shows first a retarding action to achieve a long gel time during mold filling and then increases the curing rate as a result of its co-promoting activity. Thus, it can be applied in low-temperature composite manufacturing processes to control mold filling and curing time.

10.2.2
Aromatic Amines

UPRs are composed not only of UPs and a crosslinking monomer, usually styrene; they contain, moreover, initiators (hardeners), curing promoters (accelerators) and polymerization inhibitors. In the systems with benzoyl peroxide as the initiator, tertiary aromatic amines, e.g. N,N-dimethylaniline or N,N-dimethyl-p-toluidine, are applied as the promoters. Some amino-glycols were built into the UP molecules, thus increasing the reactivity [170]. The best results were achieved when using 3,6-diaza-3,6-diphenyloctane-1,8-diol (Scheme 25):

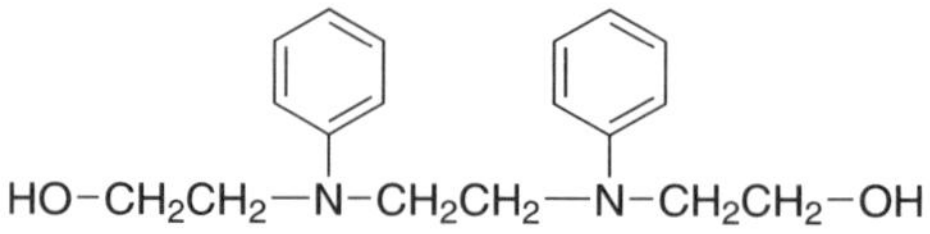

$$HO-CH_2CH_2-N(Ph)-CH_2CH_2-N(Ph)-CH_2CH_2-OH$$

Scheme 25

followed by 4,7-diaza-4,7-diphenyldecane-2,9-diol and 3,10-diaza-3,10-diphenyldodecane-1,12-diol (Scheme 26):

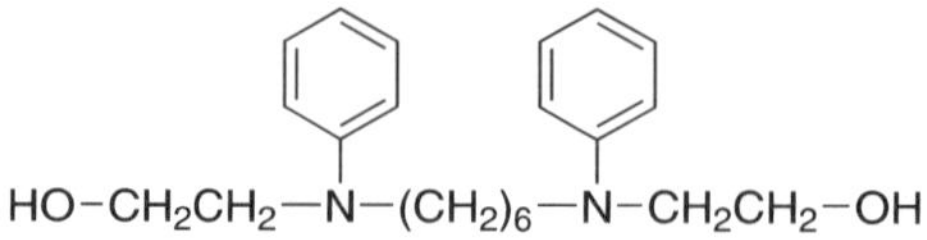

$$HO-CH_2CH_2-N(Ph)-(CH_2)_6-N(Ph)-CH_2CH_2-OH$$

Scheme 26

synthesized by the addition of ethylene oxide or propylene oxide to the corresponding amines. The increase in reactivity consisted of a reduction in gelation time, both in the cobalt- and amine-promoted system. However, the amine incorporated into the UP molecule decreased the storage stability. That property could be improved, if benzyltrimethylammonium chloride was added to the amine-modified UPR. Aminodiols were incorporated into the UP molecules as accelerators for the systems curable with benzoyl peroxide [171]. The aminodiols were synthesized by the reaction of N,N-dimethyl-p-phenylenediamine with ethylene oxide or propylene oxide (Fig. 20). The cold-curing systems consisting of benzoyl peroxide and toxic tertiary amines (N,N-dimethylaniline or N,N-dimethyl-p-toluidine) were replaced by Matynia et al. [172] with similar systems which contain non-volatile tertiary amines as the promoters. The novel promoters were synthesized by the addition reaction of aniline or p-toluidine with epoxy compounds (m-cresyl glycidyl ether, phenyl glycidyl ether, bisphenol A/epichlorohydrin and p,p'-dihydroxydiphenylmethane/epichlorohydrin

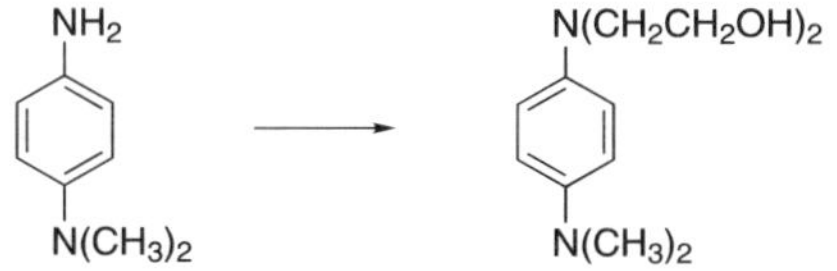

Fig. 20 Synthesis of aminodiols from N, N-dimethyl-p-phenylenediamine and ethylene oxide

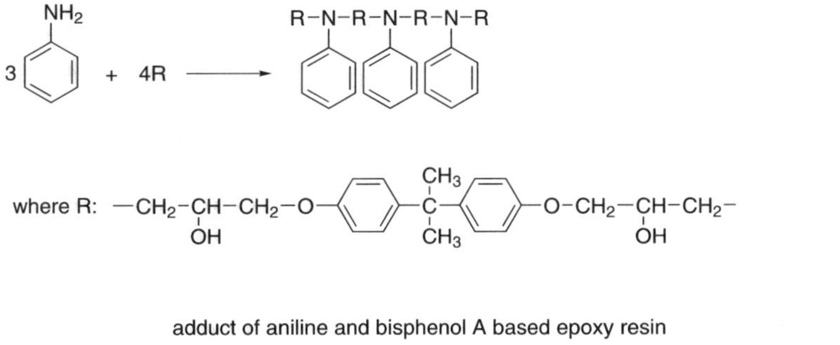

adduct of aniline and phenyl glycidyl ether

adduct of aniline and bisphenol A based epoxy resin

Fig. 21 Chemical structure of the adduct: aniline with phenyl glycidyl ether and aniline with bisphenol A-based epoxy resin. Reprinted from (1997) J Appl Polym Sci 65:1525 [172] with permission

low-molecular weight epoxy resins) (Fig. 21) and were used in combination with dibenzoyl peroxide for crosslinking UPRs.

It was found that the obtained adducts accelerated the cure of UPRs. The efficiency of the tested accelerators was characterized by the gelation time and peak exotherm temperature. The shortest gelation time was obtained for the adduct of *p*-toluidine and bisphenol A or *p,p'*-dihydroxydiphenylmethane based low-molecular weight epoxy resins. Adducts remained in the polyester resin as permanently bonded fragments and could not be extracted with methylene chloride from the cured UPR. Moreover, products extracted from the crosslinked resins were determined by the chromatographic and gravimetric methods.

10.3
Photopolymerization: Maleimides and Vinyl Ethers

Vinyl ethers add to the family of styrene-free unsaturated monomers for UPRs [173]. In a study of UV-curable UPRs, triethylene glycol divinyl ether (Scheme 27), a member of the family of large vinyl ethers, was used.

$$CH_2=CH-O-CH_2CH_2-O-CH_2CH_2-O-CH_2CH_2-O-CH=CH_2$$

Scheme 27

The unsaturated polyester, being applied in the UV-curable systems with the vinyl ether-type monomer, contained maleate ester bonds. It was shown that it (Scheme 27) copolymerizes with the maleate polyester when UV-irradiated in the presence of a commercial hydroxyphenyl ketone as the photoinitiator. Formulation of the maleate polyester was as follows: maleic anhydride (3), phthalic anhydride (2), ethylene glycol (4.5), triethylene glycol (0.5), and trimethylolpropane (0.5 mole). The maleate/vinyl ether systems may find application in very fast UV-curable coatings. The existence of a complex of electron donor-acceptors on UV-curing was ascertained. This confirmed the results of an earlier study on the photopolymerization of vinyl ether-maleate systems in the presence of different radical photoinitiators as well as using a hexafluorophosphate triarylsulfonium salt as a cationic photoinitiator [174]. A donor-acceptor mechanism of photopolymerization of vinyl ether-maleate systems makes the process less sensitive to oxygen inhibition. Moreover, triethylene glycol divinyl ether could be partially replaced by a telechelic vinyl ether polyurethane (Scheme 28):

$$CH_2=CH_2-O-[\text{polyurethane}]-O-CH_2=CH_2$$

Vectomer 2010 (M_w = 3300) from Allied Signal

Scheme 28

to increase the formulation viscosity and to additionally reduce the inhibiting effect of oxygen on photopolymerization.

Clearcoats based on combinations of various maleimides (MI) and vinyl ethers (VE) or VE blends were tested with respect to their UV-photocuring in air [175]. It was found that MI/VE binders and reactive thinners could be cured without additional photoinitiators. Moreover, the equimolar MI/VE blends could be used as photoinitiators for acrylate systems, although their reactivity compared with commercial Norrish type I initiators proved to be poor. The observed limitations, e.g. changes in the properties of cured UPRs, limited reactivity in the air and irritation effects, could be eliminated using aliphatic MI types with further increased reactivity, higher functionality and/or molecular weight and lower irritation.

Allyl ethers were used as built-in monomers in UPRs [176]. High hardness of the coatings cured in air using photoinitiators was achieved. Following polyfunctional allyl ethers were incorporated into polyester molecules: 1-allyloxy-5,6-hexanediol, trimethylolpropane monoallyl ether, pentaerythritol diallyl and triallyl ether, glycerol monoallyl ether, and diallyl ether of propoxylated cis-2-butene-1,4-diol. The allyl ether monomers with hydroxyl groups were reacted with a mixture of fumaric acid, adipic acid and 1,2-propylene glycol, thus forming a maleic unsaturated polyester with terminal and pendant allyl ether groups.

UV-curable UPRs are mainly used in coatings, waterborne ones in particular. For that purpose, unsaturated polyesters were synthesized from ethylene glycol, diethylene glycol and 1,2-propylene glycol combined with dicarboxylic acids or acid anhydrides: fumaric acid, tetrahydrophthalic anhydride, terephthalic acid and trimellitic anhydride [177]; trimethylolpropane diallyl ether was incorporated into the polyester as an air inhibitor of cure. Moreover, triethylamine was added to neutralize the unreacted carboxylic groups, whereas 2-hydroxy-2-methylphenylpropane-1-one was added as the photoinitiator. Properties of the UV-curable systems for wood coatings were investigated. The accelerated weatherability of the coatings was tested. Propenyl ethers of diols and polyols were used as comonomers in the UV photoinitiated radical or cationic copolymerization with UPs [178]. The investigated systems are suitable as UV curable coatings. The propenyl ethers were obtained by isomerization of corresponding allyl ethers, the isomerization being catalyzed with ruthenium complexes.

The matrix of thermosetting polymers is generally very brittle and susceptible to failure. Shrinkage of UPR during the crosslinking process leads to warping and cracking. These problems could be solved by the blending of UPR with thermoplastic additives, but the molecular structure developed upon curing causes an extremely broad glass transition region with a main relaxation process starting just above room temperature and extending for over 100 °C [179]. A different solution consists of the modification of UPRs by another co-reactive component, e.g. bismaleimide resin (BMI) due to

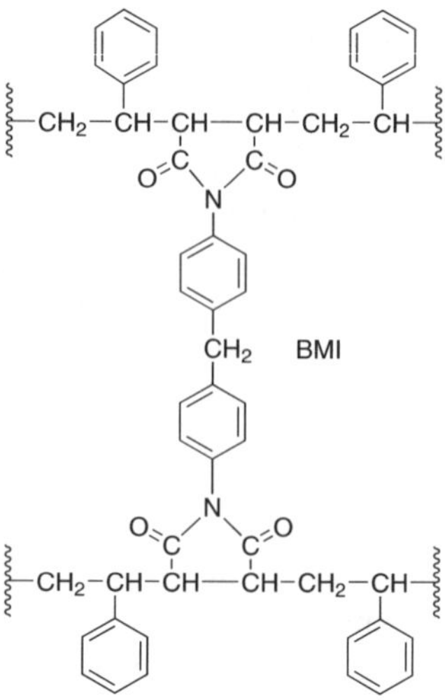

Scheme 29

its very high rigidity, excellent thermal stability, up to 300–350 °C, fatigue resistance even at high humidity and high T_g which would give an increase in the stiffness of the UP matrix. The thermosetting bismaleimide, 1,1′-(methylenedi-4,1-phenylene)bismaleimide was used as a second co-reactive monomer up to the concentration of about 20 wt %. The curing process of the bismaleimide-modified UPRs was investigated using FT-IR spectroscopy and dynamic mechanical thermal analysis [179]. BMI strongly accelerates the curing process of the UP system and enables the crosslinking to proceed even at a lower temperature, not suitable for the polymerization of UPR. Spectroscopic data and the proposed molecular structure of the UPR-BMI network (Scheme 29) proved that BMI reacted preferentially with styrene unsaturations and showed that BMI as the tetrafunctional co-reactive component increased the crosslinking density as well as the overall rigidity of the UPR network.

10.4
Copolymerization Kinetics, Thermochemistry and Rheology

The crosslinking process consists of chemical cure reactions followed by physical changes in the resin during the transformation from a liquid into a gel during the gelation process, although vitrification may take place at high conversion as the T_g increases. The formation of structural inhomogeneities (microgels) as a result of intramolecular crosslinking is one of the reasons that the gelation occurs at higher levels of conversion than that theoretically predicted [180]. The described physical changes as a result of diffusion-controlled termination and propagation could affect the kinetics of the curing process. A new mechanistic kinetic model based on a free radical polymerization mechanism and the free volume concept followed by measurement of the glass transition temperature of partially cured samples using the DSC method and dynamic mechanical analysis (DMA) was presented [180]. The combination of modeling with experimental cure data delivered a good agreement in the results and showed that a greater extent of physical trapping of radicals at a higher isothermal cure temperature occurred rendering them inactive. The effect of temperature and catalyst concentration on the curing kinetics, chemo-rheology and on the dynamic mechanical properties of UPRs were studied using DSC and static and dynamic viscometry [181]. An isothermal time-temperature-transformation (TTT) diagram was proposed and the activation energy was determined from the gel time obtained by viscometry and from the variation in the maximum exotherm temperature obtained by the DSC method. It was found that the amount of 50 wt % methylethylketone peroxide solution used as the catalyst does not significantly affect T_g and the crosslinking density, but an increase in the catalyst concentration could result in a decrease in gelation time. Moreover, an excess of the catalyst could act as a plasticizer, thus modifying the final properties of the cured resin. The curing process of the diacrylated diglycidyl ether of bisphenol A (DGEBA) with different proportions of styrene (4, 20, 40 wt %) was studied by DSC using isothermal and dynamic modes [182]. The kinetic model assumed that the curing reaction was divided into two regiments: below the vitrification regimen and during the vitrification regimen. This was proposed and confirmed by experimental data. The autocatalytic model developed by Kamal was used for the analysis of the reaction below the vitrification regimen. The diffusion control due to the low mobility of the reactive groups and molecules during the vitrification stage was incorporated into the overall rate constant according to the Rabinowitch model. Moreover, the gel influence on the radicals termination rate and the structure constraints on the reactivity of pendant vinyl groups were taken into account. The effects of the initial ratio of styrene to unsaturated $C = C$ bonds per polyester molecule on the curing kinetics of UPRs at medium-temperature reactions (70–90 °C) [183] and high-temperature reactions (100–120 °C) [184] were investigated by DSC and IR spectroscopy. Microgel-based mechanisms were proposed for the curing reactions. It was found that for high-temperature reactions the styrene content and the reaction temperature would affect the formation of the microgel structure and would affect participation of the intramicrogel and intermicrogel crosslinking reactions. For the medium-temperature reactions with the promoter-accelerated initiator systems, a predomination of intramicrogel reactions in the early stage of crosslinking was postulated. The higher molar ratio of styrene to polyester $C = C$ bonds increases the favorable styrene swelling effect and enhances the intramicrogel reactions, but the degree of $C = C$ unsaturations of UPRs has less significant effects on the curing time and the cure characteristics. The formation of a heterogeneous structure through intramolecular reactions and the phase separation process being typical for free-radical crosslinking copolymerization of UPRs with styrene was studied using optical microscopy, time-resolved light scattering goniometry and FT-IR spectroscopy [185]. It was found that phase separation started right after the primary polymers were formed, if the resin system had moved into the two-phase region. The characteristics of the primary polymers agreed well with the description of the microgels and the microgels remained unchanged before phase separation and gelation. During the phase separation or when the reaction was approaching gelation, secondary intramolecular reactions and interparticle reactions of microgels were observed. The phase separation process stopped after the resin reached its gel point, and the domain size of the separated phase remained the same throughout the rest of the reaction period. The morphological changes affect not only the cure behavior and rheological changes of the resin, but also the physical properties of the final products. Secondary intramolecular and interparticle reactions increased the styrene content in the polymer chain and the vinylene conversion.

Low temperature polymerization of UPRs and VERs is applied in important industrial processes such as Resin Transfer Molding (RTM), Vacuum-Assisted Resin Transfer Molding (VARTM), and hand lay-up. Apart from free radical initiators (mainly peroxides) and the promoters which are needed to perform curing at ambient temperatures (usually cobalt octoate, and cobalt naphthenate, when hydroperoxides and ketone peroxides are used, or aromatic tertiary amines, when benzoyl peroxide is used), inhibitors and retarders are also used. Inhibitors, e.g. hydroquinone, 1,4-benzoquinone, *tert*-butyl catechol and oxygen, increase the storage life of the resins. Retarders, e.g. 2,4-pentanedione and 1,4-benzoquinone reduce the curing rate because they slow down the rate of radical generation. Dicyclopentadiene polyester resin and the bisphenol A epoxy resin based VER are the major resins currently used for the Seemann Composites Resin Infusion Molding Process (SCRIMP), a new variety of VARTM [186]. These resins were used to study the effects of resin type, initiator, promoter, inhibitor and retarder on the reaction kinetics and rheological behavior. Investigations using DSC and rheometric dynamic analyzer (RDA) showed that when using MEKP as the initiator, 1,4-benzoquinone provides a longer induction time and a higher final conversion for UPRs. 2,4-Pentanedione seems to be a better inhibitor for VERs, but at the same time it is an accelerator for UPRs. Next, the UPR reaction is slightly more sensitive to the temperature change. Moreover, a model was proposed for the prediction of the effects of resin type, temperature and different curing agents on gel time for UPRs and VERs.

A study of the inhibitors for UPRs was published by Matynia et al. [187]. Efficient inhibiting systems consist of a quinone (benzoquinone, anthraquinone, 1,4-naphthoquinone, or 9,10-phenanthrenquinone) and a diphenol derivative [*tert*-butylhydroquinone or 4-(*tert*-butyl)pyrocatechol].

The history of the radicals was studied during the reaction of a free-radical chain growth polymerization of UP/styrene system, initiated with 2,5-dimethyl-2,5-bis(2-ethylhexanoylperoxy)hexane (DMB) and *tert*-butylperoxy-2-ethyl hexanoate (PDO) using electron spin resonance spectroscopy (ESR), DSC and RDA [188]. The effects of crosslinking and the thermoplastic additive [poly(vinyl acetate) (PVAc)] on the resin conversion and trapping of free-radicals were investigated and the interaction between the radicals behavior, reaction kinetics and rheological changes were discussed. The concentration of radicals increased during the crosslinking and changed the character from liquid-like at an initial quasi-steady-stage region to solid-like at the end of the transition region (gel time). Before any change of the conversion was observed, the radical concentration was about 75% of its final value, and could be increased when the initiator concentration was increased. What is very important, it was found that the radicals in the crosslinked systems were much more stable and their concentration was much higher than that for the linear systems. Next, more trapped radicals were generated at higher temperatures. The changed radicals concentration profiles testified that the addition of PVAc affected the reaction mechanism.

The influence of the number-average molecular weight of a standard UP on the rate of the crosslinking reaction with styrene and on the properties of the resulting materials was investigated [78]. It was found that the reaction heat did not depend on the molecular weight of the UPRs. On the other hand, it was possible only for a higher value of molecular weight to obtain a fully crosslinked material due to the higher compatibility of styrene and UP resins. Also, a slight increase in the cure rate with an increase in molecular weight was observed. This effect could be explained by an increase in the swelling of microgels by the styrene monomer for UPRs with a higher molecular weight value.

Gelation, vitrification and an autoacceleration or gel effect for the free radical chain-growth crosslinking copolymerization of UPs with styrene were studied by combining modulated temperature differential scanning calorimetry (MTDSC), Raman spectrometry (Table 27), and dynamic rheometry [189]. An important autoacceleration, assigned to an accelerating styrene consumption, occurred before vitrification. This effect observed long after the gelation of polyester resins implies that the termination reactions involving small radicals are still important at an advanced conversion. Finally, the acceleration is stopped when the propagation reaction slows down due to vitrification.

The DSC method is commonly used for the study of the curing kinetics of a UPR. The kinetic analysis of the crosslinking process initiated with MEKP was performed by means of an empirical and theoretical model based on the concept of free-radical polymerization [190]. A calculation algorithm based on the downhill simplex method and the Runge–Kutta procedure was proposed. The non-isothermal DSC analyses allowed to obtain the concentration of the initiator and the radicals as a function of cure temperature at

Table 27 Raman peak positions and assignments for the UP prepolymer and for styrene and evolution of peak intensity with cure (v = stretching; St = styrene; UP = unsaturated polyester). Reprinted from (2001) Polymer 42:2959 [189] with permission

Wave number [cm^{-1}]	Tentative assignment	Component	Cure evolution
3100–2975	Unsaturated vC – H (olefinic and aromatic)	St and UP	↓
2975–2850	Saturated vC – H	UP	↑
1729	vC = O	UP	
1659	vC = C trans	UP	↓
1650–1645	vC = C cis	UP	↓
1631	vC = C styrene	St	↓
1601	vC = C aromatic ring coupled with vC = C styrene	St and UP	↓
1581	vC = C aromatic ring	St and UP	Shift

different heating rates. The crosslinking process of the UP system containing MgO as a thickener was investigated using DSC in dynamic and isothermal modes combined with rheological measurements [191]. It was found that at higher temperatures, thermal polymerization may take place and the polymerization reaction was one-order under isothermal conditions. Moreover, the complete cure has not been reached because the vitrification or diffusion was difficult, especially when a high concentration of the thickener was applied. The Avrami theory, based on phase changes, was applied to describe the curing process of a typical UP-styrene system. Various kinetic parameters, in good agreement with the experimental data, were obtained [192]. It was demonstrated that the change of the Avrami exponent revealed different reaction mechanisms occurring at different curing stages. An investigation made in situ, during the curing, the post-curing, and second post-curing periods using DSC and FT-IR spectroscopy showed that the whole reaction enthalpy (of styrene homopolymerization and of copolymerization of the UP/styrene system) depends on the isothermal temperature [193]. It was shown that when the homopolymerization stage was achieved, the copolymerization still went on and about 5% of the styrene remained in the final solid state. Even a second post-curing period did not significantly modify the final conversion rate. The DSC method was also used for monitoring the formation of vinyl ester resin [prepared from bisphenol A epoxy resin and methacrylic acid (bisGMA) in combination with styrene] and polyurethane/simultaneous interpenetrating polymer networks (SIN) [194]. Two kinds of VERs were used: with pendant hydroxyl groups and only having negligible carboxyl group content (hydroxyl groups were capped with acetyl groups). The study of curing kinetics and the SIN structure is complicated because the hydroxyl and carboxyl groups usually contained in commercial VER or UPRs can easily react with isocyanates, giving inter-component chemical bonding between the two networks. It was found that the share of PU formation remained almost the same, whereas the copolymerization of VER was greatly affected by temperature, by the composition of SINs and by the molecular crosslinking nature of both components of SINs.

The study of the crosslinking of dimethacrylate-based VERs using scanning/isothermal DSC and DMTA describes the correlation between the reaction kinetics, as well as the relaxational behavior, cure temperature and peroxide and cobalt catalyst concentration [195]. An increase in the methylethylketone peroxide concentration increased the polymerization rate and reduced the gel time. The use of cobalt octoate as the promoter also reduced the gel time but retarded the reaction rate, except at very low concentrations. This could be explained by the dual role of the cobalt species: as a catalyst of the formation of radicals from MEKP (Scheme 30):

$$R-OOH + Co^{2+} \longrightarrow RO^{\cdot} + OH^{-} + Co^{3+}$$

$$R-OOH + Co^{3+} \longrightarrow ROO^{\cdot} + H^{-} + Co^{2+}$$

Scheme 30

and as a scavenger of the primary and polymeric radicals (Scheme 31):

$$RO^{*} + Co^{2+} \longrightarrow RO^{-} + Co^{3+}$$

$$M_n^{*} + Co^{2+} \longrightarrow M_n^{-} + Co^{3+}$$

$$M_n^{*} + Co^{3+} \longrightarrow M_n^{+} + Co^{2+}$$

Scheme 31

Differences between the values of T_g observed at low, intermediate and higher cure temperature were reported and assigned to the influence of diffusion on the polymerization rate. The T_g conversion relation of Pascault and Williams was applied to describe the dependence between T_g and the degree of cure in the VERs. Although two apparent glass transition regions were observed in a partially cured sample using the DMTA method, it was caused by the re-initiation of the cure of the sample during the DMTA experiment. Kinetics and the network structure of thermally cured VERs based on bisGMA and on epoxy-novolak based multimethacrylate oligomers (Scheme 32) were investigated using gravimetry, FT-IR spectroscopy, NMR spectroscopy, DSC, and DMTA [196].

Also in that research, it was proved that the increase in the initiator concentration raised the polymerization rate, but did not affect the final conversion. On the contrary, the increase in styrene concentration reduced the

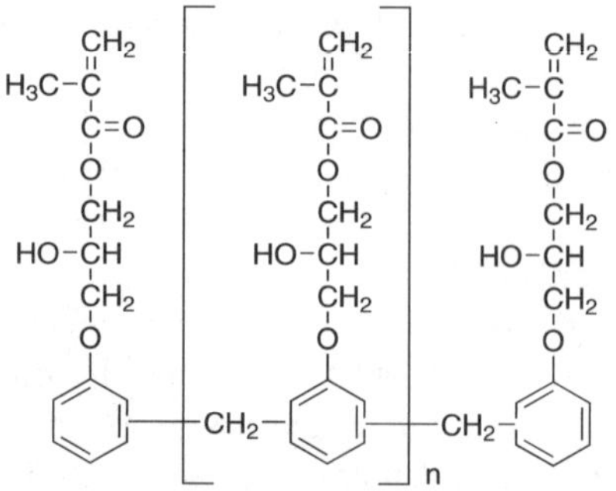

Epoxy-novolak based multimethacrylate oligomer

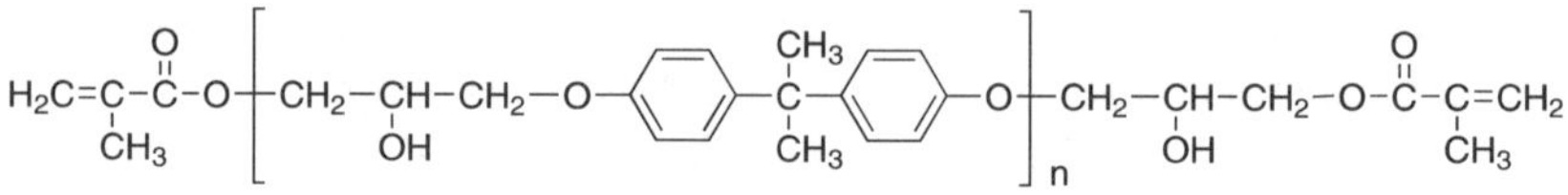

DGEBA based dimethacrylate oligomer

Scheme 32

polymerization rate due to a combination of the relative stability of the styryl radical reducing the rate of propagation, and the lower crosslink density increasing the termination rate constant. Moreover, increased styrene concentration led to the higher final conversion due to the decreased crosslink density which reduced topological constraints.

UPRs based on a mixture of maleic anhydride, phthalic anhydride and adipic acid in a molar ration of 1 : 2 : 2 were prepared in the presence of the polyesterification catalysts, i.e. lead dioxide, p-toluenesulfonic acid monohydrate and zinc acetate dihydrate, and next crosslinked with styrene by using MEKP as the initiator and cobalt naphthenate (CoNp) as the promoter [197]. Most often, the catalysts used in the polyesterification process cannot be easily separated from the polyester, thus the effect of the residual catalysts on the curing process and color of the cured polyester resin should be taken into account. It was shown that the residual catalyst could affect the curing reaction even in a small amount (Table 28), increasing the activation energy E_a, frequency factor k_0 and the reaction order x.

The polyester systems in the presence of residual catalyst behave as if having a heat rather than redox initiation mechanism, and the more organic nature of zinc acetate and p-toluenesulphonic acid produces greater differences. Moreover, the residual catalysts affect the color of the final products.

A Fourier transform infrared spectroscopy study of two different UPRs cured at two different temperatures showed that the kinetic mechanism of the beginning of the crosslinking process is dominated by near-azeotropic

Table 28 Curing exotherm (DSC) and kinetic parameters. Reprinted from (2001) Mikrochim Acta 136:171 [197] with permission

Catalyst	Peak temperature [K]			Heat released ΔH_r [J/g]	Kinetic parameters of curing		
	T_{onset}	T_{peak}	T_{final}		E_a [kJ/mol]	X	k_0 [s^{-1}]
–	309.5	355.0	385.0	233.3	62.9	1.08	1.4E+9
PbO$_2$	310.0	358.0	383.0	210.6	77.7	1.20	5.92E+11
(CH$_3$COO)$_2$Zn	313.0	357.0	384.0	202.9	91.6	1.35	5.56E+13
p-toluenesulfonic acid	312.0	353.0	382.0	193.3	89.5	1.42	7.51E+13

copolymerization, while the conversion of polyester vinylene groups becomes much more favorable than styrene later in the reaction [198]. The conversion of styrene and vinylene groups at the end of isothermal curing depends on the initial molar ratio (styrene content). The styrene dilution effect causes an increase in gelation and vitrification times as well as the time required to complete crosslinking. The conversion of styrene and vinylene groups at the end of isothermal curing depends on the initial molar ratio (styrene content). The styrene dilution effect causes an increase in gelation and vitrification time as well as the time required to complete crosslinking. The study of the relaxation mechanism occurring for UPRs cured with different amounts of styrene proved that the higher content of styrene resulted in an increased fragility of the polyester network. In addition, the styrene content does not affect the average elementary energy barrier that the molecular species (or relaxation units) must cross during the relaxation process. It changes only the number of these species [199].

The vinyl-ester/urethane hybrid (VEUH) resin consisting of the vinyl-ester resin (VER) (produced by the reaction of bisphenol-A type epoxy resin with methacrylic acid) diluted with styrene and novolak type polyisocyanate was synthesized and the reaction kinetics of the curing process of the prepared resin were studied using the DSC method and FT-IR spectroscopy [200]. The study of curing with and without incorporating the polyisocyanate showed that VEUH is crosslinked via free radical polymerization between the vinyl functions of styrene and VER, and the polyaddition reaction between the secondary hydroxyl groups of VER and isocyanate groups of the polyisocyanate, but the curing of VEUH is more complex than the described reactions. The curing process of VEUH is dominated by the radical copolymerization of styrene with VER, although both reactions are not time separated and thus the VEUH curing cannot be simplified for a peroxide-initiated free radical crosslinking. No phase separation during the curing reaction was observed.

The curing process of VER (Scheme 33) using methylethylketone peroxide (MEKP) as the initiator and cobalt hexanoate (CoHx) as the promoter was studied by means of thermal scanning rheometry (TSR) and DMTA under isothermal conditions. Some kinetic parameters, i.e. gel time, apparent activation energy (E_a) for this process were calculated [201]. The value of E_a obtained from the TSR measurements is independent of both the initiator and promoter concentration.

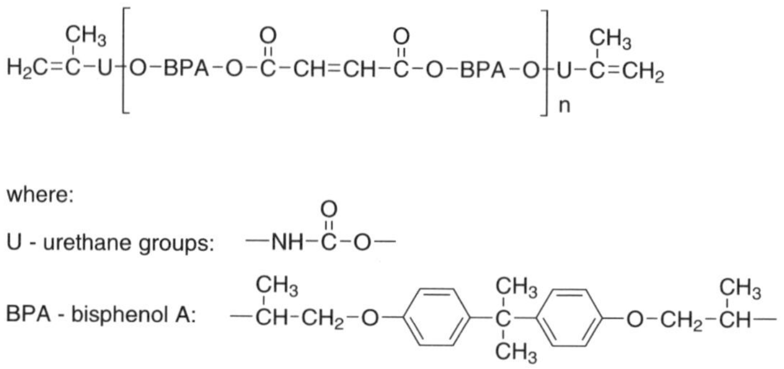

$$H_2C=\overset{CH_3}{\underset{|}{C}}-U\left[O-BPA-O-\overset{O}{\underset{\|}{C}}-CH=CH-\overset{O}{\underset{\|}{C}}-O-BPA-O\right]_n U-\overset{CH_3}{\underset{|}{C}}=CH_2$$

where:

U - urethane groups: $-NH-\overset{O}{\underset{\|}{C}}-O-$

BPA - bisphenol A: $-\overset{CH_3}{\underset{|}{CH}}-CH_2-O--\overset{CH_3}{\underset{\underset{CH_3}{|}}{C}}--O-CH_2-\overset{CH_3}{\underset{|}{CH}}-$

Scheme 33

Determination of the conversion of reaction up to the end of the gelation process:

$$\alpha = \frac{G_t^* - G_0^*}{G_\infty^* - G_0^*}$$

(where: G_t^*, G_0^*, G_∞^* are the values of the modulus at time t, at $t = 0$—uncured resin, and at the end of the gelation process) was proposed. Moreover, the DMTA data were used to obtain the vitrification times at several frequencies as a function of MEKP and CoHx wt %

Microwave heating has been widely applied in the synthesis, crosslinking and processing of polymeric materials [202], in particular in the step-growth polymerization reactions. Recently, the synthesis of unsaturated polyesters by microwave-assisted catalytic copolymerization of maleic anhydride, phthalic anhydride, epichlorohydrin, and ethylene glycol as the molecular weight regulator [molar ratio 1 : 1 : 2(-2.2) : 0.1] in the presence of 0.1 wt % LiCl as a catalyst was described by Bogdał et al. [203]. Optimum reaction temperature and time were 130 °C and 100 min, the radiation power was 90 W. The Gardner-scale color was darker than that of conventionally prepared unsaturated polyester. As expected, the reaction time was shortened thanks to the application of microwave heating. The temperature control was, however, difficult.

11
Low-Shrink and Zero-Shrink Systems

It is well known that UPRs exhibit intense shrinkage on curing at ambient temperature as a result of radical copolymerization of UP with vinyl monomers, styrene in particular. The shrinkage takes place on curing of sheet molding compounds (SMC) and recently during low temperature and low pressure processing like resin transfer molding and vacuum-assisted resin transfer molding. The shrinkage results in an uneven surface of the moldings. In order to improve the surface quality, thermoplastic "low-shrink" or "zero-shrink" additives are used.

11.1
Compositions with Thermoplastic Additives

In a paper by Cao and Lee [85], the use of a commercial carboxylated poly(vinyl acetate) as a low-profile additive for a cold-curing UPR system with a styrene monomer was described. A dual-initiator system consisting of methylethylketone peroxide and *tert*-butylperoxybenzoate with a cobalt promoter and benzoquinone inhibitor was applied. A potential increase in residual styrene content was taken into account in addition to the low-profile characteristics. Morphology of the UPR samples cured in the temperature range of 35–100 °C was studied.

In the following publication by the authors from the Ohio State University [204], the effect of the comonomers (styrene/methyl methacrylate) ratio on volume shrinkage and residual styrene content was investigated. It was found that the methyl methacrylate comonomer exhibits high reactivity and thus decreases residual styrene content. The volume shrinkage decrease mechanism was described. The effect of various thermoplastic polymers being used as low-profile additives in UPR-based sheet molding compounds and bulk molding compounds (BMC), namely thermoplastic polyurethane, poly(vinyl acetate), poly(methyl methacrylate) and polystyrene, on the mechanical properties, glass transition temperature and morphological structure was compared [205]. The molar ratio of styrene to the unsaturated bonds in unsaturated polyester was optimized.

The effects of molecular weight and chemical structure of styrene-based and vinyl acetate-based low-profile additives on the curing kinetics, compatibility, shrinkage, water absorption, surface gloss, and bulk molding compounds (BMC) pigmentability of UPRs was investigated [206]. It was found that poly(methyl methacrylate) affected the properties of wood-flour filled UPRs [207]. Poly(methyl methacrylate) acted as a low-profile additive. The composites thus prepared exhibited enhanced flexural and compressive module, whereas the flexural strength and ultimate strain were decreased. The system consisting of UP, styrene and poly(vinyl acetate) as the low-profile additive was investigated [85]. Volume shrinkage was measured of the resin cured with dual-initiator systems at lower (35 °C) and higher (100 °C) temperature. The effect of curing temperature on phase separation, polymerization kinetics, microvoid formation, and thermal expansion was studied.

Three series of UPRs were prepared from: maleic anhydride, neopentyl glycol and diethylene glycol (with various molar ratios of glycols); maleic anhydride and 1,2-propylene glycol (with and without modification by a saturated dibasic aromatic anhydride or acid, such as phthalic anhydride or isophthalic acid) and maleic anhydride, phthalic anhydride and 1,2-propylene glycol (modified by a second glycol, such as diethylene glycol, 2-methyl-1,3-propanediol or neopentyl glycol) [208]. The prepared UPRs contained thermoplastic polyurethane, poly(vinyl acetate) and poly(methyl methacrylate) as low-profile additives. The effects of glycol ratio, saturated dibasic aromatic acid modification, second glycol modification, $C=C$ unsaturation and molecular weight of the UP on the mechanical properties were investigated by an integrated approach of static phase characteristics-cured sample morphology-reaction conversion-property measurements. It was found that the mechanical properties depend also on the polarity difference between UP and low-profile additives as well as on the molar ratio of styrene consumed by the polyester unsaturated bonds reacted in the major continuous phase of styrene-crosslinked polyester as a result of phase separation phenomena. The better interfacial adhesion between UP and low-profile additives (favorable for impact strength and tensile strength) could be achieved by increasing the UP molecular weight and due to the smaller polarity difference (more compatible the styrene/UP/LPA system).

11.2
Mechanism of Decrease in Shrinkage

Thermoplastic low-profile additives are added to UPRs in order to decrease the polymerization shrinkage [209]. As the additives, saturated polyesters were used. A series of low-profile additives based on poly(vinyl acetate) were examined [210]. Poly(vinyl acetate), copoly(vinyl chloride-vinyl acetate) and copoly(vinyl chloride-vinyl acetate-maleic anhydride) differing in chemical composition and molecular weight were compared. The effects of the additives on the volume shrinkage and pigmentability were investigated. The cured UPR morphology and microvoid formation resulting in decreased shrinkage were studied. Moreover, the dependence of glass transition temperature, mechanical properties and the morphology of cured UPRs with the low-profile additives was explored [211]. In the presence of thermoplastic additives, a two-phase structure is created in UP resins during the crosslinking process. Moreover, a volume expansion occurs in the late stage of cure for resins containing high molecular weight poly(vinyl acetate). Both phenomena compensate a part of the polymerization shrinkage and reduce the final shrinkage substantially [212]. The effect of various thermoplastic additives, their molecular weight and concentration, and different UP resins on the shrinkage control of UPRs cured at a low temperature was studied by dilatometry and scanning electron microscopy. Two UPRs: a UP from maleic anhydride and propylene glycol containing 35 wt % styrene, and a flexible resin synthesized from ethylene glycol, propylene glycol, maleic anhydride and isophthalic acid were used. The thermoplastic additives, i.e. poly(vinyl acetate) with different molecular weight values ($\overline{M_w}$ = 90 000 and 190 000 g/mol, respectively), a linear polyurethane ($\overline{M_w}$ = 30 000 g/mol) and a saturated polyester ($\overline{M_w}$ = 30 000 g/mol) were applied. It was found that the thermoplastic additives are effective for shrinkage control only in the concentration range between two transitions. The concentration range depends on the additive type, as well as its molecular weight. Next, there is a close relationship between the microstructure and the microvoid formation. A co-continuous or low profile additive (LPA)-rich phase dominated structure is essential for thermoplastics to be effective as shrinkage control additives. Further investigation of the microstructure formation [213] using DSC, FT-IR, RDA and optical microscopy showed that, depending on the system miscibility and reaction kinetics, the formation of sample structure follows the same route, but may end at different stages with a different structure. The final structure depends on the volume fraction of the LPA-rich phase and the phase separation period (time period between the onset of phase separation and the gelation). A shrinkage control mechanism at low temperature cure was proposed (Fig. 22).

The partially cured UP/styrene resins with various degrees of conversion (lower than the gelation-related conversion) blended with poly(vinyl acetate) and 2-fluorotoluene were prepared and the effect of poly(vinyl acetate) on the conformation of UP microgel particles was investigated using both static and dynamic light scattering [214]. 2-Fluorotoluene was used because it is isorefractive with poly(vinyl acetate), thus only primary and partially crosslinked UPs were visible in light-scattering experiments. It was found that mixing poly(vinyl acetate) into UP/styrene resins caused an increase in the compactness of polyester coils and favored the intramolecular UP/styrene cyclization in the early stage of curing. Moreover, replacing low-molecular-weight 2-fluorotoluene with high-molecular-weight poly(vinyl acetate) in UP/2-fluorotoluene solutions caused an increase in the excluded volume of polymers. It led to a reduction of polyester microgel particle sizes and compatibility between UP and poly(vinyl acetate)/2-fluorotoluene solutions and in consequence to the decrease in anisosymmetry and differential index of refraction of UP in the solution.

The effect of the thermoplastic low-profile additives on the properties of additivated UPRs was investigated [215]. Poly(methyl methacrylate), copolymers of methyl methacrylate with butyl acrylate and terpolymers of methyl

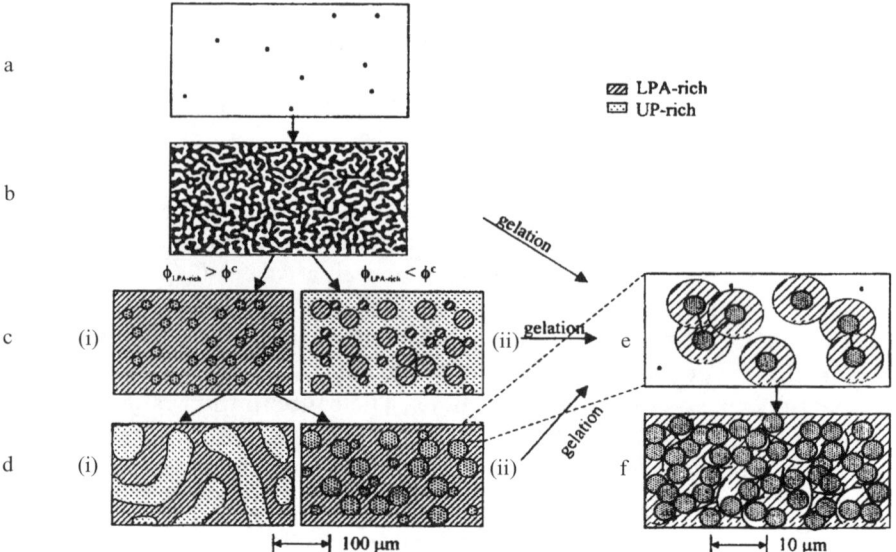

Fig. 22 Scheme of the shrinkage control mechanism of the low temperature cure: **a** induction stage; **b** spinodal decomposition; **c** coarsening to (**i**) the low profile additives (LPA)-rich phase dominated structure, or (**ii**) the UP-rich phase dominated structure; **d** coalescence and growth, (**i**) coarse co-continuous structure, or (**ii**) LPA-rich phase dominated structure; **e** gelation; and **f** microvoid formation. Reprinted from (2000) Polymer 41:697 [213] with permission

methacrylate with butyl methacrylate and maleic anhydride were used. The effect on the morphology, microvoid formation and phase separation was particularly considered. The volume shrinkage and internal pigmentability are the important factors. Curing of the investigated systems was taken into account. The SEM, DSC and DMA methods of investigation were applied [216].

12
UPR-Based Nanocomposites

Nanoparticles added to thermoplastic polymers improve the mechanical properties, increase T_g and enhance fire retardancy. Nanoparticles in UPRs bring similar effects [217]. Montmorillonite (MMT) was used to obtain polymer nanocomposites [218]. That general approach was applied to the UPR-layered MMT nanocomposites. The structure of the nanocomposites was investigated by XRD and TEM. Organophilic MMT was applied: dodecylammonium-bromide treated MMT was used. A UPR synthesized from PET glycolyzate was, together with styrene, the main component of the nanocomposites [219]. Organo-layered silicate structures ranging from intercalated to exfoliated or delaminated were obtained.

As the source of nanoparticles, organically modified montmorillonite clay was incorporated into UPRs and into diethylenetriamine crosslinked DGEBA. UPRs were used as the main component of nanocomposites which contain, moreover, sodium montmorillonite or an organically modified montmorillonite as the reinforcing agent [220]. T_g values of the composites increased from 72 °C for the neat UPR up to 86 °C for the composite with 10% organically modified montmorillonite which exhibited an increased degree of intercalation (exfoliation). The incorporation of modified montmorillonite resulted in an improvement of mechanical properties: an increase in tensile modulus, tensile strength, flexural modulus, flexural strength, and impact strength was found. Nanometer-sized Al_2O_3 particles increased considerably the fracture toughness, if the UPR matrix/alumina particles interface strength was enhanced by means of an organofunctional silane [221]. The ultrasonification method was applied to TiO_2 (titania) in order to enhance the nanoparticle dispersion in the UPR [222] which was verified by transmission electron microscopy (TEM) and scanning electron microscopy (SEM). An increase in dynamic fracture toughness was observed. The importance of the interface strength between the UPR matrix and inorganic polysilicic acid nanoparticles was ascertained using the nanoparticles modified by the reaction of silanol groups on the nanoparticles surface with phenyltrimethoxysilane, 2-(p-styryl)ethyltrimethoxysilane and 3-(trimethoxysilyl)propyl methacrylate [223]. The homogeneous product cured at 150 °C was a transparent inorganic/organic hybrid material, which exhibited reduced shrinkage on curing and enhanced dynamic mechanical properties, due to increased interfacial force between the inorganic and organic components.

In the UPR-based nanocomposites, the styrene monomer was replaced with the polar hydroxypropyl acrylate [224]. X-ray and mechanical test data indicated that mixing of the components for an extended period of time is essential to improve the physical properties of nanocomposites in the UP/Cloisite 6A system. This phenomenon was attributed to the high polarity of hydroxypropyl acrylate which may disturb the preintercalation of UPR into the galleries of montmorillonite. The UPRs with a styrene monomer were used in dispersions with polymerizable quaternary ammonium-modified montmorillonite [225]. An increase in tensile and impact strength and heat resistance (Table 29) was found when 2–5 wt % organophilic montmorillonite was added.

The effect of organophilic MMT is particularly advantageous, when the organic substituent is copolymerizable.

A different method of incorporating organophilic MMT into the molecule of UP consists of when a MMT functionalized with hydroxyl groups copolymerizes ionically with maleic anhydride, phthalic anhydride, and

Table 29 Effect of modified montmorillonite on the properties of UPR nanocomposites [225]

Modified montmorillonite (MMT), 4 wt. %	Tensile strength [MPa]	Impact strength [kJ/m^2]	Hardness [Barcol]	HDT [°C]
None	44.1	6.32	12.4	86.2
Sodium MMT	47.8	4.35	13.3	89.8
Cetyltrimethylammonium bromide MMT	56.4	8.44	13.1	103.1
Methacryloyloxyethyl-benzyl-dimethylammonium chloride MMT	71.2	9.63	13.8	110.4

epichlorohydrin. The UP thus obtained is then dissolved in styrene and crosslinked by radical copolymerization [226]. The dispersion of MMT in the UPR matrix in the form of exfoliated nanocomposites was confirmed by X-ray diffraction study. In addition to incorporation of OH groups by using methyl-di(hydroxyethyl)-alkyl (C_{14}-C_{18})-ammonium substituted MMT (Cloisite 30B), carboxylic groups were introduced by means of MMT substituted with $H_3N^+(CH_2)_{12}COOH$. The COOH groups participated in the reaction with the acid anhydride and the monoepoxide (i.e. epichlorohydrin). To compare the effect of functional groups in MMT with that of MMT without functional groups, dimethyl-dialkyl (C_{14}-C_{18})-ammonium substituted MMT (Cloisite 15A) was used. Built-in MMT increased the hardness and HDT value. It acts as a fire retardant (Table 30).

Smectic bentonite was added to UPRs of prolonged durability to provide them with thixotropic properties [227]. The quaternary ammonium salts with the structure $(R_1R_2R_3R_4)N^+X^-$ were used for bentonite modification or added directly to UPRs in order to enhance resistance to sedimentation and compatibility of resins and smectites. Different ammonium salts were tested with the model system (styrene + maleic anhydride), commercial UP resin and commercial pre-accelerated resin in relation to the structure of the substituent at the nitrogen atom. It was found that the applied ammonium salts enhanced the durability of the resin with the reactivity only slightly affected. The best result was achieved using the salts with one aromatic substituent and the other an aliphatic one.

Table 30 Properties of UPR/MMT nanocomposites [226]

Montmorillonite (MMT) (weight content)	UP		UPR (before curing)		Cured UPR			
	Acid number	Softening temp. [°C]	Styrene content [%]	Viscosity at 22 °C [mPa s]	Hardness [N/mm^2]	HDT [°C]	LOI [%]
---	---	---	---	---	---	---	---
None	40	78	35	919	116.0	63.5	24.1
Sodium MMT (2%)	18	78	36	1107	119.6	60.3	22.7
Cloisite 15A (2%)	15	83	35	1181	123.5	64.7	22.3
Cloisite 30B (2%)	18	87	36	993	127.6	65.6	27.4
Cloisite 30B (3%)	24	90	33	2583	139.7	67.6	28.7
H$_3$N$^+$(CH$_2$)$_{12}$COOH/MMT (2%)	63	88	34	6715	134.7	68.7	21.0
H$_3$N$^+$(CH$_2$)$_{12}$COOH/MMT (3%)			42	1697			

13
Major Application Processes

Manufacture of reinforced plastics is the main application of UPRs. The hand-lay-up method is the most important manufacturing method. Glass fiber fabric, mat or roving are used as the reinforcement. Apart from the traditional manual processing, many mechanized processes are applied. The processing of DMC (Dough Molding Compounds), SMC (Sheet Molding Compounds) and BMC (Bulk Molding Compounds) should be mentioned. The use of SMC needs UPRs, which are thickened with additives, like magnesium oxide or isocyanates. To improve the quality of the surface of molded pieces, the polymerization shrinkage should be diminished. To this effect, thermoplastic additives are introduced into the UPR: polystyrene, poly(vinyl acetate), polyesters, polyurethanes, a. o. Such additives are called LPAs (Low Profile Additives), LSAs (Low Shrink Additives) and ZSAs (Zero Shrink Additives).

In addition to the glass fiber reinforcement, natural fibers are used. The natural fibers involve different cellulosic materials, like jute, sizal, bast, flax, hemp, ligno-cellulosics, cotton and many other classes of natural fibers of plant origin. Natural fibers belong to the renewable materials. Actual information on this area is available from the new Journal of Natural Fibers edited by the Institute of Natural Fibers, 60-630 Poznań, Poland. As an example, the following recently published articles should be mentioned: "Study on Flax Genetic Transformation", "Alkali Extracted Stem-Exploded Wood Fiber as Reinforcing Material for Polypropylene-Based Composites", "Evaluation of Some Flax Genotypes Straw Yield", "A Review of Natural Fiber Based Composites", "Low Dielectric Constant Material from Hollow Fibers and Plant Oil".

UPRs are also used as filled materials. The fillers may contain a different amount of fillers. The content of a filler can be as high as 95 wt %. They are usually inorganic products, like sand or chalk, applied in construction. As examples, outlet pipes in sewage systems and cast seamless flooring can be mentioned. The pipes in sewage systems can replace chemical stoneware. Chalk-filled UPRs are used in the making of glass fiber reinforced sanitary fillings. Large amounts of the filled UPRs are produced as artificial marble with inorganic fillers.

Cast UPRs without fillers are used in fancy goods and for making buttons. For these applications, transparent colorless UPRs are produced. Such UPRs should be resistant to yellowing and discoloration.

UPRs are applied as a coating for wood, mainly in furniture manufacturing. For that application, two-component liquid resin systems are formulated. One component contains the peroxide initiator, whereas another one is additivated with a cobalt promoter. For the coating of particle board and other substrates, fast-curable UPR systems with UV photoinitiators are used. Recently introduced UV curable UPR-based powder coatings for heat sensitive substrates are described in a separate sub-chapter.

13.1
Composites with Glass Fibers and Natural Fibers

Glass fiber reinforced composites based on epoxy-acrylate modified UPRs were studied [228]. The authors showed that UPRs, endcapped with acrylate groups and diluted with reactive multifunctional acrylic and allylic monomers in the presence of a photoinitiator, can be photocrosslinked with UV radiation as glass fiber laminates in a rapid process. It was found that the physical properties of the photo-crosslinked laminates are well correlated with the molecular weight of the polyester, the amount of multifunctional monomer added, and the glass fiber content. A greater improvement of the tensile and flexural properties of the photocured products was observed for multifunctional acrylate or acrylether monomers added to the UPR (Table 31) than for allylic monomers.

The best mechanical properties and water resistance were obtained for the laminates containing 20–50 wt % multifunctional acryl ether monomer, 40–55 wt % glass fiber, and 1 wt % photoinitiator.

The effect of weave structure on the mode-I interlaminar fracture energy (G_{IC}) of E-glass/rubber-modified VER laminates was examined [229]. Different types of E-glass fibers were used, i.e. plain weave, twill weave, quadran 4-harness satin weave, and 8-hatness satin weave with weave index 2, 3, 4, and 8, respectively, to determine the influence of weave pattern on the delamination resistance of toughened VER laminates.

A new rapid prototyping technique, Laminated Object Manufacturing (LOM), was described [230]. It was shown that, when properly treated, LOM molds can be used in conjunction with the RTM process to produce prototype composite parts from a VER toughened with unidirectional E-glass fiber

Table 31 Effect of multifunctional monomer on mechanical strength of the photocured laminates. Reprinted from (1994) J Appl Polym Sci 51:1129 [228] with permission

Mechanical strength	Monomer	Monomer content [%]			
		10	20	30	40
Tensile strength [MPa]	Trimethylolpropane triacrylate	183	188	187	184
	Trimethylolpropane triacrylether	195	197	192	187
	Pentaerythritol tetraacrylether	191	195	194	190
Tensile modulus [10^2 MPa]	Trimethylolpropane triacrylate	89	91	90	87
	Trimethylolpropane triacrylether	97	95	94	93
	Pentaerythritol tetraacrylether	96	97	98	9

mat. The proposed method of mold fabrication reduced the man-hours and the time associated with conventional techniques.

The mechanical properties (i.e. tensile, compressive, flexural strength and modulus, impact strength, inplane shear strength and modulus, interlaminar shear strength, and Barcol hardness) of untreated woven jute fabric-reinforced polyester composites were determined [231]. Although jute-reinforced polyester composites exhibited inferior mechanical properties in comparison with conventional composites, they were characterized by higher strength than wood composites and some plastics. It was concluded that a woven jute, as an eco-friendly, non-toxic, non-health hazardous, low-cost, renewable and easily available material, could be a good substitute for wood in both indoor and outdoor applications.

13.2
Composites Reinforced with Fillers

UPRs are predominantly reinforced with glass fibers or with natural fibers. The use of powder fillers can also improve the properties of cured UPRs. A study on the filling of UPRs with fly ash was performed [232]. Fly ash is a side product of the burning of pulverized coal in power plants. It consists mainly of silicon oxide and iron oxide. It contains, moreover, aluminum oxide and unburnt carbon. The fly ash having been used exhibited a particle size below 76 μm. An isophthalic UPR was filled with 10% fly ash and cast into molds. The mechanical properties were tested in comparison with $CaCO_3$ filled UPR. Tensile and flexural strength were determined. Filling with fly ash decreases the strength, if used in an amount above 10%. The castings were found to have poor acid and solvent resistance. However, the alkali, brine, weathering, and freeze-thaw resistance was satisfactory. It is predicted that the strength may further be increased, if the adhesion between fly ash and UPR matrix is improved by applying a coupling agent. An *ortho*-phthalic UPR reinforced with fly ash/mica hybrid was prepared [233]. The effect of loading composition, the nature of filler surfaces and the surface treatment of fillers were examined in both the uncured and cured state. The tensile strength of cured composites increased with the addition of mica in the total filler. The optimum filler composition was found to be 25 phr of mica and 15 phr of fly ash.

Sand/clay UPR composites containing various amounts of styrene and sand/clay, with different particle sizes of clay were prepared [234]. The effect of a radiation-initiated (γ-irradiation at 50 kGy) polyester curing on the physicomechanical properties of the prepared composites was investigated. It was found that the compressive strength decreased with an increase in the sand/clay and styrene contents, as well as the size of clay particles, whereas the apparent porosity and water absorption of the composites increased. New bands appeared in the FT-IR spectra indicating a chemical reaction between UP and clay filler. The morphology of composites was studied using SEM.

It has been shown that recycled FR-4 epoxy laminates are advantageous fillers for UPRs [58]. Moreover, calcium carbonate and copper were used as additional fillers. The fillers affected the curing rate, decreased the flexural strength and increased the flexural modulus. The shredded laminates were used in the amount of 100 phr.

The creep of UPRs and the UPR composites filled with marble powder and powdery PVC was studied [235]. The authors proposed a jump-like character for the creep rate of materials. It was also assumed that the creep on the micron level reflected the structural inhomogeneity of UPRs and their composites. An optimized cure cycle with reduced process-induced residual stresses and the optimum temperature profile for the manufacturing of UPR composites were searched using the numerical simulation [236].

13.3
Molding Compounds

A low-profile bulk molding compound (BMC) consists of an unsaturated polyester, styrene, poly(vinyl acetate) as the low-profile additive, calcium carbonate, short-cut glass fibers, and various additives, which are contained in minor amounts. Solubility parameters of both key organic components: the polyester and poly(vinyl acetate) were determined [129]. The Hansen and the Hildebrand parameters were calculated. They may be used to predict the behavior of those materials in the presence of solvent-containing systems. It was found that the low-profile additive significantly modifies the solubility parameter values. The relationship between morphology and paint solvents interactions of a BMC was studied [237]. The existence of a poly(vinyl acetate)-filler free polyester skin of about 0.1 μm thickness and the existence of heterogeneously distributed porosities were also discussed with special reference to a protective effect towards solvent diffusion.

A reversible thickening method of UPRs was developed. The method consisted of the use of thermally breakable functional groups derived from diketogulonic acid and a salt-forming diamine [238]. Diketogulonic acid itself was prepared by hydratation of dehydroascorbic acid, being the product of oxidation of L-ascorbic acid with H_2O_2 (Fig. 23).

Then the standard UP with terminal COOH groups was reacted with a diamine [$A(NH_2)_2$: 1,6-hexanediamine or p-xylilenediamine] followed by the reaction with DKGA (Fig. 24) and the urethanization of terminal CH_2OH groups of DKGA with phenylisocyanate and a diisocyanate (MDI). The thickening of the thus modified UP which contained the terminally breakable diketo groups and amine salt groups was compared with the conventional one being carried out with MgO in the SMC and BMC manufacturing technology.

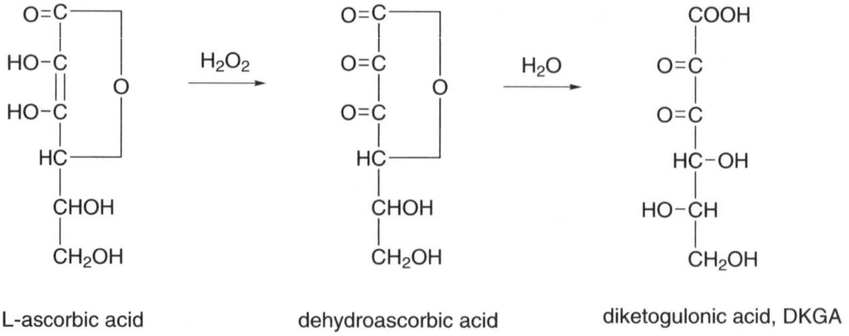

Fig. 23 Synthesis of diketogulonic acid. Reprinted from (1996) Polymer 37:2179 [238] with permission

Polyester-based SMCs, their basic composition (various resin types), fillers, and modifiers as well as the equipment and methods used to prepare and to process SMCs are reviewed [239] with particular reference to short-stroke polyester presses and to conditions of preparation and the properties of moldings.

Chemical recycling of cured UPRs and sheet molding compounds may become a feasible process of waste utilization. The processes are based on the breaking of ester bonds in the crosslinked UPR. The breaking of ester bonds consists of hydrolysis at 225 °C for several hours [240]. After filtration, isophthalic acid, styrene/fumaric acid copolymer and propylene glycol are recovered. According to another process, cured UPR is hydrolyzed with butanone/water at 220–275 °C for 6 hours [241]. The separation into an aqueous and organic layer resulted in the recovery of phthalic acid and styrene oligomer, correspondingly. Hydrolysis of ground SMC with methanolic KOH in dioxane under reflux was described [242]. Recently, a utilization of cured UPR by pulverization followed by glycolysis with 1,2-propylene glycol at a temperature in the range of 70–80 °C was presented [243]. The glycolyzate was reacted with maleic anhydride and then dissolved in styrene.

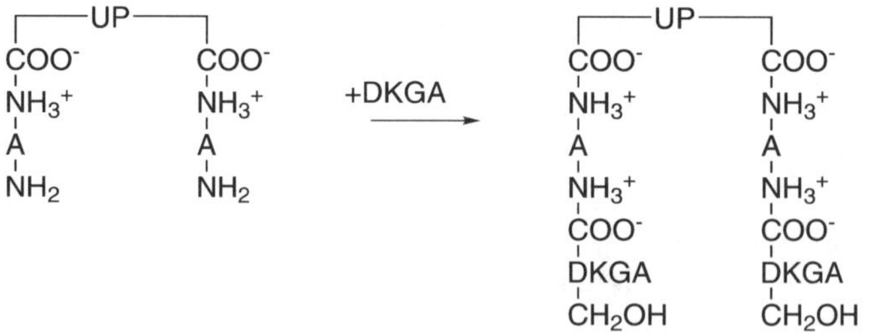

Fig. 24 Modification of UPRs with diketogulonic acid. Reprinted from (1996) Polymer 37:2179 [238] with permission

Thus, a UPR with acceptable mechanical properties after curing by radical copolymerization was eventually obtained. The recycling of cured UPR was developed whereby the resin was mechanically disintegrated and the thus obtained fine powder was glycolyzed with propylene glycol. The glycolyzate was then reacted with maleic anhydride to form a UP [243]. The UP was dissolved in styrene and the resultant UPR was cured by radical copolymerization. Tensile strength, tensile modulus and impact strength were determined. The idea was based on the previous results by Kinstle et al. [240], Tesoro et al. [244] and Patel et al. [242] who described the recycling of cured SMC. Those processes began with hydrolysis of the ground material.

Waste polyester molding compounds/bulk molding compound articles as well as glass-reinforced cold-cured polyester laminates scrap, and waste printed-circuit boards made of epoxy-glass copper-coated laminates were used as fillers for making polyester molding compounds [245]. Partial, up to 10–15 wt %, replacement of a powdered chalk by the recyclates raised the viscosity, accelerated the chemical thickening caused by MgO, and slightly delayed the exothermic peak of the crosslinking process and lowered the heat of copolymerization, but did not significantly affect the properties of the cured resin.

13.4
Powder Coatings from UPs

Solid UPs with solid monomers and photoinitiators are used after having been powdered as UV curable powder coatings [246]. Both main components are solvent-free and are melted after being heated with infrared irradiation. The powder coating is first applied with a gun onto an electrostatically charged surface. The powder particles undergo an electrostatic charging in the gun. The molten powder coating is spread out on the surface of the object to be coated. The components of the powder coating copolymerize (are hardened) within a few seconds of the UV irradiation. In contradiction to the thermal curing, the UV hardening does not need a high temperature or a long application time. Thus, the UV curable coatings can be applied to the elevated temperature sensitive substrates, such as wood, particle board and thermoplastics.

Powder coatings have to be resistant to agglutination after prolonged storage at an elevated or ambient temperature under a compression. The agglutination resistance concerns both the powder and the solid monomer component. The solid powdered components should exhibit a T_g value of around 50 °C. The UP component is a fumaric acid polyester with a glycol like ethylene, 1,4-butylene and neopentyl glycol or 1,4-cyclohexane dimethanol. The solid unsaturated monomer may comprise a vinyl-ester type DGEBA diacrylate which is reacted with a diisocyanate, e.g. isophorone diisocyanate (IPDI).

The urethanization of DGEBA diacrylate with IPDI results in an increase in the T_g value.

The meltable components of the powder coating are mixed as a melt in an extruder. To be extruded in melt, the components should exhibit a melting temperature in the range of 90–110 °C. The extruded composition is then cooled down, powdered and screened. The given melting temperature enables the powder coating to be spread out on the coated object.

References

1. Penczek P (1966) Polimery 11:351
2. Yang YS, Pascault JP (1997) J Appl Polym Sci 64:133
3. Yang YS, Pascault JP (1997) J Appl Polym Sci 64:147
4. Piotrowska Z, Sowińska W, Kiełkiewicz J, Kozłowski A (1990) Polimery 35:9
5. Piotrowska Z, Kiełkiewicz J (1995) Polimery 40:669
6. Fekete F (1963) Plastics Technol 9(3):42
7. Matynia T, Gawdzik B, Zarêbska E (1998) Polimery 43:738
8. Matynia T, Księżopolski J, Gawdzik B (2002) J Appl Polym Sci 84:716
9. Matynia T, Księżopolski J, Pawłowska E (1998) J Appl Polym Sci 68:1423
10. Matynia T, Księżopolski J (1999) J Appl Polym Sci 73:1815
11. Matynia T, Księżopolski J (2000) J Appl Polym Sci 77:3077
12. Matynia T, Pawłowska E, Księżopolski J (1999) Polimery 44:739
13. Abdel-Azim A-A (1995) Polym Bull 35:229
14. Agrawal JP, Bhale VC (1997) React Funct Polym 34:145
15. Hoa LTN, Pascault JP, My LT, Son ChPN (1993) Eur Polym J 29:491
16. Penczek P (1970) Polimery 15:421
17. Batog AE, Tkachuk BM, Aldoshin VA (1995) Polimery 40:624
18. Nalampang K, Johnson AF (2003) Polymer 44:6103
19. Kicko-Walczak E (1990) Polimery 35:24
20. Bicak N, Karagoz B, Tunca U (2003) J Polym Sci Pol Chem 41:2549
21. Vlad S, Oprea S, Stanciu A, Ciobanu C, Bulacovschi V (2000) Eur Polym J 36:1495
22. Skrifvars M, Schmidt HW (1995) J Appl Polym Sci 55:1787
23. Mormile P, Musto P, Petti L, Ragosta G, Villano P (2000) Appl Phys B 70:249
24. Sanchez EMS, Zavaglia CAC, Felisberti MI (2000) Polymer 41:765
25. Yang YS (1996) J Appl Polym Sci 60:2387
26. Abbate M, Martuscelli E, Musto P, Ragosta G (1996) Angew Makromol Chem 241:11
27. Gawdzik B, Matynia T, Chmielewska E (2001) J Appl Polym Sci 82:2003
28. Wardzińska E, Rudnik E, Penczek P (1995) Polimery 40:636
29. Frydrych A, Ostrysz R, Penczek P (1999) Polimery 44:745
30. Meixner J, Fisher W, Mueller M, Rebuscini C (1996) German Patent 19 501 176
31. Tsuchiya S, Oshima A, Hayashi H (1982) US Patent 4 360 622
32. Kuang J (1966) Reguxing Shuzhi 11:23; Chem Abs 126:132064d
33. Penczek P, Abramowicz D, Rokicki G, Ostrysz R (2004) Fatipec Congr 2:617
34. Hegemann G (2003) Macromol Symp 199:333
35. Fischer W, Kniege W (1986) US Patent 4 625 008
36. Gude F, Bellut H (1989) German Patent 3 723 891
37. Kleine HW (1987) German Patent 3 619 797

38. Tarnawski W, Motak A (1999) Polimery 44:777
39. Ostrysz R, Kłosowska Z, Jankowska F (1964) Polish Patent 76 005
40. Ostrysz R (1969) Polimery 14:203
41. Ostrysz R (1970) Polimery 15:406
42. Spychaj T (2000) Handbook of thermoplastic polyesters, Fakirov S (ed), Chapter 27, Chemical Recycling of PET. Wiley-VCH, Weinheim
43. Paszum D, Spychaj T (1997) Ind Eng Chem Res 36:1373
44. Nowaczek W, Królikowski W, Pawlak M, Kłosowska-Wołkowicz Z, Mieczkowski W (1992) Polish Patent 153 520
45. Rebeiz KS, Fowler DW, Paul DR (1992) J Appl Polym Sci 44:1649
46. Gawdzik B, Matynia T, Maciąga-Dembińska D, Kurczak M (2000) Przem Chem 79:197
47. Pepper TP (1995) US Patent 5 380 793
48. Matynia T, Gawdzik B, Andrachiewicz M, Gadaczowski S, Pazgan A (1999) Przem Chem 78:303
49. Vaidya UR, Nadkarni VM (1987) J Appl Polym Sci 34:235
50. Campanelli JR, Kamal MR, Cooper DG (1994) J Appl Polym Sci 54:1731
51. Chen JY, Ou CF, Hu YC, Lin CC (1991) J Appl Polym Sci 42:1501
52. Yang LS, Cai G, Armstead D (1999) US Patent 5 880 225
53. Kao CY, Cheng WH, Wan BZ (1997) Thermochim Acta 292:95
54. Suh DJ, Park OO, Yoon KH (2000) Polymer 41:461
55. Pimpan V, Sirisook R, Chuayjuljit S (2003) J Appl Polym Sci 88:788
56. Vaidya UR, Nadkarni VM (1989) J Appl Polym Sci 38:1179
57. Mansour SH, Ikladious NE (2003) J Elastom Plast 35:133
58. Hong SG, Su SH (1997) Angew Makromol Chem 246:125
59. Lu M, Kim S (2001) J Appl Polym Sci 80:1052
60. Tawfik ME (2003) J Appl Polym Sci 89:3693
61. Rebeiz KS, Banko AS, Nesbit SM, Craft AB (1995) J Appl Polym Sci 56:757
62. Scott TF, Cook WD, Forsythe JS (2002) Eur Polym J 38:705
63. Cook WD (1992) Polymer 33:2152
64. Scott TF, Cook WD, Forsythe JS (2003) Polymer 44:671
65. Matynia T, Penczek P (1976) Plaste Kautsch 23:403
66. Starr B, Burts E, Upson JR, Riffle JS (2001) Polymer 42:8727
67. Gawdzik B, Matynia T (2001) J Appl Polym Sci 81:2062
68. Auad ML, Frontini PM, Borrajo J, Aranguren MI (2001) Polymer 42:3723
69. Egan D, Weber C, Fielder P (1996) Reinf Plastics 10:50
70. Schultze U, Skrifvars M, Reichelt N, Schmidt HW (1997) J Appl Polym Sci 64:527
71. Pilati F, Toselli M, Messori M, Credali U, Tonelli C, Berti C (1998) J Appl Polym Sci 67:1679
72. Messori M, Toselli M, Pilati F, Tonelli C (2001) Polymer 42:9877
73. Abbate N, Martuscelli E, Musto P, Ragosta G, Leonardi M (1996) J Appl Polym Sci 62:2107
74. Martuscelli E, Musto P, Ragosta G, Scarinzi G, Bertotti E (1993) J Appl Polym Sci 31:619
75. Cherian AB, Thachil ET (2003) J Elastom Plast 35:367
76. Auad ML, Borrajo J, Aranguren MI (2003) J Appl Polym Sci 89:274
77. Karger-Kocsis J, Gryshchuk O, Jost N (2003) J Appl Polym Sci 88:2124
78. Eisenberg P, Lucas JC, Williams RJJ (1997) J Appl Polym Sci 65:755
79. Penczek P, Lewandowska T (1968) Polimery 13:59
80. Szayna A (1963) Ind Eng Chem, Prod Res Devel 2:105

81. Jedliński Z, Penczek P (1964) Plaste Kautsch 11:580
82. Yuan CD, Xu YU, Wang YJ, Liu DH, Chai YJ, Cao TY (2000) J Appl Polym Sci 27:3049
83. Klaban J, Luňak S, Kitzler J (1985) Plasty Kaučuk 22:6
84. Xu MX, Liu WG, Yao KD (1996) Angew Makromol Chem 240:163
85. Cao X, Lee LJ (2003) Polymer 44:1893
86. Valette L, Hsu CP (1999) Polymer 40:2059
87. Ramís X, Cadenato A, Morancho JM, Salla JM (2001) Polymer 42:9469
88. Wojturska J (2004) Thesis, Rzeszów University of Technology
89. Tang DY, Qin CL, Cai WM, Zhao L (2003) Mater Chem Physics 82:73
90. Dinakaran K, Alagur M (2003) Polym Adv Technol 14:544
91. Guhanathan S, Harikaren R, Sarojadevi M (2004) J Appl Polym Sci 92:817
92. Ajithkumar S, Patel NK, Kansara SS (2000) Eur Polym J 36:2387
93. Wang GY, Wang YL, Hu ChP (2000) Eur Polym J 36:735
94. Alvarez-Castillo A, Costaño VM (1994) Polym Bull 32:447
95. Galina H, Potoczek M (1999) Polimery 44:730
96. Penczek P, Kicko-Walczak E (1981) Österr Kunststoff-Z 12:106
97. Kicko-Walczak E, Penczek P (1984) Kunststoffe 74:464
98. Marsh G (1999) Reinf Plastics 10:36
99. Anon (1999) Reinf Plastics 10:19
100. Heilmann H (1997) Reinf Plastics 10:44
101. Skrifvars M, Säämänen A (1997) Reinf Plastics 3:46
102. Prado C, Ibarra I, Periago JF (1997) J Chromatogr A 778:225
103. Rahman F, Langford KH, Scrinshaw MD, Lester JN (2001) Sci Total Environ 75:1
104. Brzozowski ZK, Szymańska E, Bratychak MM (1997) React Funct Polym 33:217
105. Wolf R (1986) Kunststoffe 76:943
106. Hörold S (1999) Polym Degr Stab 64:427
107. Penczek P, Kicko-Walczak E (1987) Kunststoffe 77:415
108. Galip H, Hasipoğlu H, Gündüz G (1999) J Appl Polym Sci 74:2906
109. Snyder CA (1985) Plast Compd 8:1
110. Kłosowska-Wołkowicz Z (1988) Polimery 33:428
111. Shen KK, Sprague RW (1982) Plast Compd 5:67
112. Shih YF, Jeng RJ (2002) Polym Degrad Stab 72:67
113. Penczek P, Ostrysz R, Krassowski D (2000) Flame Retardants 2000. Interscience Commun, London, p 105
114. Atkinson PA, Haines PJ, Skinner GA (2001) Polym Degr Stab 71:251
115. Kicko-Walczak E (2003) Macromol Symp 202:221
116. Kicko-Walczak E (2003) Macromol Symp 199:343
117. Kicko-Walczak E (2003) Polimery 48:351
118. Cusack PA (1995) Polimery 40:650
119. Kicko-Walczak E (1999) Polym Degr Stab 64:439
120. Kicko-Walczak E (2000) Polimery 45:808
121. Baudry A, Dufay J, Regnier N, Mortaigne B (1998) Polym Degr Stab 63:441
122. Mortaigne B, Bourbigot S, Le Bras M, Cordellier G, Baudry A, Dufay J (1999) Polym Degr Stab 64:443
123. Evans SJ, Haines PJ, Skinner GA (1997) Thermochem Acta 291:43
124. Kłosowska-Wołkowicz Z, Penczek P, Piechocki J (1986) Polimery 31:465
125. Agrawal JP, Sarwade DB, Makashir RR, Dendage PS (1998) Polym Degr Stab 62:9
126. Penczek P, Kłosowska-Wołkowicz Z, Kicko-Walczak E, Urbańska J (1978) Plaste Kautsch 25:17
127. Bélan F, Bellenger V, Mortaigne B, Verdu J (1997) Polym Degrad Stabil 56:301

128. Bélan F, Bellenger V, Mortaigne B (1997) Polym Degrad Stabil 56:93
129. Deslandes N, Bellenger V, Jaffiol F, Verdu (1998) J Appl Polym Sci 69:2663
130. Valea A, Gonzales ML, Modragon I (1999) J Appl Polym Sci 71:21
131. Marais S, Métayer M, Nguyen TQ, Labbé M, Saiter JM (2000) Eur Polym J 36:453
132. Marais S, Métayer M, Nguyen TQ, Labbé M, Perrin L, Saiter JM (2000) Polymer 41:2667
133. Gu X, Raghavan D, Nguyen T, VanLandingham MR, Yebassa D (2001) Polym Degrad Stab 74:139
134. Valiente N, Lalot T, Brigodiot M, Maréchal E (1998) Polym Degrad Stabil 61:409
135. Evans SJ, Haines PJ, Skinner GA (1997) Thermochim Acta 291:43
136. Bansal RJ, Mittal J, Singh P (1989) J Appl Polym Sci 37:1901
137. Agrawal JP, Sarwade DB, Makashir PS, Mahajan RR, Dendage PS (1998) Polym Degrad Stabil 62:9
138. Baudry A, Dufay J, Regnier N, Mortaigne B (1998) Polym Degrad Stabil 61:441
139. Shih YF, Jeng RJ (2002) Polym Degrad Stabil 77:67
140. Auad ML, Aranguren M, Borrajo J (1997) J Appl Polym Sci 66:1059
141. Fan LH, Hu CP, Zhang ZP, Ying SK (1996) J Appl Polym Sci 59:1417
142. Xu MX, Liu WG, Yao KD (1996) Angew Makromol Chem 240:163
143. Larez V, Cristobal J, Perdomo Mendoza GA (1993) J Appl Polym Sci 47:121
144. Tawfik SY (2001) J Appl Polym Sci 81:3388
145. Kharas GB, Kamenetsky M, Simantirakis J, Beinlich KC, Rizzo A-MT, Caywood GA, Watson K (1997) J Appl Polym Sci 66:1123
146. Galcéra T, Fradet A, Maréchal E (1995) Eur Polym J 31:733
147. Grobelny J (1995) Polymer 36:4215
148. Youssef MH, Mansour SH, Tawfik SY (2000) Polymer 41:7815
149. Grobelny J (1997) Polymer 38:751
150. Grobelny J (1999) Polymer 40:2939
151. Newman R, Patterson KH (1996) Polymer 37:1065
152. López-González MMC, Callejo Cudero MJ, Barrales-Rienda JM (1997) Polymer 38:6219
153. Hietalahti K, Root A, Skrifvars M, Sundholm F (1997) J Appl Polym Sci 65:77
154. Hietalahti K, Skrifvars M, Root A, Sundholm F (1998) J Appl Polym Sci 68:671
155. Zhang YH, Feng YA, Lu WK (2000) J Chrom A 904:87
156. Podzimek Š, Hyršl J (1994) J Appl Polym Sci 53:1351
157. Mortaigne B, Feltz B, Laurens P (1997) J Appl Polym Sci 66:1703
158. Bharadwaj RK, Mehrabi AR, Hamilton C, Trujillo C, Murga M, Fan R, Chavira A, Thompson AK (2002) Polymer 43:3699
159. Pucić I, Ranogajec F (1999) Radiat Phys Chem 54:95
160. Chen JS, Yu TL (1998) J Appl Polym Sci 69:871
161. Orak S (2000) Cem Conc Res 30:171
162. Stein DL (1998) US Patent 5 710 213
163. Sanchez J, Stein DL (1999) US Patent 5 981 787
164. Blot E, Stock C (1999) US Patent 5 907 003
165. Diloy BJL (2001) US Patent 6 214 898
166. Gawdzik B, Księżopolski J, Matynia T (2003) J Appl Polym Sci 87:2238
167. Kalenda P (1995) Eur Polym J 31(11):1099
168. Martín JL (1999) Polymer 40:3451
169. Li L, Lee J (2002) Polym Compos 23:971
170. Duliban J (2001) Macromol Mater Eng 286:624
171. Kucharski M, Duliban J, Chmielszukiewicz E (2003) J Appl Polym Sci 89:2973

172. Matynia T, Gawdzik B (1997) J Appl Polym Sci 65:1525
173. Zhang L, Liu L, Chen Y (1999) J Appl Polym Sci 74:3541
174. Decker C, Decker D (1997) Polymer 38:2229
175. Pietschmann N (2002) Macromol Symp 187:225
176. Rokicki G, Szymańska E (1998) J Appl Polym Sci 70:2031
177. Jung DJ, Lee SJ, Che WJ, Ha CS (1998) J Appl Polym Sci 69:695
178. Martysz D, Antoszczyszyn M, Urbala M, Krompiec S, Fabrycy E (2003) Progr Org Coatings 16:302
179. Martuscelli E, Musto P, Ragosta G, Scarinizi G (1996) Polymer 37:4025
180. Zetterlund PB, Johnson AF (2002) Polymer 43:2039
181. de la Caba K, Guerrero P, Eceiza A, Mondragon I (1996) Polymer 37:275
182. Auad ML, Aranguren MI, Eliçabe G, Borrajo J (1999) J Appl Polym Sci 74:1044
183. Huang YJ, Chen ChJ (1993) J Appl Polym Sci 47:1533
184. Huang YJ, Chen ChJ (1993) J Appl Polym Sci 48:151
185. Hsu CP, Lee LJ (1993) Polymer 34:4496
186. Yang H, Lee LJ (2001) J Appl Polym Sci 79:1230
187. Matynia T, Gawdzik B, Osypiuk J (1998) Int J Polym Mater 41:215
188. Tollens FR, Lee LJ (1993) Polymer 34:29
189. Van Assche G, Verdonck E, Van Mele B (2001) Polymer 42:2959
190. Marín JL, Cadenato A, Salla JM (1997) Thermochim Acta 306:115
191. Lu M, Shim M, Kim S (2001) Eur Polym J 37:1075
192. Lu MG, Shim MJ, Kim SW (1998) Thermochim Acta 323:37
193. Delahaye N, Marais S, Saiter JM, Metayer M (1998) J Appl Polym Sci 67:695
194. Fan LH, Hu ChP, Ying SK (1996) Polymer 37:975
195. Cook WD, Simon GP, Burchill PJ, Lau M, Fitch TJ (1997) J Appl Polym Sci 64:769
196. Scott TF, Cook WD, Forsythe JS (2002) Eur Polym J 38:705
197. Simitzis JCh, Zoumpoulakis LTh, Sulis SK, Mendrinos LN (2001) Mikrochim Acta 136:171
198. de la Caba K, Guerrero P, Eceiza A, Mondragon I (1997) Eur Polym J 33:19
199. Bureau E, Chebli K, Cabot C, Saiter JM, Dreux F, Marais S, Metayer M (2001) Eur Polym J 37:2169
200. Jost N, Karger-Kocsis J (2002) Polymer 43:1383
201. Martin JS, Laza JM, Morrás ML, Rodríguez M, León LM (2000) Polymer 41:4203
202. Bogdat D, Penczek P, Pielichowski J, Prociak A (2003) Adv Polym Sci 163:193
203. Pielichowski J, Bogdał L, Wolff E (2003) Przem Chem 82:938
204. Cao X, Lee LJ (2003) Polymer 44:1507
205. Huang YJ, Horng JC (1998) Polymer 39:3683
206. Ma CCM, Hsieh CT, Kuan HC, Tsai TY, Yu SW (2003) Polym Eng Sci 43:989
207. Acha BA, Aranguren MI, Marcivich NE, Reboredo MM (2003) Polym Eng Sci 43:997
208. Huang YJ, Chen LD (1998) Polymer 39:7049
209. Boyard N, Vayer M, Sinturel C, Erre R, Delaunay D (2003) J Appl Polym Sci 88:1258
210. Huang YJ, Chen TS, Huang JG, Lee FH (2003) J Appl Polym Sci 89:3336
211. Huang YJ, Chen TS, Huang JG, Lee FH (2003) J Appl Polym Sci 89:3347
212. Li W, Lee LJ (2000) Polymer 41:685
213. Li W, Lee LJ (2000) Polymer 41:697
214. Sun MC, Chang YF, Leon Yu T (2001) J Appl Polym Sci 79:1439
215. Dong JP, Huang JG, Lee FH, Roan JW, Huang YJ (2004) J Appl Polym Sci 91:3388
216. Dong JP, Huang JG, Lee FH, Roan JW, Huang YJ (2004) J Appl Polym Sci 91:3369
217. Xu WB, Bao SB, Shen SJ, Wang W, Hang GP, He PS (2003) J Polym Sci B Polym Physics 41:378

218. Giannelis EP (1998) Appl Organomet Chem 12:675
219. Suh DJ, Lin YT, Park OD (2000) Polymer 41:8557
220. Iceoglu AB, Yilmazer U (2003) Polym Eng Sci 43:661
221. Zhang M, Singh RP (2004) Mater Lett 58:408
222. Evora VMF, Shukla A (2003) Mater Sci Eng A 361:358
223. Hsu YG, Wang CP (2003) J Polym Res – Taiwan 10:201
224. Kim HG, Oh DH, Lee KE (2004) J Appl Polym Sci 92:238
225. Zhang YH, Cai QY, Jiang ZJ, Gong KC (2004) J Appl Polym Sci 92:2038
226. Kędzierski M, Penczek P (2004) Polimery, 49:801
227. Oleksy M, Galina H (2000) Polimery 45:541
228. Shi W, Rånby B (1994) J Appl Polym Sci 51:1129
229. Suppakul P, Bandyopadhyay S (2002) Compos Sci Technol 62:709
230. Tari MJ, Bals A, Park J, Lin MY, Hahn HT (1998) Compos Part A 29A:651
231. Gowda TM, Naidu ACB, Chaya R (1999) Compos Part A 30:277
232. Saroja Devi M, Murugesan V, Rengaraj K, Anand P (1998) J Appl Polym Sci 69:1385
233. Şen S, Nugay N (2001) Eur Polym J 37:2047
234. Ismail MR, Ali MAM, El-Millicy AA, Afifi MS (1999) J Appl Polym Sci 72:1031
235. Peschanskaya NN, Hristova J, Yakushev PN (2001) Polymer 42:7101
236. Gopal AK, Adali S, Verijenko VE (2000) Compos Struct 48:99
237. Deslandes N, Bellenger V, Jaffiol F, Verdu J (1998) Compos Part A 29A:1481
238. Chiu YY, Saito R, Lee LJ (1996) Polymer 37:2179
239. Królikowski W, Spaay A (1999) Polimery 44:716
240. Kinstle JF, Forskey LD, Valke R, Campbell RR (1983) Am Chem Soc Div Polym Chem 32:446
241. Tesoro G, Wu Y (1992) Am Chem Soc PSME Prepr 67:459
242. Patel SH, Gonzalves KF, Stivala SS, Reich L, Trivedi DH (1993) Adv Polym Technol 12:35
243. Yoon KH, DiBenedetto AT, Huang SJ (1997) Polymer 38:2281
244. Tesoro G, Chum H, Power A (1992) Compos Inst, 47th Annu Conf SPI, Session 4C:1
245. Nowaczek W (1999) Polimery 44:758
246. Fink D, Brindöpke G (1995) Eur Coatings J 9:606

Crosslinked Polyolefin Foams: Production, Structure, Properties, and Applications

M. A. Rodríguez-Pérez

Departamento de Física de la Materia Condensada,
Universidad de Valladolid, 47011 Valladolid, Spain
marrod@fmc.uva.es

1	Introduction	99
2	Production	101
2.1	Commercial Processes	101
2.2	Foaming of New Polymers	104
2.3	Improvements of Formulations	105
2.4	Experimental Characterization and Modeling of Foaming	107
3	Structure	107
3.1	Characterization of the Cellular Structure	108
3.2	Characterization of the Polymer Morphology	109
4	Physical Properties	110
4.1	Mechanical Properties	110
4.1.1	Mechanical Properties at Low Strain Rates	111
4.1.2	Impact	114
4.1.3	Creep and Gas Diffusion	116
4.1.4	Dynamic Mechanical Behavior	117
4.2	Thermal Properties	118
4.2.1	Thermal Conductivity	118
4.2.2	Thermal Expansion	119
4.3	Recycling	120
5	Applications	120
6	Conclusions	122
	References	123

Abstract The purpose of this paper is to review the most significant developments of the last 10 years in the field of crosslinked polyolefin foams. The methods to produce the foams, the relationships between structure and properties, and the main applications of these materials are briefly reviewed. Topics of possible future research are proposed.

Keywords Polyolefin foams · Polyethylene foams · Crosslinking · Foaming agents · Physical properties of foams

Abbreviations

AAGR	Average annual growth rate
AZD	Azodicarbonamide
CPOF	Crosslinked polyolefin foams
C	Constant that accounts for the cell shape
C_1, C_2, C_3	Fitting constants in the Gibson and Ahsby equations
Φ	Mean cell size
δ	Mean cell wall thickness
σ_c	Collapse or yield stress
ρ	Density of the foam
ρ_s	Density of the solid polymer
DCP	Dicumyl peroxide
EVA	Ethylene vinyl acetate copolymer
ESI	Ethylene styrene interpolymer
D_{ef}	Effective diffusion coefficient
f_s	Fraction of material in the struts or edges
T_N	Fraction of radiant energy sent forward by a solid membrane of thickness δ
r	Fraction of incident energy reflected by each gas-solid interface
t	Fraction of energy transmitted through a solid membrane
HDPE	High-density polyethylene
K	Foam bulk modulus
L	Foam thickness
p_0	Initial gas pressure
LDPE	Low-density polyethylene
LLDPE	Linear low-density polyethylene
PO	Polyolefin
PP	Polypropylene
ρ/ρ_s	Relative density
ω	Refractive index of the plastic
ν	Poisson ratio
P	Polymer permeability
ε	Strain
σ	Stefan–Boltzman constant
λ	Foam thermal conductivity
T	Temperature
λ_g	Thermal conductivity by conduction through the gas phase
λ_s	Thermal conductivity by conduction through the solid phase
λ_r	Thermal conductivity by radiation
λ_{gas}	Thermal conductivity of the gas
λ_{solid}	Thermal conductivity of the solid
α	Thermal expansion of the foam
α_g	Thermal expansion of the gas
α_s	Thermal expansion of the solid
V_{gas}	Volume fraction of gas
V_{poly}	Volume fraction of solid
E_s	Solid polymer Young modulus
E	Foam Young modulus

1
Introduction

Polymer foams are two-phase materials in which a gas is dispersed in a continuous macromolecular phase [1–4]. These materials are important items in the economy, and, because of technical, commercial, and environmental issues, they represent an interesting dynamic in 21st-century society. The plastic foam industry is a major segment of the U.S. plastics industry, accounting for about 10% of total commodity resin consumption. This market [5] was estimated at more than 3.34 billion kg in 2001 and it is expected to grow at an average annual growth rate (AAGR) of about 2.8%, thus increasing to 3.83 billion kg by 2006. In terms of sales, polyolefin (PO) foams are ranked fourth, behind polyurethane, polystyrene, and poly(vinyl chloride) foams; however, polyolefinic materials have the highest AAGR of any resin group. A growth rate of 3.8% is forecast for POs in 2006, largely based on increasing demand for CPOF in market sectors such as construction, packaging, buoyancy, automotive, medical, and the new and growing field of leisure and sports items.

From a technical point of view a variety of properties, such as light weight, buoyancy, chemical resistance, skin friendliness, no water absorption, cushioning performance, shock absorption, and thermal insulation, have all ensured the success of these materials.

Olefins are defined as unsaturated aliphatic hydrocarbons. The commercial materials currently used to produce PO foams are low-density polyethylene (LDPE), high-density polyethylene (HDPE), linear low-density polyethylene (LLDPE), polypropylene (PP), and ethylene vinyl acetate copolymers (EVA). The density of a cellular material profoundly affects its properties. PO foams are divided into two density categories: high and low density. The borderline is at approximately 240 kg/m^3 [3]. High-density materials are mainly used in wires and cables and for structural purposes. Low-density foams are

Table 1 Market for polymeric foam in the USA by resin family (million kg). AAGR (average annual grow rate)

	2001	2006	AAGR % 2001–2006
Polyurethane	1782	2124	3.6
Polystyrene	861	914	1.2
Poly(vinyl chloride)	520	592	2.6
Polyolefin	135	162	3.8
Others	41	46	2.3
Total	3339	3838	2.8

mainly used for energy absorption and thermal insulation. This review focuses on these low-density foams. The majority of these foams have densities lower than approximately 100 kg/m^3, the relative density (ρ/ρ_s), defined as the density of the foam ρ divided by the density of the solid polymer ρ_s, is lower than 0.1, i.e., the volume of gas is higher than 90%.

PO foams are classified into three product groups depending on the production process and final shape [3, 6]. Extruded PO foams were introduced into the market by Dow Chemical Company in 1958. Crosslinked PE foams emerged in the Japanese market in the mid-1960s. PO foam moldings were introduced in the early 1970s by BASF AG in Germany [3, 6]. The production routes are based on different technologies for the noncrosslinked and crosslinked materials. This review considers mostly CPOF, though most of the

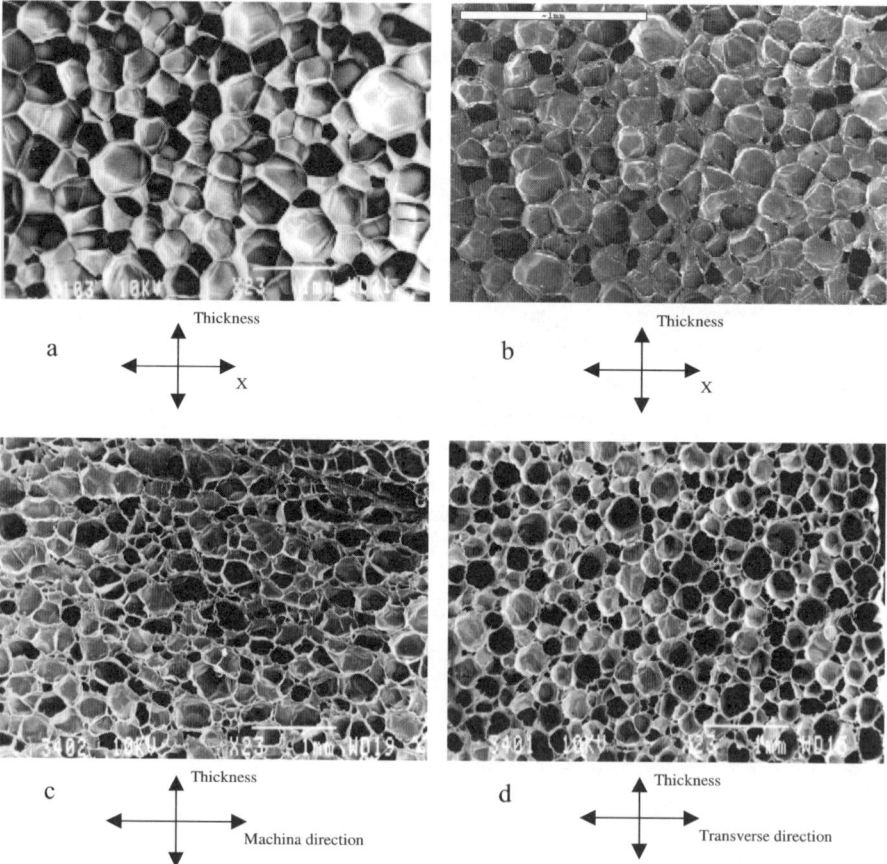

Fig. 1 Micrographs of foams produced from **a** high-pressure nitrogen solution process, **b** compression molding process, **c** and **d** semicontinuous process, crosslinking by irradiation (micrographs of the same sample in two perpendicular directions)

results on the structure property relationships could be directly applied to the noncrosslinked materials.

The structure is based on closed cells, which are polyhedra of different sizes and shapes (Fig. 1). The solid phase is a crosslinked PO, which is stretched during foaming, and, depending on the production route, this polymer could contain a nonnegligible amount of foaming agent residues. In addition, this solid phase crystallizes in very thin faces (for a LDPE foam of 15 kg/m^3 the cell wall thickness is approximately 1 μm). These features make crosslinked poyolefin foams remarkable materials in which interesting relationships between structure and properties should be expected [1, 2].

Due to the commercial and technical aspects mentioned in this introduction, the interest in these materials has increased notably in the last decade.

2
Production

2.1
Commercial Processes

The processes are characterized by three steps: sheet formation, crosslinking, and foaming. There are several ways to perform and combine the steps, yielding a number of possible routes [3, 6–8]. Figure 2 shows the processes that are commercially used and the main steps of each one.

The raw materials are the base polymer, the crosslinking agent (for chemically crosslinked foams), the foaming agent, and several additional additives; some of them are included as processing aids; others are added to impart specific properties to the foam (carbon black, UV stabilizers, etc.).

In processes where foam expansion is accomplished by heating, cells need to be stabilized by crosslinking the polymer. Crosslinking extends the rubbery plateau of the polymer melt, widening the temperature range in which a stable foam can be produced (Fig. 3) [3, 6–8].

Two main crosslinking procedures are used, physical crosslinking by irradiation (electrons or gamma irradiation) (process c in Fig. 2) and chemical crosslinking by using crosslinking agents such as dycumil peroxide (processes b, d, and e in Fig. 2).

Azodicarbonamide (AZD) meets most of the requirements for a successful foaming agent [3, 6–8]. It decomposes in a narrow temperature range (approximately 210–220 °C); however, the compound can be modified with an activator (ZnO for instance) to decompose at temperatures as low as 130 °C. The gas it releases consists of nitrogen (65%) and carbon monoxide (32%). This compound is used in processes b–e in Fig. 2. The other foaming agent commercially used is gas nitrogen (process a in Fig. 2). This method

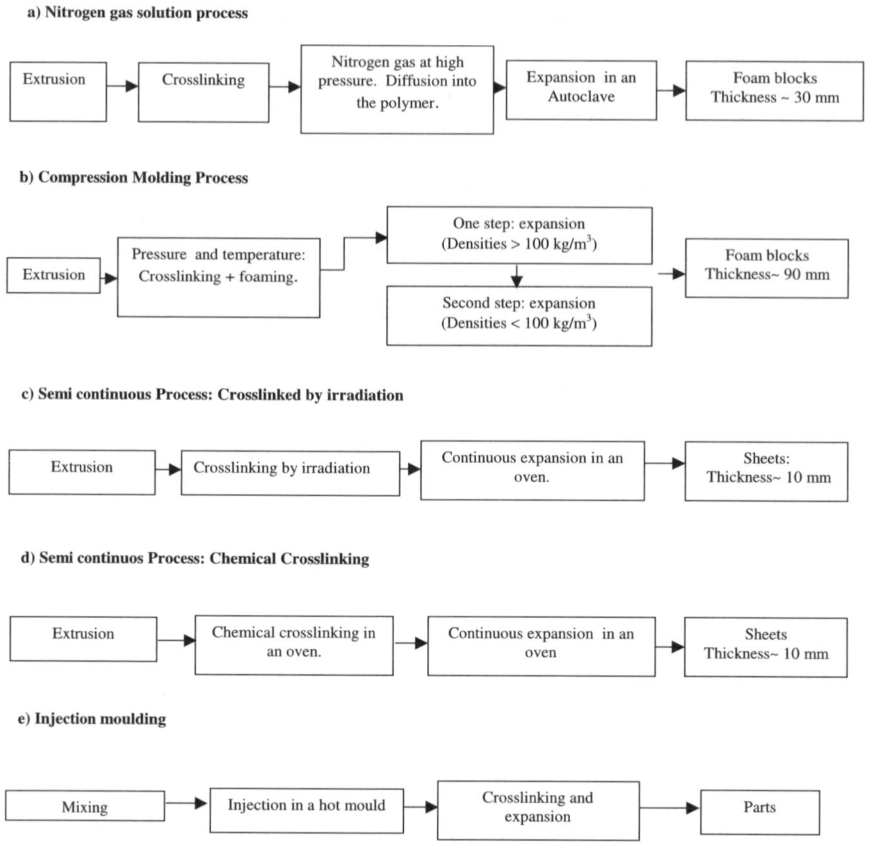

a) Nitrogen gas solution process
Extrusion
Crosslinking
Nitrogen gas at high pressure. Diffusion into the polymer.
Expansion in an Autoclave
Foam blocks Thickness ~ 30 mm
b) Compression Molding Process
Extrusion
Pressure and temperature: Crosslinking + foaming.
One step: expansion (Densities > 100 kg/m3)
Second step: expansion (Densities < 100 kg/m3)
Foam blocks Thickness~ 90 mm
c) Semi continuous Process: Crosslinked by irradiation
Extrusion
Crosslinking by irradiation
Continuous expansion in an oven.
Sheets: Thickness~ 10 mm
d) Semi continuos Process: Chemical Crosslinking
Extrusion
Chemical crosslinking in an oven.
Continuous expansion in an oven
Sheets Thickness~ 10 mm
e) Injection moulding
Mixing
Injection in a hot mould
Crosslinking and expansion
Parts

Fig. 2 Schematic diagram showing the main steps of the most important commercial technologies used to produce crosslinked closed-cell polyolefin foams

Solid
Fluid
Optimum melt viscosity for expansion
Crosslinked polyethylene
Polyethylene
Melt Viscosity
Tm
Range of temperatures for optimum expansion
Temperature

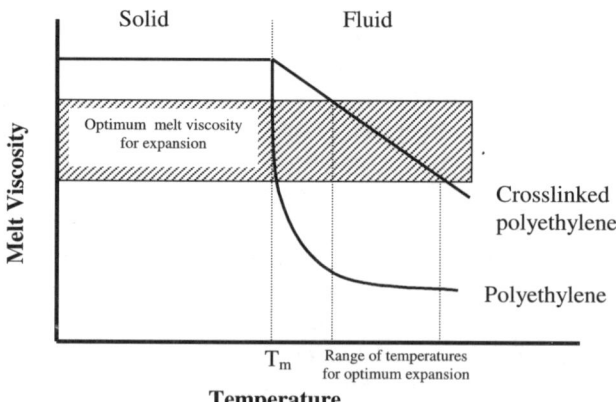

Fig. 3 Melt viscosity of polymer after crosslinking

Table 2 Characteristic temperatures of the materials used in the production of CPOF

Base material (T_m)	Crosslinking agent (T_d)	Foaming Agent (T_f)
Low-density polyethylene 110 °C		
EVA copolymers	Dicumyl peroxide	Azodicarbonamide
≈ 80 to 100 °C	Start at 130 °C,	Start at 150 °C
	Finish at 200 °C	Finish at 200 °C
High-density polyethylene 130 °C	Radiation, room temperature	
Polypropylene 170 °C		

is based on the diffusion of gas nitrogen on a solid sheet using high pressures and temperatures above the melting point of the base polymer. After the diffusion step the sheet is expanded by proper control of pressure and temperature [3, 6–8].

A strong impression of the complexity of this technology is obtained by considering the melting and decomposition temperatures of the various chemical materials used in each process step (Table 2).

The following principles of temperature control must be established [9]:

1. Form a sheet at a temperature higher than T_m but below T_d or T_f.
2. Start crosslinking above T_d but below T_f.
3. Foam at temperatures higher than T_m and T_d.

The advantage of using radiation for crosslinking comes in freeing the temperature need for gas release from the relatively close temperature levels of starting peroxide crosslinking. The overlap of the temperature ranges for accomplishing peroxide cure and for decompose AZD are observed in the table.

Different processes result in foams with different cellular structures. Figure 1 shows the cellular structure of foams produced from three different technologies. Foams produced from a nitrogen gas solution process are isotropic and homogeneous with polygonal cells [10, 11]. Foams produced from a compression molding technique show a microstructure characterized by a small average cell diameter, with the produced blocks being nonhomogeneous; the cellular structure changes along the block thickness (cells are smaller close to the surface of the blocks) [12]. Foams produced from compression molding or semicontinuous processes present some residues of foaming agent in the final structure. For a 30 kg/m^3 density foam, foaming residues account for 10% of the solid phase. The foams produced from the semicontinuous process are anisotropic with cells elongated in the machine direction (Fig. 1c) [13–16].

Early reviews on this topic considered in detail the different routes to produce crosslinked closed-cell PO foams [3, 7–9]. Recent research has concentrated on the foaming of new polymers, on the improvement of formulations, and on the experimental characterization and modeling of foaming.

2.2
Foaming of New Polymers

In the last few years, some new materials have emerged as base materials for CPOF [17, 18]. Examples are ethylene–styrene interpolymers, metallocene polyethylenes, polypropylene foams, and blends of different POs.

Ethylene–Styrene Interpolymer Foams

The properties of these random copolymers vary with the copolymerization styrene content. Interpolymers having a styrene content greater than 65% are amorphous and semirigid, while those containing a lower styrene content are crystalline, soft, and resilient. Interpolymers have both aromatic and aliphatic functionality. Hence, they are compatible with a variety of other thermoplastics as LDPE and polystyrene [19]. In addition, ESI resins with a high styrene content exhibit good vibration damping because their glass transition temperatures are near room temperature [18, 20]. On the other hand, they are relatively expensive; consequently, there has been an interest in blending these materials with cheaper LDPE, EVA, HDPE, and PP [21–23]. Applications that have been analyzed are midsoles for shoes and soccer shin guards.

Metallocene PE

The unique features of metallocene POs make these resins attractive for a wide range of applications [24, 25]. In particular, foams from these POs have been recently commercialized [26]. Foaming of metallocene LDPE and LLDPE have been performed successfully by using DCP as foaming agent [27–30] and silane grafted metallocene POs [31].

Polypropylene Foams

The interest in producing crosslinked PP foams has been enormous. However, due to a high melting temperature, near the range of the AZD decomposition (Table 2), and the inherent difficulties associated with the crosslinking of this polymer, mainly PP/PE blends have been commercially produced [32].

When a few years ago high-melt-strength PP became commercially available, several efforts were made to produce 100% noncrosslinked PP foams [18, 33]. Moreover, attempts have also been conducted to produce crosslinked PP foams. On the one hand, using a PP random copolymer with low ethylene content, electron irradiation, and multifunctional nomomers as crosslinking promoter, PP foams and PP/LLDPE blend foams have been produced and characterized [34]. On the other hand, PP films and sheets for use in automotive applications have also been manufactured, using organosilane crosslinking of selected PP copolymers [35]. The Zotefoams company [36], using the high-pressure nitrogen solution process, has introduced a 100% PP foam. However, no details of the process are given in this paper.

Blends

Blending is one of the easiest ways to modify the properties of CPOF. The physical properties of foams manufactured from blends of LDPE and EVA have been studied as a function of the LDPE content in the blends [15, 16]. The properties of the blends usually lie between the limits of the foamed constituents, although the relationship is not always linear, which mainly depends on the blend morphology.

2.3
Improvements of Formulations

Crosslinking

Crosslinking is a critical step for a successful foaming process. Methods to improve crosslinking efficiency have been analyzed empirically. One possibility is to add crosslinking promoters [28, 37], such as triallyl-1,3,5-benzenetricarboxylate or triallyl cyanurate. Another method is by using POs that are silane grafted to enhance physical properties and processability [38].

The effect of gamma irradiation on the crosslinking degree and foaming has also been considered [31, 39]. Cardoso et al. [40] compared foams crosslinked by irradiation and materials crosslinked by using peroxides. The foams, produced by radio-induced crosslinking, showed a smooth and homogeneous surface.

Foaming Agents

As was explained previously, commercial processes by chemical foaming use AZD [6]. This chemical compound and the technology behind it are well established, and therefore few trials exist to discover alternatives. In fact, comparative studies on the thermal behavior between AZD and other foaming agents showed that the older and lower cost AZD-based agent performed well and was shown to be better than most of the bicarbonate/citric acid-based agents. [41, 42]

Formulations

The interrelationships of base polymer type, crosslinking blowing agent concentration and physical properties of resultant foams have been investigated [43–50]. A small addition of crosslinked LLDPE, which has a low density of crosslink points to LDPE, enhances the mechanical behavior at high temperatures, which reduces heterogeneous deformation during foaming. As a result, foams with a uniform cell size distribution can be obtained [44]. It has been found that the expansion ratio decreases with increasing molecular weight and increases with blowing agent content for HDPE foams. Moreover, increasing both the molecular weight and amount of blowing agent decreases the cell size [46]. The expansion ratio of the LDPE foams decreases with increasing DCP content, which is due to the enhancement of the elastic modulus. The crystallization temperature (T_c) of the foams is also responsible for the expansion ratio. The degree of shrinkage decreases with increasing T_c because immediate crystallization prevents shrinkage [45].

Open-Cell Foams

One of the challenges in increasing the range of properties and applications of these materials is to produce the foams with open cells. Open-cell foams would have better acoustic absorption and better recovery after mechanical testing. In the last few years several papers and patents have been published on this topic. Two different approaches have been proposed.

On the one hand, the cell opening is produced by pressing the foam up to very high strains (higher than 95%). This compression causes the rupture of cell faces, producing a cellular structure with interconnected cells (Fig. 4). This procedure can be performed at room temperature or at a temperature below the glass transition temperature of the matrix [51]. The inconvenient aspect of this method is a reduction of the mechanical properties due to the plastic deformation of the polymeric matrix.

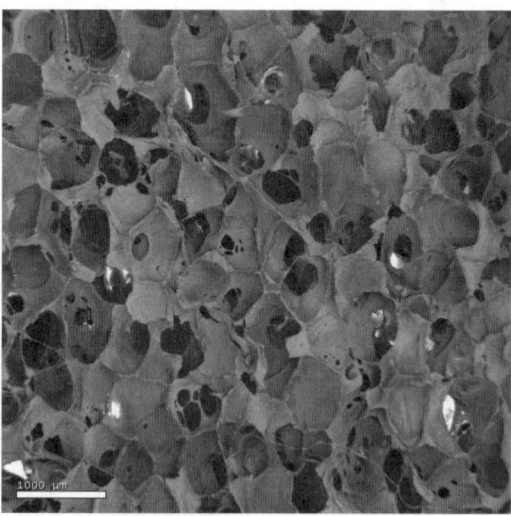

Fig. 4 Cellular structure of an open-cell polyethylene foam 30 kg/m³ density

Other methods are able to produce the cell opening without pressing the foams. For example, several patents propose to invert the relationships between the decomposition temperature of the crosslinking agent and the decomposition temperature of the blowing agent [52, 53], which results in an open-cell material with a similar structure to that shown in Fig. 4, but without plastic deformation of the polymeric matrix.

2.4
Experimental Characterization and Modeling of Foaming

Most of the developments in the field of foaming are based on empirical knowledge. However, a few papers have considered more fundamental aspects of foaming.

For example, Tai et al. [54] investigated the reaction kinetics of foaming. The crosslinking reaction of an LDPE/DCP system is controlled by a first-order decomposition of DCP. The addition of ZnO into AZD not only lowers the decomposition temperature, but also accelerates its decomposition. The reaction kinetics of the LDPE/AZD/ZnO system was found to be similar to those of the AZD/ZnO system. The numerical results of this paper could be used as a framework for future prediction and simulation of these processes.

Tatibouët et al. [55] proposed an interesting experimental setup to follow the reactions taking place during injection molding. The ultrasonic quasistatic technique can mimic adequately the conditions prevailing during the molding process (pressure, temperature, time) and monitor the blowing process through sound velocity, attenuation, and specific volume measurements.

Mahapatro et al. [56] proposed a theory to predict the foam density of a given formulation. The model is based on the gas pressure in a Kelvin foam structure (see below) and a rubber-elastic analysis of the biaxial stretching of the cell faces. The theory predicts with good agreement the final foam density.

3
Structure

The structure of PO foams can be studied at at least three different levels (Fig. 5). The cell size distribution and anisotropy of the cellular structure is observed in low-magnification micrographs (Fig. 5a). From these micrographs important parameters such as mean cell size (ϕ), anisotropy ratio, (ratio between the cell size in different directions), possible residues of foaming agent, and overall cell shape are usually obtained. To obtain a more detailed characterization of the cellular structure, it is necessary to determine the mean cell wall thickness (δ) and the mass fraction in the edges

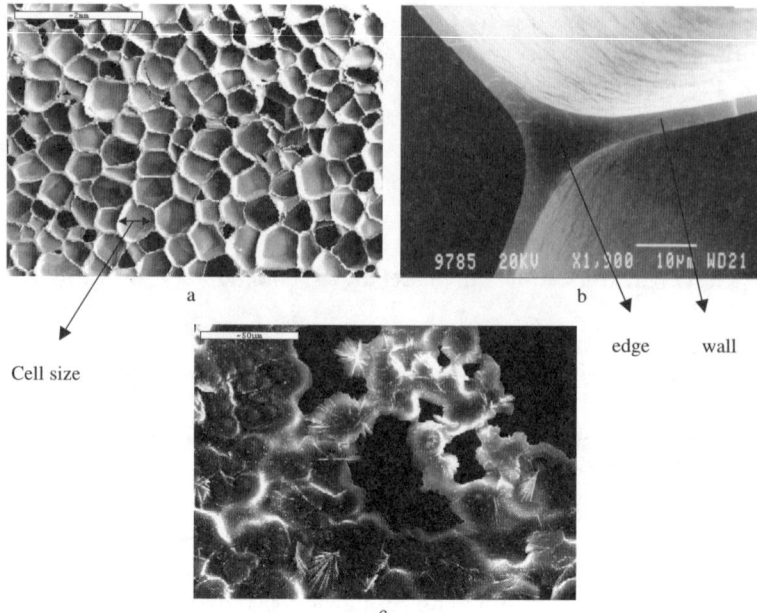

Fig. 5 The three levels of the structure in polyolefin foams. **a** General view of the cellular structure. **b** The polymer phase is built from cell walls and cell edges. **c** This micrograph corresponds to a cell wall after etching [11]

(f_s) (Fig. 5b). Finally, it is convenient to consider a third structural level, the morphology of the base polymer (Fig. 5c).

3.1
Characterization of the Cellular Structure

The most characterized parameter is cell size [11, 13, 57, 58], which can be determined using various methods [59]. An interesting contribution is given by Sims and Khunniteekool [60], who analyzed micrographs of closed-cell foams by different techniques. Excellent agreement for mean cell size data was established between image analysis and the intercept method, in which the number of cells per unit length is measured. It was concluded that rapid accurate determination of apparent cell size was more readily obtained by the intercept counting method. Moreover, they found that characterization of cell size distribution was most accurately performed using an image analysis method.

The other parameters that characterize the cellular structure, i.e., mean cell wall thickness measured directly in the micrographs, fraction of mass in the struts (obtained by the Kuhn method [57]), cell shape, and anisotropy, have been analyzed in several papers [10, 11, 58]. The most difficult characteristic to evaluate is cell shape. Because of the closed-cell structure it is not

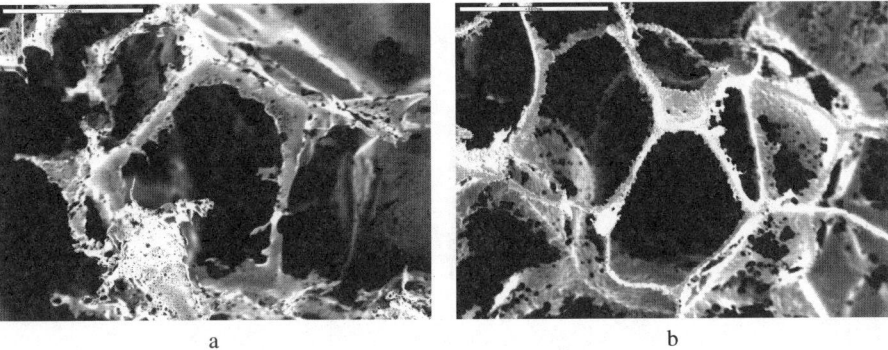

Fig. 6 Examples of cell shapes of closed-cell LDPE foams of **a** 18 kg/m³ and **b** 30 kg/m³ density produced by a high-pressure nitrogen solution process

possible to observe this shape directly with a microscope. Two interesting ways of determining it have been proposed [10, 11]. First, it can be determined using the equation

$$C\delta = \frac{\rho}{\rho_s} \Phi(1 - f_s). \tag{1}$$

The constant C, which can be obtained by fitting the experimental data to the equation, accounts for the cell shape. For example, C takes a value of 3.46 for pentagonal dodecahedrons and a value of 3.35 for tetracaidecahedra. The other way to estimate the cell shape is by doing a permanganate etching to the foam. Thin cell walls are removed, and the shape of the cells can be observed directly with a microscope (Fig. 6). Pentagonal dodecahedrons and tetracaidecahedra are usually observed.

Another interesting finding, which as we will see below has important consequences for the physical properties, is that cell walls in low-density foams are not completely flat; in fact, they are partly buckled or wrinkled [62].

3.2
Characterization of the Polymer Morphology

The polymer in the walls and edges of a foam could have different morphological characteristics than that of continuous polymers produced by extrusion or injection molding. Several reasons support this hypothesis.

1. The cell wall thickness of very low-density foams could be less than 1 μm. These sizes are smaller than the typical spherulite dimensions of a LDPE. As a consequence, a possible influence of the cellular structure on the way in which the foam crystallizes could be expected.
2. The polymer in the foam is crosslinked.
3. The polymer is stretched during foaming.

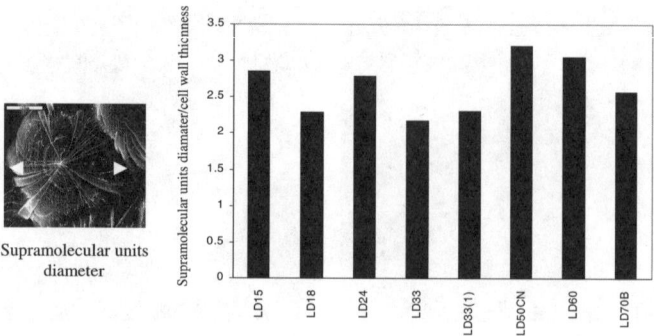

Fig. 7 Ratio between supramolecular units diameter and cell wall thickness for a collection of polyethylene foams of different densities. LDPE foams of different nominal densities (LD15 means LDPE foam of 15 kg/m^3 density). LD50CN and LD70B were black foams. The ratio is a constant for all the foams

All these features suggest a different morphology of the polymer in the foam compared to that of the solid sheet from which the material is produced. This idea has been suggested by several authors [11, 60–63], and, although it has not been completely solved, several experimental studies have considered this topic.

Using scanning electron microscopy on etched cell walls, Almanza et al. [11, 60] found that the microstructure is not spherulitic and, in addition, that cellular structure and polymer morphology are not independent variables. In fact, the diameter of the supramolecular units that form the structure is conditioned by the cell wall thickness. Figure 7 shows this result. A possible explanation lies in the restriction of space the crystals have during their growth in a cell wall. In a similar way, Zipper et al. [63] analyzed PP-structural foam moldings by site-resolved X-ray scattering. The authors showed differences in the microstructure of the foams compared to that of compact injection molded parts.

4
Physical Properties

4.1
Mechanical Properties

PO foams' main applications are packaging and protection against impact. For this reason, an important part of the research on these materials is focused on their mechanical properties. This part of the paper is divided into four sections, each one dedicated to the different types of mechanical situations in which these materials are used: behavior at low and high strain rates (impact), creep response, and dynamic mechanical behavior. Most of the investigations considered compression experiments, which is the common situation in real applications.

4.1.1
Mechanical Properties at Low Strain Rates

The compressive stress–strain curve for PO foams shows three different regions, which are associated to different deformation mechanisms [1] (Fig. 8). At low strains, below approx. 5%, the stress–strain relationship is almost linear. Linear elasticity is controlled by cell wall bending and cell face stretching, and the plateau is associated with the collapse of the cells by elastic buckling. When the cells are almost completely collapsed, opposing cell walls touch, and further stress compresses the solid itself, giving the final region of rapidly increasing stress.

Foam mechanical properties are often explained using Gibson and Ashby's approach [1]. Their simplified microstructural model is built from cubic cells. For closed-cell foams Gibson and Ashby predict three contributions for the Young modulus of a foam:

$$E = E_s \left[C_1 f_s^2 \left(\frac{\rho}{\rho_s} \right)^2 + C_2(1 - f_s) \left(\frac{\rho}{\rho_s} \right) \right] + \frac{p_0(1 - 2\nu)}{(1 - \rho/\rho_s)}, \qquad (2)$$

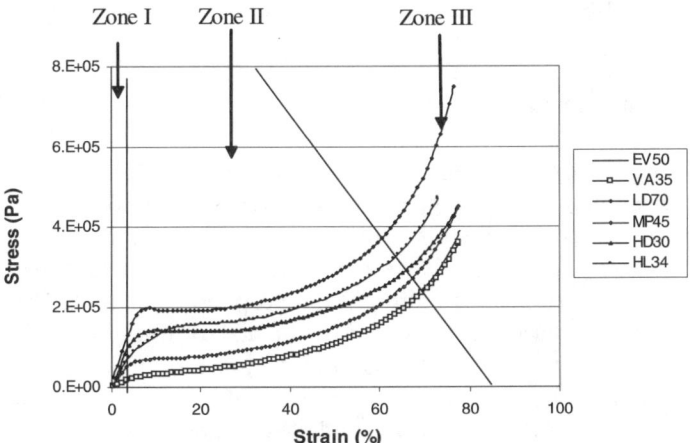

Fig. 8 The three different regions of the compressive stress–strain curve for foams of different densities and chemical compositions. EV50: EVA foam 50 kg/m^3 density, VA35: EVA foam 35 kg/m^3 density, LD70: LDPE foam 70 kg/m^3 density, MP45: LDPE foam (metallocene PE) 45 kg/m^3 density, HD30: HDPE foam 30 kg/m^3 density, HL34, 50% LDPE, 50% HDPE foam, 34 kg/m^3 density

which represents contributions from the bending of cell edges, stretching of cell faces, and compression of the gas. v is the Poisson ratio, p_0 is the initial gas pressure, and E is the Young modulus of the polymer. The constants C_1 and C_2 are fitting constants.

A similar equation is proposed for the collapse or yield stress:

$$\sigma_c = E_s \left(C_3 \left(\frac{\rho}{\rho_s} \right)^2 + \frac{p_0 - p_{at}}{E_s} \right), \tag{3}$$

where p_{at} is the atmospheric pressure.

The postcollapse stress has also been modeled. Gibson and Ashby's slight modification of Rusch analysis [64] gives the following equation:

$$\sigma = \sigma_c + \frac{p_0 \varepsilon}{1 - \varepsilon - \rho/\rho_s}, \tag{4}$$

where ε is the compressive strain. The assumptions are zero lateral expansion of the foam, isothermal gas compression, and an incompressible polymer. The polymer contribution σ_c to stress is assumed to be constant. The variable $\varepsilon/(1 - \varepsilon - R)$ is referred to as gas volumetric strain. Mills and Gilchrist [65] repeated the analysis for foam with a nonzero, but constant, Poisson ratio v. They found that the slope of the graph of applied stress against the gas volumetric strain was $p_0(1 - 2v)$.

None of these equations considers the possible effect of microstructural parameters such as cell size, cell shape, cell wall thickness, etc. For this reason, there has been an interest in improving the oversimplified model of Gibson and Ashby both from an experimental and a theoretical point of view.

Experimental Investigations

The following trends have been proposed [66–69]. The density of the foam is the dominating parameter in the mechanical response followed by f_s and matrix Young modulus [66]. Foams with open cells are less stiff and strong than those with closed cells. In anisotropic foams, strength and stiffness in the direction of elongated cells are higher than perpendicular to it. The temperature dependence is mainly determined by the matrix Young modulus [67, 68]. Nonrecovered strain, collapse stress, and Young modulus seems to be approximately constant as a function of cell size [69].

PO foams, in contrast to PS or rigid PVC foams, are considered multi-impact materials. This means that the foams can be used for more than one impact. The modulus, collapse stress, and cushioning properties of CPOF changes after preloading. Increase in the amount of strain increases the loss in mechanical properties. The greater the foam stiffness, the larger the reduction in properties. [70, 71].

Fatigue loading can occur, for example, during transit of packaging. Fatigue test on crosslinked closed-cell PO foams showed [72], first, that local temperature rises under cyclic loading and, second, that the static compressive stress increased with increasing test speed. The maximum stress in the fatigue loading cycle decreased by 20–30% from the initial value. This could be due both to the temperature increase or changes in the polymer structure.

Theoretical Approach

A more realistic model of the foam's microstructure is the Kelvin model (Fig. 9). In this model a collection of tetracaidecahedra are used as a basis to model the cell morphology.

By using this microestructural model, Kraynik et al. [74] predicted, using finite element analysis, the Young modulus. Kraynik assumed that the cell faces remained flat. The values lay above the experimental data (Fig. 10).

Mills and Zhu [73] used the same microstructural model assuming 60% of the polymer in the cell faces and compression in the (001) direction. Cell edges were bent and compressed axially, while cell faces acted as membranes. The predicted Young's moduli were slightly low (Fig. 10) because compressive face stresses were ignored, but the Poisson ratio was correctly predicted. The predicted value of collapse stress for polyethylene foams was close to the experimental value.

The foam bulk modulus also can be predicted by using the Kelvin model [62]

$$K = \frac{2E}{9(1-v)} \frac{\rho}{\rho_s} + p_{at}. \tag{5}$$

The experimental data are lower by a factor of four than the predicted values. As was pointed out in , this difference is explained by the fact that the faces in low-density foams are partly buckled or wrinkled as a result of processing [62].

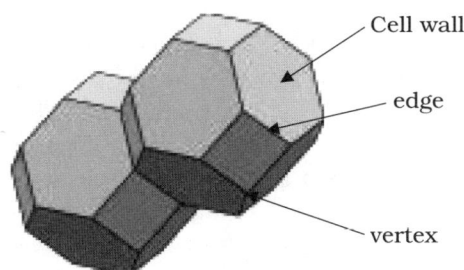

Fig. 9 The Kelvin model, which is currently used to model the foam properties

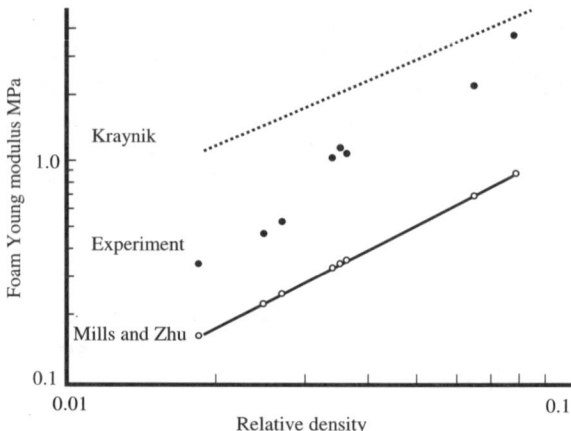

Fig. 10 Comparison between the predictions of different models and the experimental values of the Young modulus [73]

4.1.2
Impact

The last review on this topic was published by Mills in 1994 [75]. In recent years interest has been focused on:

1. The prediction of impact curves from experiments at low strain rates.
2. The estimation of cushion curves from a few impact tests.
3. The behavior of foams subjected to multiple impacts.
4. The analysis of impact with different geometries.

Gruenbaum and Miltz [76] found that the amount of energy absorbed in low strain compression tests and impact tests is almost the same, while the force applied to the product is about 30–50% higher in the dynamic test. Therefore, there is a need for reliable data under impact conditions.

Two methods for the prediction of force-deformation curves of closed-cell plastic foams at any strain rate from a limited number of experiments were described [77]. Both methods use constitutive equations and experimentally determined parameters. The modified Boltzman integral model uses data obtained in a limited number of stress-relaxation experiments, while the reference model uses a very limited number of stress-relaxation and one force-deformation curve data. Both models predict well the force-deformation curves, the reference model providing somewhat better predictions.

Cushion curves are used in the industry to select the most appropriate foams and geometry for a given packaging application. In these diagrams, the maximum deceleration during the impact is presented as a function of the static stress used in the experiment (Fig. 11). Loveridge and Mills [78] proposed a method for the prediction of cushion curves from a single-impact stress–strain curve. The method is valid if there is a master curve for the increasing stress part of the stress–strain curve. For viscoelastic PO foams the divergence from the master curve is less than 10%.

As was mentioned, PO foams are considered multi-impact materials. It has been found [79, 80] that strains in the 80–90% region cause some permanent buckling of the cell walls, but the majority of the deformation recovers within 24 h. With HDPE foam the performance can deteriorate by 30% in a single severe impact, but this is still better than the PS foam widely used. Recovery occurs by the viscoelastic straightening of the buckled faces, but it is incomplete due to some plastic deformation in the structure, so the faces remain slightly buckled [75, 77]. There is no gas diffusion in the time scale of an impact (ms). Consequently, the gas inside the cells plays an important role in the recovery of low-density PO foams after impact testing.

The effect of temperature was analyzed by Marcondes et al. [81]. The materials were tested for shock and vibration under four different temperatures (– 17 °C, 3 °C, 23 °C, and 43 °C). The results show that the properties of expanded polystyrene (in the glassy state in this range of temperatures) were least influenced and those of expanded polyethylene (between the glass transition and melting temperatures) were most influenced by changes in temperature.

Mills and Gilchrist [82] studied oblique impacts in PP bead foams. In these tests, the material was compressed and sheared. This strain combination could occur when a cycle helmet hit a road surface. The results were compared with simple shear tests at low strain rates and with uniaxial com-

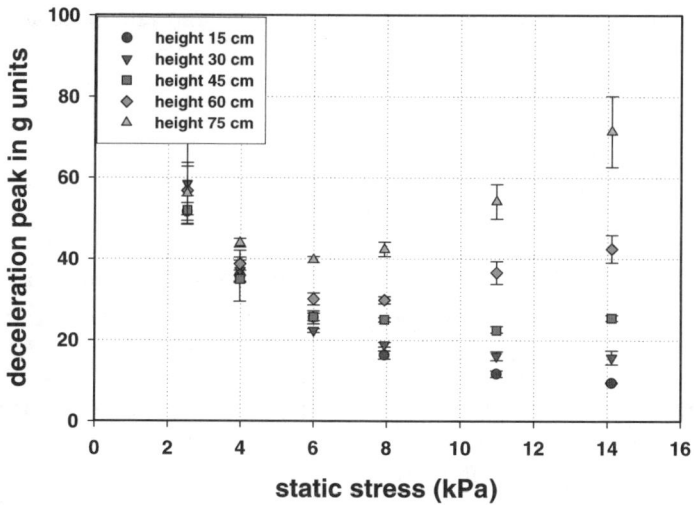

Fig. 11 Cushion curves for a LDPE foam 45 kg/m^3 and 50 mm thickness. The deceleration values are measured in the first impact and for different drop heights

pressive tests at impact strain rates. The observed shear hardening was greatest when there was no imposed density increase and practically zero when the angle of impact was less than 15°. The shear hardening appeared to be a unique function of the main tensile extension ratio and was a polymer contribution, whereas the volumetric hardening was due to the isothermal compression of the cell gas. Foam material models for finite element analysis needed to be reformulated to consider the physics of the hardening mechanisms, so their predictions were reliable for foam impacts in which shear occurred.

Velasco et al. [83] performed indentation rebound tests on PO foams of different densities and chemical compositions. Elastic moduli were calculated and found to increase when foam density and polymer crystallinity increased.

4.1.3
Creep and Gas Diffusion

Creep is dominated by the polymer viscoelasticity if the stress is less than the collapse stress, but at higher stresses gas compression takes an increasing proportion of the load. Gas diffusion is a creep mechanism operating on a time scale that depends on the size of the foam block [84]. For a sample ($20 \times 20 \times 20$ mm^3) of EVA foam with a density of 275 kg/m^3, experimental testing has shown a 50% air loss on a time scale of 10 h [85].

The gas loss was analyzed by fitting the creep data at fixed times to the postcollapse equation (Eq. 4) (Fig. 12). The slope of the different curves accounts for the gas pressure at each time [84, 85].

Modeling of creep was performed following two different approaches. On the one hand, using a discrete model, Mills et al. [84] calculated an effective diffusion coefficient given by:

$$D_{\text{ef}} = \frac{6Pp_a}{\phi\rho}. \tag{6}$$

Therefore, polymer permeability P, density, and mass fraction in the edges are the key parameters for this property. Bhatt et al. [31, 39] suggested that increasing the crosslinking degree reduces the gas loss during mechanical testing, which is probably due to a reduction in polymer permeability.

On the other hand, Pilon et al. [86] presented an engineering model for practical predictions of the effective diffusion coefficient of gases through closed-cell foams. The analysis suggests that the effective diffusion coefficient through the foam can be expressed as a product of a geometric factor and the gas diffusion coefficient through the foam membrane. Comparison of model predictions with experimental data available in the literature shows satisfactory agreement.

Recovery after creep is a slow process [84, 85]. For LDPE, EVA, and PP foams, subjected to creep for 10^6 s, it appears that 100% recovery from

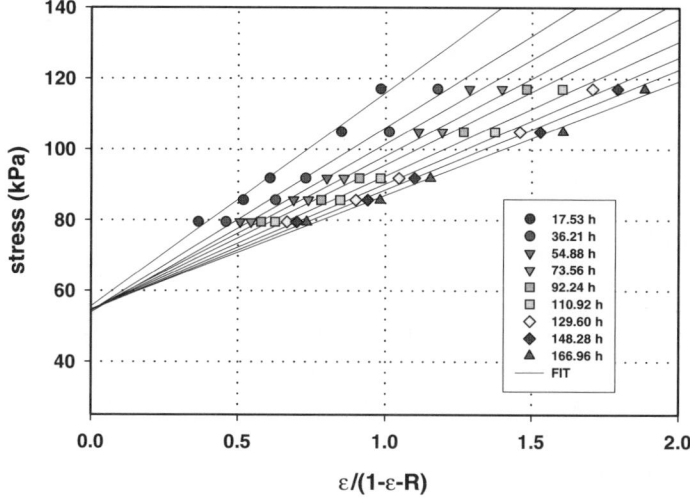

Fig. 12 Isochronous stress vs. volumetric strain curves for a LDPE foam of 60 kg/m^3 density for different creep times (hours). The slope of the linear fits is related to the pressure of the gas inside the cells

the high creep strains will eventually occur, but this will take longer than 10^6 s. The deformation mechanism in creep and recovery occurs in a different order. If a significant percentage of the cell air has diffused out of the foam during creep, the recovery will be slow, because there will be a weak viscoelastic recovery, hindered by the slow reentry of the air to the foam.

4.1.4
Dynamic Mechanical Behavior

The applications of DMA for CPOF foams have been analyzed [87–90]. It has been shown that the storage modulus decreases significantly when the temperature increases. This is typical behavior for semycristalline polymers between the glass transition of the amorphous phase and the melting of the crystalline phase. Consequently, the mechanical properties of all PE and PP foams are strongly temperature-dependent above room temperature. The loss factor (tan δ) of PE foams presents the typical relaxations of polyethylene; the β and α relaxation can be observed in the temperature range between −20 and 80 °C. The deformation mechanisms that control the response of these materials are the cell edge bending and cell face stretching and the gas pressure inside the cells, a contribution which has a higher importance at low densities and high temperatures. It has also been shown that the dynamic mechanical response is nonlinear. Taking into account that the main mechanism that controls the foam response comes from the matrix polymer, this technique can be used to obtain information of the matrix polymer morphology.

4.2
Thermal Properties

4.2.1
Thermal Conductivity

Because of the small proportion of solid in the foam and the consequent large volume of gas, which has a much lower thermal conductivity, the resultant conductivity of the foams is much less than that of the solid materials from which the foams are produced. The other reason for the low conductivity is the negligible contribution of convection because of the small cell size (lower than 2–3 mm) [91].

It is well known [91–93] that heat transfer through a foam is a consequence of four mechanisms: convection in the gas phase, conduction along the struts and cell walls of the solid polymer, conduction through the gas within the cells, and thermal radiation. Many of the early works in this field postulated that the effective conductivity of the foam could be expressed by a superposition of the different mechanisms taken separately.

$$\lambda = \lambda_s + \lambda_g + \lambda_r, \tag{7}$$

where λ is the thermal conductivity, λ_s is the conductivity due to conduction in the solid phase, λ_g is the conductivity through the gas phase, and λ_r is the conductivity by radiation. This concept is an accurate approximation except for cases in which low emissivity boundary layers are used. In that case, the actual effect on the foam conductivity will be far less than predicted by assuming that radiation acts independently of the other heat transfer mechanisms [91].

A good approximation to the conductivity by conduction can be obtained by using the equation [91, 94]:

$$\lambda = \lambda_s + \lambda_g \tag{8}$$
$$= \lambda_{gas} V_{gas} + \lambda_{solid} \left(\frac{2}{3} - \frac{f_s}{3}\right) V_s,$$

where λ_{gas} is the thermal conductivity of the gas that fills the cells, λ_{solid} is the thermal conductivity of the matrix polymer, V_{gas} is the volume fraction of gas, and V_s is the volume fraction of polymer.

The radiation contribution has also been analyzed [10] using the models of Roseland [91], Glicksman [91], Boets and Hoogendoorn [95], and Williams and Aldao [96]. The best fit was obtained using the Williams and Aldao equation [96]:

$$\lambda_r = \frac{4\sigma T^3 L}{1 + \left(\frac{L}{\Phi}\right)\left(\frac{1}{T_N} - 1\right)} \tag{9}$$

$$T_N = \frac{(1-r)}{(1-rt)}\left\{\frac{(1-r)t}{(1+rt)} + \frac{(1-t)}{2}\right\}$$

$$r = \left\{\frac{\omega-1}{\omega+1}\right\}^2 \quad t = \exp(-a\delta),$$

where σ is the Stefan–Boltzman constant, T is the temperature, L is the foam thickness, Φ is the cell size, T_N is the net fraction of radiant energy emitted by a solid membrane of thickness δ, r is the fraction of incident energy reflected by each gas–solid interface, t is the fraction of energy transmitted through the solid membrane, ω is the refractive index of the plastic, and a is the absorption coefficient of the plastic.

Two important parameters, cell size and foam thickness, are taken directly into account for the model. The predicted increase of the thermal conductivity with the cell size is the same as that shown in the experimental results. The model predicts an increase of the thermal radiation term for thicknesses lower than 10 mm, with the thermal radiation being almost constant for foams of higher thicknesses. Experimental results confirm this behavior. Moreover, black foams have a lower conductivity than white foams of the same density and cell size due to the higher extinction coefficient of the black cell walls [69]. Thermal conductivity increases linearly with temperature for PO foams in a temperature range between 24 and 50 °C.

4.2.2
Thermal Expansion

The thermal expansion coefficient of PE foams [13, 15] (a) increases slightly between 0 and 40 °C, (b) is approximately equal to the PE value for foam densities > 80 kg/m^3 but increases at lower densities, (c) is reduced if the Young modulus of the PE increases, and (d) can be anisotropic, with lower values in the direction in which cells are elongated. These trends have been explained qualitatively in terms of the gas and polymer contributions to the total thermal expansion coefficient [13, 15]. The Kelvin microstructural model can be used to predict the thermal expansion coefficient [62]

$$\alpha = \alpha_s + (\alpha_{gas} - \alpha_s)\frac{p_0}{K}, \tag{10}$$

where K is the foam bulk modulus (Eq. 5), α_s is the thermal expansion of the plastic, and α_{gas} is the thermal expansion of the gas.

The model underestimates the foam thermal expansion coefficient because it assumes that the cell faces are flat. The measured bulk modulus, which is considerably smaller than the theoretical value, was used to estimate the linear thermal expansion coefficient of the LDPE foams, obtaining good agreement [62].

4.3
Recycling

One of the main disadvantages of CPOF compared to noncrosslinked PO foams is that, due to the crosslinked matrix, these materials are not recyclable by melting of the polymer phase. Nowadays, the environmental impact is reduced by using the foam scrap to produce other materials. For example, Tamboli et al. [97] used finely ground waste of crosslinked foam as filler in high-density polyethylene (HDPE). Waste foam powder concentration was increased up to 40% by weight basis. The overall changes in mechanical properties are similar to the crosslinking effect. It seems that waste foam particles act as a point of entanglement with different chains of polyethylene.

5
Applications

Table 3 shows the main applications of PO foams and the key properties that are used in each application [3, 26, 32].

It is interesting to note that properties such as light weight, energy absorption, or thermal insulation, related to the cellular structure, are used in many of the applications of these materials. In addition, other characteristics that are connected to the polyolefinic character of the base material, such as low water absorption and low water transmission, chemical resistance, nontoxicity, skin friendliness, and thermoformability are also used.

A growing sector is the sports and leisure markets [5]. The foam works under very exigent conditions in many of these applications; therefore, much care must be taken in the design of these materials. For this reason, several recent studies have considered this aspect.

Mills and Gilchrist [98] performed impact tests on body protectors. Dubois et al. [23] considered the use of ethylene styrene interpolymers (ESI) to produce extruded sheets, bun foams, and injection molded foams (IMF) for footwear parts providing properties that enhance and/or outperform current foams of EVA. Ankrah and Mills [89] evaluated football ankle protectors against a kick from a studded boot. An anatomically correct test rig was used to evaluate materials and designs. Mills and Gilchrist [99] analyzed the optimization of the foam used for shock absorption in bicycle and motorcycle

Table 3 Main applications of CPOF

Type of Application	Main properties	Examples
Cushion packaging	Multi drop-energy absorption Low weight Vibration damping Surface protection Thermal insulation	Corner pads, pads and saddles, encapsulation, case inserts, overlap sheeting, bracing and blocking, micro electronics components packaging, thermal bags
Sports a leisure	Buoyancy Low water absorption Energy absorption Low weight Nontoxicity Skin friendliness Easy cleaning and drying Wide color choice	Body protection, camping mats, swimming pool covers, life vests, swim aids, football ankle protectors, Helmets, midsoles in running shoes
Construction	Low thermal conductivity Sealing against water, dust, air, etc. Sound reflection Vibration damping	Tube insulation, parquet underlay, air duct insulation, sealing strips, gymnastic walls, floating floors, sealant backer, expansion joint filler, closure strips
Industrial uses	Compressibility Oil resistance Buoyancy	Gaskets, flotation collars, ship fenders, tape backing, vibration pads
Automotive/transportation	Weight savings Sound reflection Thermal insulation Sealing behavior (noise, air, and water) Cost savings Soft feel Crash protection Thermoformability	Headliner, watershields, motor undershield, gaskets, dashboard, doorpanels
Health	Nontoxicity Chemically resistance Hygiene Adhesive compatibility Safe disposability/environmentally neutral Wide range of mechanical properties	Plaster, electrocardiogram pads, orthopedic shoe inlays, covering for orthopedical supports

Table 3 (continued)

Type of application	Main properties	Examples
Shoes	High level of comfort, cushioning and sock Absorption Light weight Nontoxicity Good aesthetics	Insoles, midsoles, shoe lining, needle pouched insoles
Adhesive coating	Ease of adhesive coating after surface treatment or adhesive modification Ease of postfabrication Smooth surface Regular cellular structure Wide range of mechanical properties	Pressure-sensitive adhesive coated applications, tapes, strips, pad, joints and gaskets

helmets. Verdejo and Mills [100] performed measurements of plantar pressure distribution in running shoes. The peak plantar pressure increased on average by 100% after a 500-km run. Scanning electron microscopy shows that structural damage (wrinkling of faces and some holes) occurred in the foam after a 750-km run. Fatigue of the foam reduces heelstrike cushioning and is a possible cause of running injuries.

6
Conclusions

The commercial processes used to produce CPOF are well established and based on time-tested concepts and empirical methods. Different technologies result in foams with different cellular structures, and, as a consequence, different properties should be expected. In the last few years, investigations in the area of foaming have focused mainly on the foaming of new resins such as ESI, metallocene polyethylenes, or PP and on the improvement of formulations. In addition, to increase the range of properties and applications of PO foams, open-cell materials have also been produced.

Experimental and theoretical studies have led to significant gains in knowledge on the structure (both cellular structure and polymer morphology) of foams and on the physical mechanisms that control the different properties of foams. Models have been developed to predict the Young and bulk modulus, creep and gas diffusion coefficient, thermal expansion, and thermal conductivity. In this area, the microstructural model based on the Kelvin cell has played an important role.

CPOF foams are being used in more performance applications (especially in the sports and leisure market). Due to this reason, the characterization, selection and design of the foams and the foaming of new materials is nowadays more important for this market sector.

Although knowledge in this field is continuously growing, several topics require more research. Areas for future research are:

1. A more fundamental knowledge of the reactions taking place during foaming could be helpful in the control and optimization of the different processes, especially when a new polymer is being foamed.
2. The characterization of polymers in the cell walls of foams is a topic that has not been deeply studied. The use of micro Raman spectroscopy or X-ray diffraction with beams of a few microns in size would provide more reliable knowledge of the polymer morphology in a given foam.
3. The deformation mechanisms are not fully understood. In situ microcomputed tomography combined with mechanical testing experiments would yield more information on this important aspect.
4. The recyclability of CPOF is one of the major drawbacks for these materials. Finding ways to reuse or recycle these foams is of great interest for this market sector.

References

1. Gibson LJ, Ashby MF (1998) Cellular Solids: Structure and Properties, 2nd edn. Pergamon, Oxford
2. Cunningham A, Hilyard NC (1994) Physical Behavior of Foams: An Overview. In: Hilyard NC, Cunningham A (eds) Low Density Cellular Plastics: Physical Basis of Behaviour. Chapman and Hall, London
3. Park CP (1991) Polyolefin Foams. In: Klempner D, Frisch KC (eds) Handbook of Polymeric Foams and Foam Technology. Hanser, Munich
4. Khemani KC (1997) Polymeric Foams: An overview. In: Khemani KC (ed) Polymeric Foams: Science and Technology, ACS Symposium Series
5. Business Communications Company (2004) RP-120X Polymeric Foams – Updated Edition. Norwalk, CT
6. Eaves DE (1998) Cell Polym 7:297
7. Puri RR, Collington KT (1988) Cell Polym 7:57
8. Puri RR, Collington KT (1988) Cell Polym 7:219
9. Trageser DA (1977) Radiat Phys Chem 9:261
10. Almanza OA, Rodríguez-Pérez MA, de Saja JA (2000) J Polym Sci Part B Polym Phys 38:993
11. Almanza OA, Rodríguez-Pérez MA, de Saja JA (2001) Polymer 42:7117
12. Martínez-Díez JA, Rodríguez-Pérez MA, de Saja JA, Arcos y Rábago LO, Almanza OA (2001) J Cell Plast 37:21

13. Rodríguez-Pérez MA, Alonso O, Duijsens A, de Saja JA (1998) J Polym Sci Part B Polym Phys 36:2587
14. Rodríguez-Pérez MA, Diez S, de Saja JA (1998) Polym Eng Sci 38:831–838(1998)A
15. Rodríguez-Pérez MA, Duijsens A, de Saja JA (1998) J Appl Polym Sci 68:1237
16. Rodríguez-Pérez MA, de Saja JA (1999) Cell Polym 18:1
17. Kyung WS, Park CP, Myron J, Maurer J, Tusim MH, Genova RD, Bross R, Sophiea DP (2000) Adv Mater 23:1779
18. Mapleston P (1998) Mod Plast Int 18:43
19. Park CP, Clingerman GP (1997) Plast Eng March 1997
20. Liu IC, Tsiang RC (2003) Polym Compos 24:304
21. Ankrah S, Verdejo R, Mills NJ (2002) Cell Polym 21:237
22. Chaudhary BI, Barry RP, Tusim MH (2000) J Cell Plast 36:397
23. Dubois R, Karande S, Wright DP, Martínez F (2002) J Cell Plast 38:149
24. Synclair KB (1993) Proc 8th Polyolefins Int Conf SPE, p 1
25. Anon (1995) J Plast Eng 16:20
26. Zotefoams (2003) Product Guide, Croydon, UK
27. Kim DW, Kim KS (2001) J Cell Plast 37:333
28. Kim DW, Kim KS (2002) J Cell Plast 38:471
29. Abe S, Yamaguchi M (2001) J Appl Polym Sci 79:2146
30. Heck RL (1998) Cell Polym 17:31
31. Bhatt CU, Royer JR, Hwang CR, Khan SA (1999) J Poly Sci Part B Polym Phys 37:1045
32. Sekisui Alveo (2003) Product Guide. Roermond, The Netherlands
33. Djoumaliisky S, Christova D, Touleshkov N, Nedkov E (1998) J Macromol Sci A A35:1147
34. Tokuda S, Kemmotsu T (1995) Radiat Phys Chem 46:905
35. Fritz HG, Bolz U, Lu R (1998) Int Polym Process 13:129
36. Zotefoams (1999) High Perform Polym Oct 1999, p 2
37. Sims GLA, Sipaut CS (2001) Cell Polym 20:255
38. Bambara JD, Kozma ML, Hurley RF (1999) US Patent 5883144
39. Bhatt CU, Hwang CR, Khan SA (1998) Radiat Phys Chem 53:539
40. Cardoso ECLm, Lugao AB, Andrade E, Silva LG (1998) Radiat Phys Chem 52:197
41. Dixon D, Martin PJ, Harkin-Jones E (2000) J Cell Plast 36:310
42. Sims GLA, Sirithongtaworn W (1997) Cell Polym 16:271
43. Sims GLA, Khunniteekool C (1996) Cell Polym 15:1
44. Yamaguchi M, Susuki KI (2001) J Polym Sci Part B Polym Phys 39:2159
45. Abe S, Yamaguchi M (2001) J Appl Polym Sci 79:2146
46. Zhang Y, Rodrigue D, Ait-Kadi A (2003) J Appl Polym Sci 90:2111
47. Gendron R, Vachon C (2003) J Cell Plast 39:117
48. Kotzev G, Touleshkov N, Christova D (2002) Macromol Symp 181:507
49. Kotzev G, Touleshkov N, Christova D, Nedkov E (2000) J Cell Plast 36:29
50. Kotzev G, Touleshkov N, Christova D, Nedkov E (1998) J Macromol Sci A A35:1127
51. Hiroo I (1989) US Patent 4877814
52. Park CP (2003) US Patent 6541105
53. Akitaka S, Aizawa T (1984) US Patent 4424181
54. Tai HJ, Wang JB (1997) J Cell Plast 3:304
55. Tatibouët J, Gendron R, Haïder L (2003) Polym Test 23:125
56. Mahapatro A, Mills NJ, Sims GLS (1998) Cell Polym 17:252
57. Kuhn J, Ebert HP (1992) Int J Heat Mass Transfer 35:1795

58. Mills NJ, Zhu HX (1999) The Compression of Closed-Cell Polymer Foams. In: Sadoc F, Rivier N (eds) Foams and Emulsions. NATO ASI Series. Kluwer, Dordrecht, p 175
59. Rhodes MB (1994) Characterization of Polymeric Cellular Structures. In: Hilyard NC, Cunningham A (eds) Low Density Cellular Plastics: Physical Basis of Behaviour. Chapman and Hall, London
60. Sims GLA, Khunniteekool C (1994) Cell Polym 13:137
61. Rodríguez-Pérez MA, de Saja JA (2002) J Macromol Sci Phys Vol B41:761
62. Almanza OA, Masso-Moreu Y, Mills NJ, Rodríguez-Pérez MA (2004) J Polym Sci Part B Polym Phys 42:3741
63. Zipper P, Djoumaliisky S (2002) Macomol Symp 181:421
64. Rusch KC (1970) J Appl Polym Sci 14:1263
65. Mills NJ, Gilchrist A (1997) Cell Polym 16:87
66. Almanza OA, Arcos y Rábago LO, Rodríguez-Pérez MA, González A, de Saja JA (2001) J Macromol Sci Phys B40:603
67. Ramsteiner F, Fell N, Forster S (2001) Polym Test 20:661
68. Clutton EQ, Rice GN (1991) Prog Rubber Plast Tech 7:38
69. Rodriguez-Perez MA, Gonzalez-Pena JI, Witten N, de Saja JA (2002) Cell Polym 21:165
70. Ozkul MH, Mark JE (1994) Polym Eng Sci 34:798
71. Clutton EQ, Rice GN (1991) Prog Rubber Plast Tech 7:38
72. Sombatsompop N, Saengjun B, Tareelap N, Sudaprasert T (1999) Cell Polym 18:197
73. Mills NJ, Zhu H (1999) J Mech Phys Solids 47:669
74. Kraynik AM, Neilsen MK, Reinelt DA, Warren WE (1999) Foam Micromechanics Structure and Rheology of Foams, Emulsions and Cellular Solids. In: Sadoc F, Rivier N (eds) Foams and Emulsions. NATO ASI Series. Kluwer, Dordrecht, p 259
75. Mills NJ (1994) Impact Response. In: Hilyard NC, Cunningham A (eds) Low Density Cellular Plastics: Physical Basis of Behaviour. Chapman and Hall, London
76. Gruenbaum G, Miltz J (1983) J Appl Polym Sci 28:135
77. Miltz J, Ramon O, Mizrahi S (1989) J Appl Polym Sci 38:281
78. Loveridge P, Mills NJ (1993) Proc Cellular Polymers II Conf, Rapra Technology, paper 21
79. Totten TL, Burgess GJ, Singh SP (1990) Packag Tech Sci 3:117
80. Mills NJ, Hwang AMH (1989) Cell Polym 8:259
81. Marcondes J, Hatton K, Graham J, Schueneman H (2003) Packag Tech Sci 16:69
82. Mills NJ, Gilchrist A (1999) Cell Polym 18:157
83. Velasco JI, Martínez AB, Arencón D, Almanza O, Rodríguez-Pérez MA, de Saja JA (2000) Cell Polym 19:115
84. Mills NJ, Gilcrist A (1997) J Cell Plast 33:264
85. Mills NJ, Rodriguez-Perez MA (2001) Cell Polym 20:79
86. Pilon L, Fedorov AG, Viskanta R (2000) J Cell Plast 36:451
87. Rodríguez-Pérez MA, de Saja JA (1999) Polym Test 19:831
88. Rodríguez-Pérez MA (2002) Cell Polym 21:117
89. Ankrah S, Verdejo R, Mills NJ (2002) Cell Polym 21:237-264
90. Pritz T (1994) J Sound Vibrat 178:315
91. Glicsksman LR (1994) Heat Transfer in Foams. In: Hilyard NC, Cunningham A (eds) Low Density Cellular Plastics: Physical Basis of Behaviour. Chapman and Hall, London
92. Leach AG (1993) J Phys D Appl Phys 26:733
93. Collishaw PG, Evans JRG (1983) J Mater Sci 29:486

94. Almanza O, Rodríguez-Pérez MA, de Saja JA (1999) Cell Polym 18:6
95. Boetes R, Hoogendoorn CJ (1987) Proc Int Conf Heat Mass Trans 24:14
96. Williams JR, Aldao CM (1983) Polym Eng Sci 23:32
97. Tamboli SM, Mhaske ST, Kale DD (2004) J Appl Polym Sci 91:110
98. Mills NJ, Gilchrist A (1995) Cell Polym 14:461
99. Mills NJ, Gilchrist A (1991) Accid Anal Prev 23:153
100. Verdejo R, Mills NJ (2004) J Biomechanics 37:1379

Crosslinking of Vinylidene Fluoride-Containing Fluoropolymers

A. Taguet · B. Ameduri (✉) · B. Boutevin

Laboratoire de Chimie Macromoléculaire, Ecole Nationale Supérieure de Chimie de Montpellier, Unité Mixte de Recherche 5076, 8 rue de l'Ecole Normale, 34296 Montpellier Cedex5, France
aurelie.taguet@enscm.fr, bruno.ameduri@enscm.fr, bernard.boutevin@enscm.fr

1	Introduction	129
1.1	Introduction to Fluoropolymers	129
1.2	PVDF	131
1.3	Copolymers Based on VDF	131
2	Generalities	133
2.1	Different Crosslinking Agents	133
2.2	Compounding	134
2.3	Press-cure and Post-cure Steps for Crosslinking	134
3	Crosslinking of VDF-Based Fluoroelastomers	136
3.1	Crosslinking with Amines and Diamines	136
3.1.1	Dehydrofluorination of the Fluoropolymer	136
3.1.2	Second Step: the Michael Addition of the Amine	145
3.1.3	The Different Amines and Diamines	147
3.1.4	Formation of Two Networks During Post Cure	167
3.2	Crosslinking with Bisphenols	170
3.2.1	Crosslinking Mechanism	170
3.2.2	^{19}F NMR Study	171
3.2.3	Oscillating Disc Rheometer (ODR) Response	173
3.2.4	Limitations of the Bisphenol-Cured Fluoroelastomers	175
3.3	Crosslinking with Peroxides	175
3.3.1	Reaction Conditions	176
3.3.2	Importance of the Coagent	179
3.3.3	Influence of the Nature and the Amount of the Peroxide	181
3.3.4	Mechanism of Crosslinking	184
3.4	Radiation Crosslinking	186
3.4.1	Crosslinking Mechanism by Electron Beam Radiation	187
3.4.2	Influence of Irradiation Parameters on the Properties of Crosslinked Fluoropolymers	190
3.5	Crosslinking with a Thiol-ene System	196
4	Comparison of Physical and Mechanical Properties	197
5	Applications	203
6	Conclusion	204
References		206

Abstract Fluoropolymers are well-known for their good properties in terms of chemical, thermal and electrical stabilities, inertness to acids, bases, solvents and oils, and high resistance to ageing and oxidation. Polyvinylidene fluoride (PVDF) is useful as a homopolymer endowed with interesting characteristics. It contains a high crystallinity rate, but is base sensitive. In addition, VDF can be co- or terpolymerized with several fluorinated monomers, rendering them suitable as elastomers and various examples of synthesis of VDF-copolymers are also presented. This review also focusses on binary and tertiary systems containing VDF. Several curing systems for these VDF-containing copolymers have been investigated, especially diamines and their derivatives, aromatic polyhydroxy compounds, peroxides with coagents, such as triallylisocyanurate, radiations, and thiol-ene systems. The best vulcanizate properties are obtained by a two-step process. First, the material is press cured at different times and temperatures, then, it is post cured in air or under nitrogen at higher temperature and time, and under atmospheric pressure. Poly(VDF-*co*-HFP) copolymers can react with primary, secondary or tertiary monoamines, but they are mainly crosslinked by diamines such as hexamethylene diamine (HMDA), their carbamates (HMDA-C), and derivatives. A mechanism of crosslinking is identified by Infrared and ^{19}F NMR spectroscopies, and was evidenced to proceed in three main steps. First, a VDF unit undergoes a dehydrofluorination in the presence of the diamine, then the Michael addition occurs onto the double bonds to form crosslinking, while HF is eliminated from crosslinks in the presence of HF scavengers. The crosslinking mechanism with bisphenols takes place also in three main steps (dehydrofluorination, then substitution of a fluorine atom by a bisphenol, and elimination of HF). The most efficient crosslinking bisphenol is bisphenol AF.

A fluoropolymer crosslinked with peroxide/coagent systems needs to be functionalized or halogenated to insure a free radical attack from peroxide. The peroxide is introduced with a coagent that enhances the crosslinking efficiency, and the most efficient one is triallylisocyanurate (TAIC). The crosslinking mechanism of the peroxide/triallylisocyanurate system proceeds in three main steps. The crosslinking reaction occurs from a macroradical arising from the functional or halogenated polymer which is added onto the three double bonds of the TAIC. A fourth way to crosslink VDF-based fluoropolymers deals with high energy radiation, such as X and γ (^{60}Co or ^{137}Cs)-rays, and charged particles (β-particles and electrons). Three different reactions are possible after irradiation of a PVDF, and the one that leads to crosslinking is the recombination between two macroradicals. The irradiation dose used on the VDF-based copolymer has an influence on the thermal and mechanical properties.

Finally, a crosslinking system also used to vulcanize hydrogenated elastomers concerns a thiol-ene system which requires a mercapto function born by the VDF-based polymer. Crosslinking occurs via a non-conjugated diene. The mechanical properties (tensile strength, elongation at break, hardness, elongation modulus, compression set resistance ...) of the three main crosslinking systems of fluoroelastomers are compared. Finally, the main applications of crosslinked VDF-based fluoropolymers are summarized which include tubing in the aircraft building industry, sealing, tube or irregular-profile items of any dimension, films with good adhesion to metallic or rigid surfaces, multilayer insulator systems for electrical conductors, captors, sensors, and detectors, and membranes for electrochemical applications.

Keywords Crosslinking · VDF-containing copolymers · Amines · Bisphenols · Peroxides/Triallylisocyanurate

Abbreviations

BTPPC	benzyltriphenylphosphonium chloride
CTFE	chlorotrifluoroethylene
DBU	1,8-diazabicyclo[5-4-0]-undec-7-ene
DETA	diethylene triamine
DMAC	dimethylacetamide
d.o.g.	degree of grafting
DSC	differential scanning calorimetry
DTA	differential thermal analysis
EDA (-C)	ethylene diamine (carbamate)
HBTBP	hexamethylene-N,N'bis(*tert*-butylperoxycarbamate)
HFP	hexafluoropropene
HMDA	hexamethylene diamine
HMDA-C	hexamethylene diamine carbamate
HPFP	1H-pentafluoropropene
MBTBP	methylene bis-4-cyclohexyl-N,N'(*tert*-butylperoxycarbamate)
ODR	oscillating disc rheometer
PMVE	perfluoro(methyl vinyl ether)
PVDF	polyvinylidene fluoride
$t_{1/2}$	half life
TAC	triallylcyanurate
TAIC	triallylisocyanurate
TFE	tetrafluoroethylene
THF	tetrahydrofurane
VDF	vinylidene fluoride

1
Introduction

1.1
Introduction to Fluoropolymers

Fluorinated polymers are particularly interesting and attractive compounds because of their properties. Indeed, the electronegativity of the fluorine atom implies strong C – F bonds (about 110 kcal mol^{-1}), and a higher strength of the C – C bonds in fluorinated compounds (97 kcal mol^{-1}). It also supplies to fluoropolymers strong Van der Waals forces between hydrogen and fluorine atoms [1–3], and it confers a lot of good properties to the fluorinated polymers such as:
- Chemical, thermal, electric stabilities [4–6];
- Inertness to acids, bases, solvents and oils;
- Low dielectric constant;
- Low refractive index;
- No flammability;

- High resistance to ageing, and to oxidation;
- Low surface tension.

Fluorinated polymers range from a wide scope of thermoplastics, elastomers, plastomers, thermoplastic elastomers [7–14], and can be semi-crystalline or totally amorphous. Hence, fluorinated polymers have been used in many applications: building industries (paints and coatings resistant to UV and to graffiti), petrochemical and automotive industries, aerospace and aeronautics (use of elastomers as seals, gaskets, O-rings used in extreme temperature for tanks of liquid hydrogen for space shuttles), chemical engineering (high-performance membranes), optics (core and cladding of optical fibers), treatment of textile, stone protection (especially for old monuments), microelectronics [8–14], and for cable insulation.

As a matter of fact, the performance of fluoropolymers, especially insolubility and fusibility can be improved by crosslinking. Indeed, the crosslinking reaction takes advantage of the base-sensitive characteristic of the VDF-based polymer [15]. The crosslinking is a chemical reaction between the polymer backbone and an ex-situ agent, both of which possess the same functions in order to couple covalently the polymeric chains together, to produce a network structure, and to increase the molecular weight.

Sulfur has been the predominant curing agent in the rubber industry since it was first used with rubber in 1840[16]. Many efforts have been devoted over the 40 years of existence of fluoroelastomers toward the development of practical crosslink systems.

Fluoroelastomers are now usually cured by nucleophiles such as diamines [17–31], or bisphenols [3, 32–38], or with peroxides [3, 35, 39–43], by chemical reactions when the polymers based on VDF contain a cure-site monomer, such as a thiol function [44], or by radiation, such as an electron beam [45–52].

The cure chemistry of VDF-based fluoroelastomers is connected with the strong polarity of the C–F bond and specific polarization of molecules, which determine their selective ability to split off hydrogen fluoride under the influence of internal factors.

The first part of the review presents the generalities of the crosslinking of VDF-based fluoroelastomers, and especially the two steps of the cure (the press cure and the post cure). The second part deals with the crosslinking involving different agents: first, the aliphatic and aromatic amines and diamines, then the bisphenol-cure, third the peroxide-cure, fourth the crosslinking by irradiation, and finally the thiol-ene system-curing. Then, the third part compares all the crosslinking systems, by considering the main mechanical properties, and finally, the last part concerns the applications of the crosslinked VDF-based fluoroelastomers.

1.2
PVDF

Among fluoropolymers, polyvinylidene fluoride (PVDF) is a semi-crystalline and thermoplastic polymer, with a glass transition temperature of – 40 °C [53, 54]. This polymer exhibits interesting thermal, chemical and physical properties, especially when it is co- or ter-polymerized with a fluorinated alkene [14, 35, 50, 55–59]. Its main drawback is its sensitivity to base that can degrade it by creating insaturations. PVDF homopolymer is a long chain macromolecule endowed with a high crystallinity rendering it unsuitable as an elastomer, and unsuitable for curing. Therefore, copolymers of VDF with various comonomers can fall into three categories: (i) when the amount of comonomers in the copolymer is small, the resulting materials are thermoplastics with a lower crystallinity than that of the PVDF [60]; (ii) with a slightly higher content of comonomer, thermoplastic elastomers are obtained; (iii) for a higher proportion of comonomers, the produced copolymers are elastomeric and amorphous with low intermolecular forces [35, 50, 57, 58, 61–65]. In the case of the poly(VDF-*co*-HFP) copolymer, when the molar percentage of VDF is higher than 85%, the copolymer is a thermoplastic, whereas for a smaller content, the copolymer is an elastomer [35, 50, 66].

1.3
Copolymers Based on VDF

VDF has been involved in radical copolymerization with many monomers [14, 60–67], listed in Table 1 [44, 68–94].

Most common co- or termonomers of VDF [14, 50] are hexafluoropropene (HFP) [67, 80, 82, 83, 95–97], tetrafluoroethylene (TFE) [78, 80, 81, 98, 99], chlorotrifluoroethylene (CTFE) [67, 78, 79, 100–102], trifluoroethylene (and in that case, interesting piezoelectrical materials have been obtained) [75], perfluoro(methyl vinyl ether) (PMVE) [84, 103], perfluoro(alkoxyalkylvinylether) [104, 105] enabling the resulting copolymers to exhibit very low T_g, and 1*H*-pentafluoropropene (HPFP) [106, 107]. Interestingly, functional fluoromonomers (also called cure site monomers) useful for further crosslinking, have been successfully used, bearing OH [86], CO_2H [71, 89], $Si(OR)_3$ [94] functions, or bromine [87] and iodine atoms mentioned in Sect. 3.3.1. Table 1 supplies a non-exhaustive list of fluoromonomers that were copolymerized with VDF, and their reactivity ratios r_i, when assessed.

Although it is difficult to compare their reactivities [since (i) the copolymerization were not carried out under similar conditions, (ii) certain articles do not mention if the kinetics of copolymerization were realized at low monomer conversion, and (iii) various kinetic laws were used], it was worth examining a reactivity series of fluorinated monomers with VDF. The traditional method involving the determination of the reactivity of a macroradical

Table 1 Monomer reactivity ratios for the radical copolymerization of VDF (A) with other fluoroalkenes (B) (and vinyl acetate and ethylene)

Monomer B	r_A	r_B	$r_A r_B$	$1/r_A$	Ref.
$H_2C=CH_2$	0.05	8.5	0.42	20.00	[68]
$H_2C=CHOCOCH_3$	−0.40	1.67	−0.67	−2.5	[69]
	0.50	2.0	1.00	2.0	[70]
$H_2C=C(CF_3)CO_2H$	0.33	0	0	3.03	[71]
$FCH=CH_2$	0.17	4.2–5.5	0.71–0.94	5.88	[72]
	0.20–0.43	3.8–4.9	0.76–2.11	2.33–5.00	[73]
$H_2C=CFCF_2ORF$	0.38	2.41	0.92	2.63	[74]
$F_2C=CFH$	0.70	0.50	0.35	1.43	[75]
$F_2C=CHCF_3$	9.0	0.06	0.54	0.11	[76]
$F_2C=CHC_6F_{13}$	12.0	0.90	10.80	0.08	[77]
$CFCl=CF_2$	0.73	0.75	0.55	1.37	[78]
	0.17	0.52	0.09	5.88	[79]
$CFBr=CF_2$	0.43	1.46	0.63	2.33	[78]
$CF_2=CF_2$	0.23	3.73	0.86	4.35	[78, 80]
	0.32	0.28	0.09	3.13	[81]
$CF_3-CF=CF_2$	6.70	0	0	0.15	[82]
	2.45	0	0	0.40	[80]
	2.90	0.12	0.35	0.34	[83]
$F_2C=CFOCF_3$	3.40	0	0	0.29	[84]
$F_2C=CFOC_3F_7$	1.15	0	0	0.86	[84]
$F_2C=CFO(HFP)OC_2F_4SO_2F$	0.57	0.07	0.04	1.75	[85]
$CF_2=CFCH_2OH$	0.83	0.11	0.09	1.02	[86]
$CF_2=CF(CH_2)_2Br$	0.96	0.09	0.09	1.00	[87]
$CF_2=CF(CH_2)_3OAc$	0.17	3.26	0.59	5.56	[88]
$F_2C=CF(CH_2)_3SAc$	0.60	0.41	0.25	4.07	[44]
$CF_2=CFCO_2CH_3$	0.30	0	0	3.33	[89]
$F_2C=C(CF_3)COF$	7.60	0.02	0.15	0.13	[90]
$F_2C=C(CF_3)OCOC_6H_5$	0.77	0.11	0.08	1.30	[91]
$F_2C=CFOC_6H_4R$ [1]	n.d.[3]	n.d.	n.d.	n.d.	[92, 93]
$F_2C=CFC_3H_6Si(OR)_3$ [2]	n.d.	n.d.	n.d.	n.d.	[94]

[1] $R = Br, SO_2X (X = Cl, F)$
[2] $R = CH_3, C_2H_5$
[3] n.d. = not determined

to several monomers was used. Indeed, it is common to compare the value $1/r_A = k_{AB}/k_{AA}$, as the ratio of rate constants of co-propagation (k_{AB}) to that of homo-propagation (k_{AA}). Thus, the higher the $1/r$ value, the higher the co-propagation reactivity of the radical. On the basis of the data in Table 1, the increasing order of relative reactivities of monomers to $\sim$ VDF• macroradicals is as follows:

$F_2C=CHC_6F_{13} < F_2C=CHCF_3 < HFP < PMVE < PPVE < F_2C=CFC_2H_4Br < VDF < F_2C=CFCH_2OH < F_2C=C(CF_3)OCOC_6H_5 < TrFE < CTFE$ (recent value) $\approx$ BrTFE $< H_2C=CFCF_2ORF < F_2C=CFCO_2CH_3 < TFE < F_2C=CFC_3H_6SCOCH_3 < F_2C=CFC_3H_6OAc < H_2C=CHF \approx CTFE$ (old value) $< H_2C=CH_2$, although numerous kinetics still deserve to be investigated.

2
Generalities

In order to improve their properties, poly(VDF-*co*-HFP) copolymers or poly(VDF-ter-HFP-ter-TFE) terpolymers can be crosslinked by bisnucleophiles, such as diamines or bisphenols, or by irradiation. On the other hand, poly(VDF-ter-HFP-ter-termonomer containing an iodine or bromine atom) terpolymer can be crosslinked by peroxide/coagent systems. Those three main ways of crosslinking exhibit two main crosslinking mechanisms (ionic and radical mechanisms) and different properties.

2.1
Different Crosslinking Agents

Several curing systems have been investigated or developed for the crosslinking of fluoroelastomers. Some of them are [35, 108]:

- high energy radiation [19, 45–52];
- peroxide with or without coagent [3, 35, 39–43];
- dithiols in combination with amines [19];

Table 2 Improvement of mechanical properties of bisphenol, and peroxide-cured poly(VDF-*ter*-HFP-*ter*-TFE)terpolymer with post cure step [35]

Properties	Bisphenol		Peroxide	
	Press cure[1]	Post cure[2]	Press cure[1]	Post cure[2]
Modulus at 100% strain, (MPa)	5.0	7.9	5.0	7.9
Tensile strength at break, (MPa)	10.0	13.8	9.7	15.9
Elongation at break, (%)	225	175	165	150
Compression set, (%) (200 °C, 70 h)				
O-rings	63	25	50	27
Pellets	85	20	52	20

[1] press cure at 177 °C, for 10 min
[2] post cure at 232 °C, for 24 h

- aromatic polyhydroxy compounds [3, 32–38];
- diamines and their derivatives [17–31];
- thiol-ene systems [44, 50].

Each curing system exhibits a different crosslinking mechanism, and results in different mechanical properties and crosslinking densities. Indeed, Table 2 [35] shows different mechanical properties for bisphenol and peroxide cured systems.

The comparison of the mechanism is comprehensively described in Sect. 3, while Sects. 4 and 5 are devoted to the study of mechanical properties and the applications of each system.

2.2
Compounding

In order to improve the properties of the raw elastomer, many materials that facilitate mixing or processing may be compounded with the vulcanizing agent [28, 55], (i) accelerators and accelerator activators to increase the rate of vulcanization and to improve product properties; (ii) fillers to enhance physical properties and/or to reduce costs; (iii) softeners to process or to plasticize the product; (iv) antioxidants and other materials which slow down decomposition of the product by oxidation; (v) heat and/or radiation; (vi) pigments and blowing agents.

For the main additional materials, the proportions (in part per hundred of polymer) are [55]:

- Raw polymer 100;
- Curing agent 1–6;
- Basic metallic oxide 6–20;
- Filler > 60.

2.3
Press-cure and Post-cure Steps for Crosslinking

The best vulcanizate properties are obtained by a two step-process [35, 58, 109]. Fluoroelastomers and additives are generally molded in a press and then post cured in an oven [28].

First, the materials are press cured at different times and temperatures, depending on the size of the product, the structure of the polymer, the curing systems, and on end-use requirements (paints, O-rings, membranes, seals) [28]. Press-cure conditions vary from 4 min at temperatures approaching 200 °C for thin cross sections, to 30 min at 150–170 °C for thick sections [28, 110]. The purpose of this step is to develop sufficient crosslinks in the sample to prevent the formation of bubbles due to the release of trapped air during the early stages of the subsequent oven cure [111].

Then, the second step (post cure or oven cure) is carried out in air or under nitrogen at higher temperature than that of the press cure, and under atmospheric pressure [35, 58].

This post-cure step is required to reach the best vulcanizate properties (tensile strength, modulus at 50 or 100% elongation, compression set resistance, elongation at break) [28, 40, 108, 111]. Table 3 [35] shows the improvement of compression set resistance with post curing, for four samples containing poly(VDF-ter-HFP-ter-TFE) terpolymer crosslinked with a peroxide [2,5-bis(t-butylperoxy)-2,5-dimethylhexyne] in the presence of triallylisocyanurate [35, 108, 112, 113]. Table 2 [35] presents the improvement of some mechanical properties of bisphenol and peroxide cured systems with post cure.

An improvement in compression set resistance is observed after post cure under nitrogen compared to that realized under air (Table 3). The C=C double bond of the polymeric backbone undergoes an oxidation from the oxygen of the air atmosphere, that prevents a good compression set resistance. Table 2 shows a 50% increase in modulus at 100% elongation (M_{100}) and tensile strength at break, and a 50% decrease in elongation at break.

During the step of crosslinking of fluoroelastomers, water is formed, and post cure removes this water, whose presence prevents the full development of the diamine cure and causes reversion of the bisphenol cure [3, 23, 40, 114]. Indeed, during press cure, water is formed from the reaction between the acid acceptor and HF, caused by dehydrofluorination.

For thick sections, the temperature of the post-cure oven is usually raised in several steps to prevent fissuring. Generally, 12–24 h reaction time at a temperature of 200–260 °C is used [28, 35, 58, 110]. Typically, 200 °C is suffi-

Table 3 Compression set resistance measured at 204 °C for 70 h, of a peroxide cured-poly(VDF-ter-HFP-ter-TFE)terpolymer, after press cure in air at 177 °C for 15 min, and after post cure in air or under nitrogen at 232 °C for 24 h [35]

Compound[1]	Compression set %		
	Press cured	Post cured	
	air	air	N_2
A	71	38	20
B	70	37	25
C	59	27	12
D	52	21	9

[1] compounds A and C contain: 100 parts of polymer, 3 parts of peroxide II, 3 parts of TAIC, and 3 parts of PbO. In compound D PbO is replaced by 2 parts of MgO, and 2 parts of ZnO. And in compound D, PbO is replaced by 6 parts of $Ca(OH)_2$, and 3 parts of MgO. Moreover, the bromo cure site in A and B differed from that in C and D

cient for amines [59, 111], whereas bisphenol and peroxide cures need higher temperatures (230 to 260 °C).

All these results suppose a difference in the crosslinking mechanism of bisphenol, peroxide and diamine cured systems, that are the most important crosslinking agents for VDF-based fluoroelastomers. The crosslinking mechanisms and the properties of the resulting crosslinked polymers are the subject of the following parts of this article.

3
Crosslinking of VDF-Based Fluoroelastomers

3.1
Crosslinking with Amines and Diamines

Curing by diamines, originally introduced in the late 1950s, was the predominant way of crosslinking raw fluoroelastomers until the late 1960s, when bisphenol curing was introduced [55, 58]. The polyamine system is the best for general use because of easier processing [96]. Indeed, it only needs the presence of hydrogen atoms in the polymer backbone. Moreover, the mechanism of crosslinking can be a simple addition to this backbone [96].

The diamine curing system generally results in relatively poor processing, safety concerns, thermal and ageing resistance, and compression set resistance. However, this cure system has demonstrated specific properties, such as excellent adhesion to metal [115].

The curing of elastomer with an amine or a diamine usually takes place in the three following steps [24, 114, 116, 117]:

1. an elimination of HF (dehydrofluorination) from VDF segments adjacent to HFP in the main chain to generate internal double bonds;
2. a Michael addition of the diamine onto the resulting double bonds to form crosslinks;
3. an elimination of HF from the crosslinks, during post cure to form further double bonds.

These steps are detailed below.

3.1.1
Dehydrofluorination of the Fluoropolymer

The dehydrofluorination of a solution of Viton poly(VDF-*co*-HFP) copolymer treated with several amines, or heated at high temperature can be monitored by measurement of hydrogen fluoride elimination (titration of the HF in the solution) [5, 15, 19, 116], infrared study [19, 24, 118], viscosity [19], solubility and determination of the gel content [5].

3.1.1.1
Evidences of Dehydrofluorination

Solutions of Viton in tetrahydrofurane were treated with primary, secondary and tertiary monoamines for periods of several weeks at room temperature. The reaction was followed by the measurement of HF elimination by a titration of the hydrogen fluoride in the solution.

Figure 1 [19] shows the evolution of the quantity of HF in the solution of THF as a function of time for primary, secondary and tertiary monoamines. All of the monoamines used caused dehydrofluorination of the polymer to some degree. Tertiary amines are the least efficient, primary amines are by far the most active.

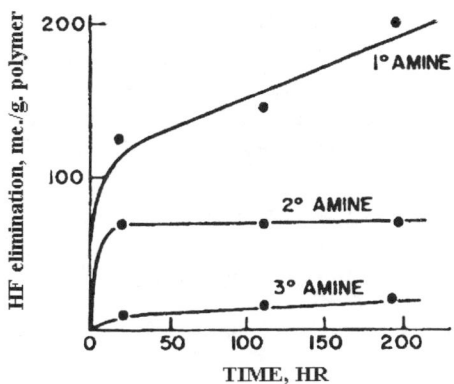

Fig. 1 Amount of elimination of HF from primary, secondary and tertiary amines cured VitonA (Reprinted with permission of Lippincott et Peto) [19]

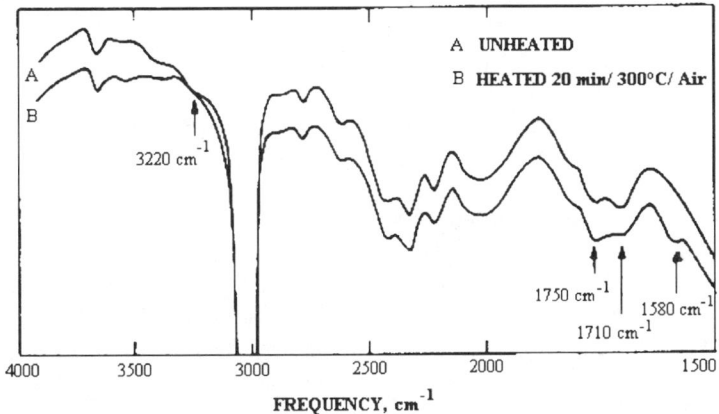

Fig. 2 Infrared spectra of an uncured poly(VDF-*co*-HFP) copolymer before (A) and after (B) heating at 300 °C for 20 min under air (Reprinted with permission of ACS) [118]

Figure 2 [24, 118] shows the infrared spectrum of uncured poly(VDF-co-HFP) copolymer (FKM gum), before and after a thin film of polymer is heated in air at 300 °C. Two new bands centered at 1580 and 1750 cm^{-1} appeared after heating, which are assigned to the conjugated double bonds and to the $-CH=CF_2$ end groups, respectively. Unsaturation is likely to be caused by elimination of HF from PVDF block of the FKM chain, in particular from the head-to-tail position of the structure.

So, in the presence of a base or under heating, the VDF-based fluoroelastomers are submitted to dehydrofluorination.

3.1.1.2
Consequences of the Dehydrofluorination

The conjugated double bonds evidenced by infrared measurements allowed us to interpret a new mechanism. Figure 2 [118] exhibits the presence of isolated double bonds (1710 cm^{-1}), and the presence of conjugated double bonds (1580 cm^{-1}). It is proposed that the initial double bond (1710 cm^{-1}) activates the elimination of HF from neighboring atoms leading to conjugated double bonds (1580 cm^{-1}). This process would lead to the formation of a brown color [20, 24, 118].

Such a conjugate site would then be expected to react with a double bond, in an adjacent chain by a Diels–Alder reaction, leading to a fluorinated cyclohexene which should readily lose HF to form an aromatic ring (Scheme 1) [19, 119].

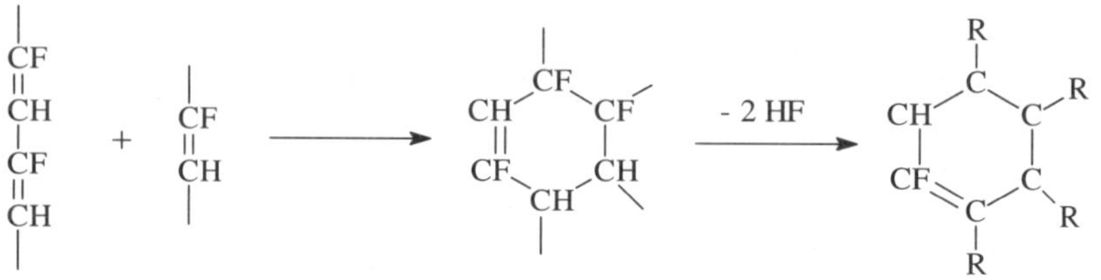

Scheme 1 Diels–Alder reaction during post-curing forms aromatic ring with loss of HF [19]

The observed absorption at 1580 cm^{-1} could be ascribed to such a site.

The evolution of the solubility of a raw poly(VDF-co-HFP) copolymer heated in air at 250 °C is shown in Table 4 [24]. Indeed, there is an initial rapid decrease in solubility, and then it proceeds to rise slowly. This type of variation of solubility, together with the formation of a swollen gel, indicates the simultaneous occurrence of crosslinking and chain scission in the polymer [5, 114].

During heating or attack with a base, the polymer undergoes a dehydrofluorination, creating conjugated double bonds that can be involved in a Diels–

Table 4 Fraction soluble in acetone of a poly(VDF-co-HFP)copolymer heated in air at 250 °C [24]

Time (hr)	Fraction soluble in acetone at 28 °C	Volume fraction of polymer in swollen gel fraction at equilibrium in acetone at 28 °C
3	0.78	Not determined
24	0.52	0.02
42	0.47	0.02
48	0.48	0.02
137	0.53	0.01

Alder reaction. But, at higher temperature or in the presence of a stronger base, it also creates degradation such as oxidation or scissions that can be evidenced by the measurements of the decrease in the intrinsic viscosity, caused by the decrease in molecular weight [19]. In order to avoid any degradation, the created double bonds can become the site of the addition of several agents like diamines, bisphenols or peroxides, that can increase the mechanical and chemical properties. The formation of the scissions in the network are explained in Sect. 3.1.4.

3.1.1.3
Sites of Dehydrofluorination

In VDF-based fluoropolymers, and especially poly(VDF-co-HFP) copolymer, dehydrofluorination occurs on special sites.

Paciorek et al. [23] studied the crosslinking of amines on several fluorocompound models. The model of addition of butylamine onto 1,5,5-trihydro-4-iodoperfluorooctane and 4-hydroperfluoroheptene-3, in diethylether at room temperature, is the only one known. It proceeds according to the following scheme:

$$C_2F_5CFICH_2C_3F_7 + H_2NC_4H_9 \rightarrow$$
$$C_2F_5C(=NC_4H_9)CH_2C_3F_7 + C_2F_5C(NHC_4H_9) = CHC_3F_7$$

The reaction occurs mainly on the carbon adjacent to the iodine atom, because dehydrofluorination is the main process under the selected conditions.

From ^{19}F NMR characterization, Schmiegel [3, 15, 32, 33, 67] showed that a polymer based on VDF units with HFP, TFE, PMVE co- or ter-monomers in a solution of DMAC can undergo dehydrofluorination from the n-Bu$_4$N^{+-}OH in specific sites.

Figure 3 [33] represents the 294.1 MHz ^{19}F NMR spectra of poly(VDF-*co*-HFP) copolymer before (top) and after (bottom) treatment with hydroxylic base in DMAC at 20 °C. Peaks A and B are assigned to CF_3 group, peaks C, D, E, F, G, H, I, J, K, and L are attributed to CF_2 of VDF, peaks M and N are

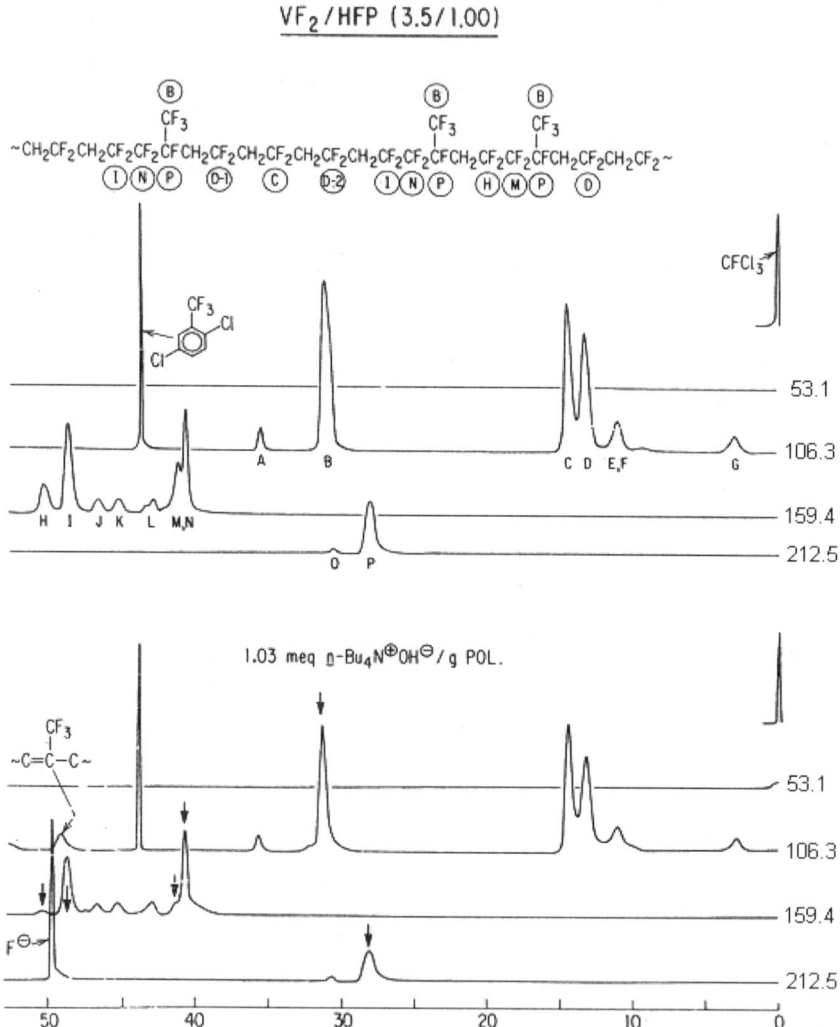

Fig. 3 ^{19}F NMR spectra of a poly(VDF-*co*-HFP) copolymer before (*top*) and after (*bottom*) treatment with hydroxylic base (2,5-trifluorobenzotrifluoride internal standard). Changes in peak intensities are indicated (Reprinted with permission of Verlag Chemie) [33]

assigned also to CF_2 of the HFP, and finally peaks O and P are assigned to CF. The small resonances A, G and O correspond to HFP inversions, whereas F, J, K, and L are attributed to VDF inversions. The spectrum at the bottom exhibits selective intensity reduction of resonance B, H, I, M, N and P after add-

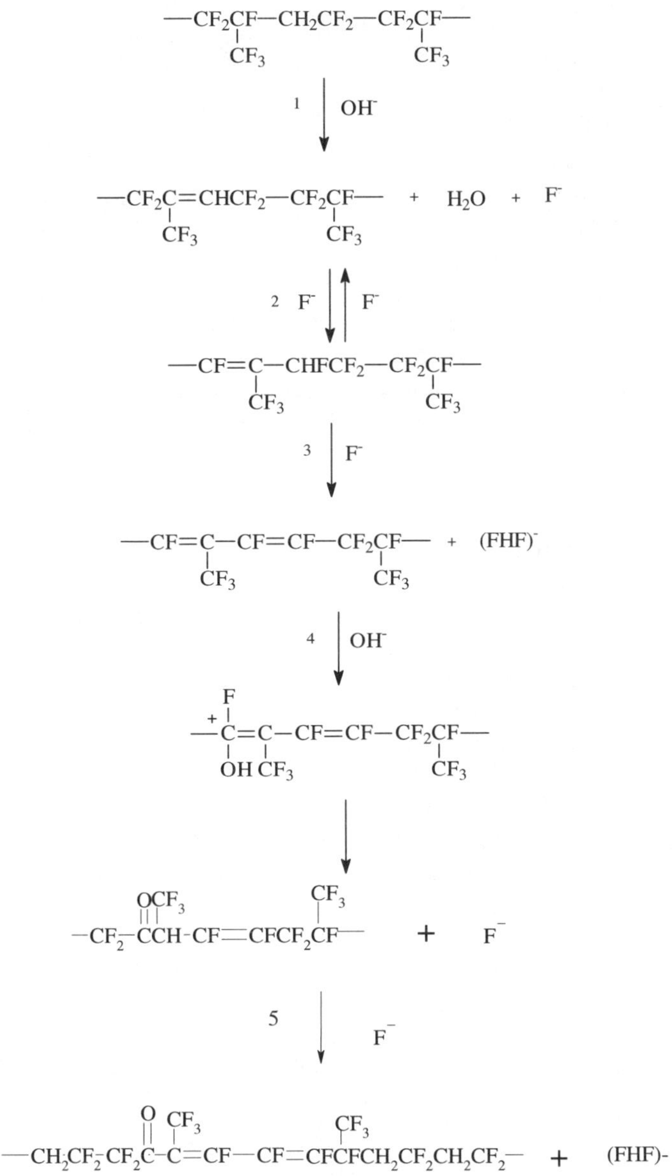

$$—CF_2CF(CF_3)—CH_2CF_2—CF_2CF(CF_3)—$$

$$\xrightarrow{1,\ OH^-}$$

$$—CF_2C(CF_3)=CHCF_2—CF_2CF(CF_3)— + H_2O + F^-$$

$$\xrightarrow{2,\ F^-\ \rightleftarrows\ F^-}$$

$$—CF=C(CF_3)—CHFCF_2—CF_2CF(CF_3)—$$

$$\xrightarrow{3,\ F^-}$$

$$—CF=C(CF_3)—CF=CF—CF_2CF(CF_3)— + (FHF)^-$$

$$\xrightarrow{4,\ OH^-}$$

$$\overset{+}{-}\underset{OH\ CF_3}{C}(F)=C—CF=CF—CF_2CF(CF_3)—$$

$$\downarrow$$

$$-CF_2-\underset{\overset{\|}{O}CF_3}{C}CH-CF=CFCF_2\underset{CF_3}{CF}— + F^-$$

$$\xrightarrow{5,\ F^-}$$

$$—CH_2CF_2CF_2\underset{\overset{\|}{O}}{C}—\underset{CF_3}{C}=CF—CF=CFCFCH_2CF_2CH_2CF_2— + (FHF)-$$

Scheme 2 Dehydrofluorination mechanism of poly(VDF-co-HFP) copolymer in the presence of base [32, 33]

ition of $Bu_4N^+OH^-$. A peak assigned to CF_3 groups of $C=C(CF_3)-C$ appears also at −55 ppm. These observations can be accommodated to the highly selective dehydrofluorination of isolated VDF units, i.e. HFP-VDF-HFP structures [3, 32–34]. The concentration of this site in a 3.5 poly(VDF-*co*-HFP) copolymer is about 0.6 mol/kg. The same results were observed in poly(VDF-*co*-TFE) and poly(VDF-*co*-PMVE) copolymers, and poly(VDF-ter-HFP-ter-TFE) and poly(VDF-ter-PMVE-ter-TFE) terpolymers [33, 67]. For example, in poly(VDF-*co*-TFE) copolymer, dehydrofluorination occurs on VDF units having a TFE-VDF-TFE triad, or in a poly(VDF-ter-HFP-ter-TFE) terpolymer, it occurs on the HFP-VDF-TFE structure.

A reaction scheme of dehydrofluorination of poly(VDF-*co*-HFP) copolymer in the presence of a base was given by Schmiegel (Scheme 2) [32, 33].

First, the attack of hydroxide creates a double bond on VDF units in the VDF-HFP diad. Then, a fluoride ion rearrangement of the initial double bond occurs. The resulting allylic hydrogen is abstracted by fluoride, followed by an elimination of a second fluoride. So, a bifluoride and a formally conjugated non-coplanar diene are formed. Then, a nucleophilic attack by the hydroxide on the diene forms an enone and a subsequent attack of the fluoride ion onto the highly acidic hydrogen of the tertiary carbon atom. The final product is the dienone [32, 33].

3.1.1.4
Role of the Acid Acceptor

An acid acceptor of metal oxide type is a necessary ingredient of all VDF-based polymer curing formulations. No cure is obtained without any metal oxide (magnesium oxide), and the state of cure developed is directly related to the amount of MgO [111, 114, 120].

Figure 4 [111] represents the evolution of the tensile strength and the modulus versus the quantity of MgO, for a trimethylamine hydrochloride cured poly(VDF-*co*-HFP) copolymer.

Indeed, there is evidence from infrared that MgO contributes to the elimination of HF from the polymer during irradiation, and probably also in the course of the chemical cures.

Figure 5 [116] shows the variation of the amount of fluoride ions at 200 °C with MgO content. The presence of MgO does not prevent HF elimination; it merely reduces its rate of evolution from the elastomer, a 15% addition giving a result comparable with that of the raw polymer alone.

The reaction between MgO and HF is given in the following scheme [114]:

$$MgO + 2HF \rightarrow MgF_2 + H_2O$$

Several metal oxides can be used as HF scavengers for VDF-based polymers. The relative efficiencies of a number of basic oxides, hydroxides and carbonates as HF acceptors at approximately 275 °C are illustrated in Fig. 6 [116]. It

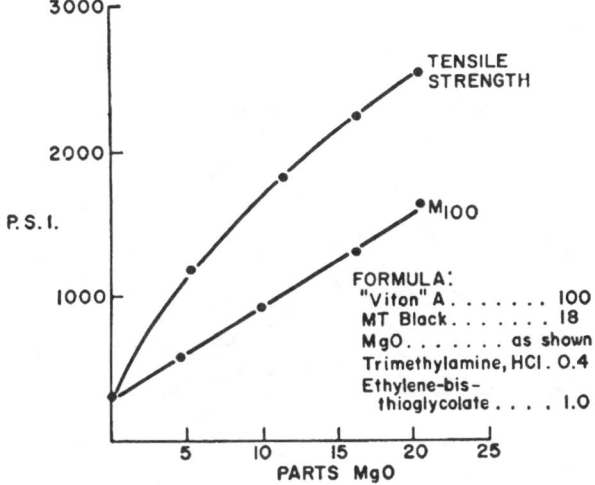

Fig. 4 Effect of MgO on the mechanical properties of a formula comprising VitonA cured with dithiol (Reprinted with permission of Lippincott and Peto) [111]

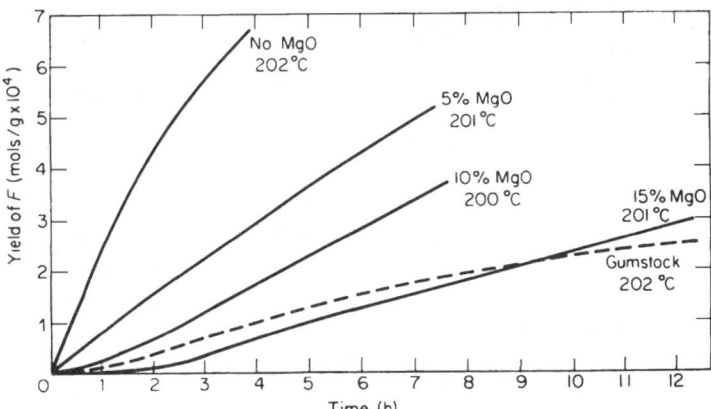

Fig. 5 Evolution of the yield of fluoride atom of a VDF-based fluoropolymer heated at 200 °C versus time and amount of MgO (acid acceptor) (Reprinted with permission of Wiley) [116]

is apparent that there are many variations in the efficiencies of the different compounds. The decreasing order of efficiencies is as follows:

$CaO \gg Li_2O \approx B_2O_3 \approx BeO > Al_2O_3 > MgO > TiO_2$

The hydroxides are significantly better acceptors than their analogous oxides. The decreasing order of efficiencies is [116]:

$Ca(OH)_2 > Mg(OH)_2 > LiOH > Al(OH)_3$

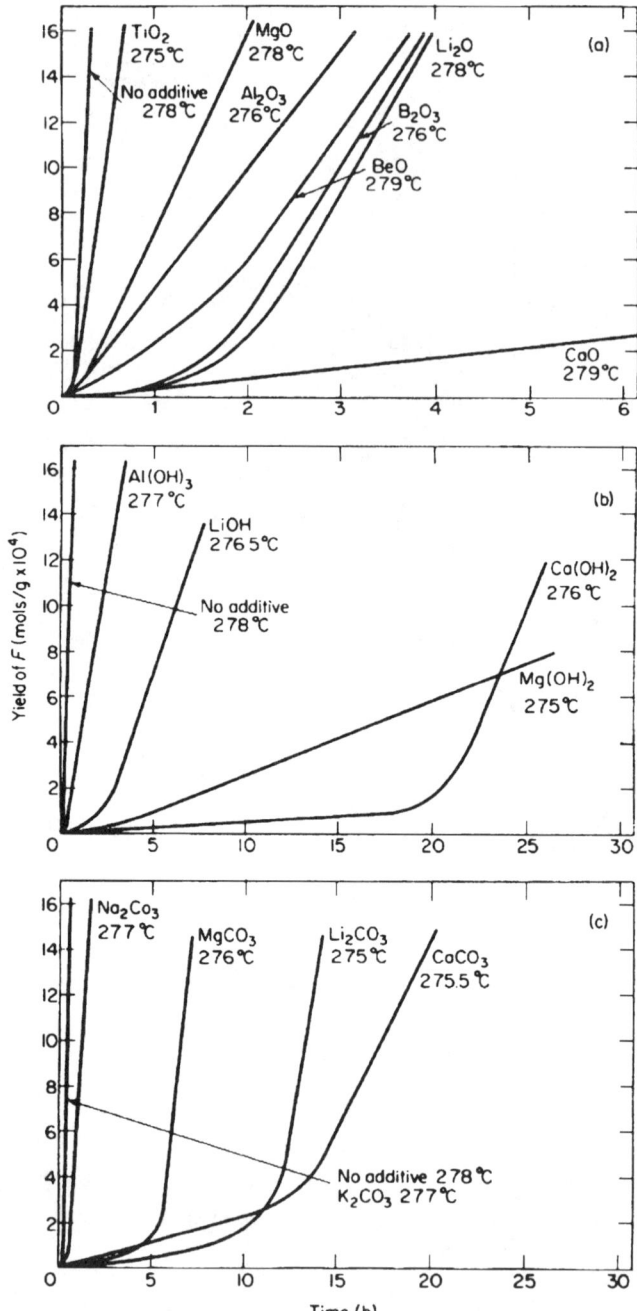

Fig. 6 Comparison of the efficiency of acid acceptors: metal oxide at 275 °C (**a**), hydroxide acceptors at 275 °C (**b**), and carbonate acceptors at 275 °C (**c**) (Reprinted with permission of Wiley) [116]

Finally, the decreasing order of efficiencies for carbonates is [116]:

$CaCO_3 > Li_2CO_3 > MgCO_3 > Na_2CO_3 > K_2CO_3$

The most commonly used acid acceptor is MgO.

Thus, dehydrofluorination of a VDF comonomer in the diad is the first step of a crosslinking mechanism with diamine. The second step consists in the addition of the amine or the diamine onto that unsaturation.

3.1.2
Second Step: the Michael Addition of the Amine

After the dehydrofluorination of the poly(VDF-co-HFP) copolymer, the amine can add across the unsaturated center. Addition can be carried out with primary and secondary diamine, and less readily with primary and secondary monoamines. Vulcanization of VDF-based fluoroelastomers is induced by secondary and tertiary monoamines [20].

Paciorek et al. [20] studied the treatment of Viton-A [poly(VDF-co-HFP) copolymer] and Kel-F [poly(VDF-co-CTFE) copolymer] with different primary, secondary and tertiary mono- and diamines. It appears that Kel-F elastomer required specific crosslinking conditions according to the nature of the (di)amine, at room temperature for primary mono- and diamines, at 50–60 °C for secondary mono- and diamines, at 90–100 °C for tertiary diamines, and at 180–190 °C for tertiary monoamines.

3.1.2.1
Mechanism with Monoamines

The general mechanism of grafting of a primary or a secondary monoamine onto a model compound is given in Scheme 3 [22]. The different steps of this mechanism are identified by infrared spectroscopy. The primary monoamine (butylamine), like a secondary monoamine, dehydroiodinates the model compound creating $CF = CH$ double bonds. Then, the amine can add onto the unsaturation thanks to Michael addition. Finally, as has been shown by Pruett et al. [17], a structure containing $- NH - CF(X)$-group readily eliminates hydrogen fluoride, leading to $- N = C(X) -$, or $C = C(N) - X$.

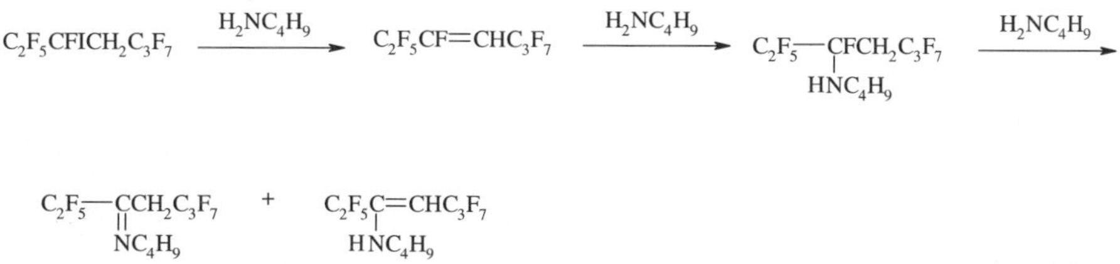

$$C_2F_5CFICH_2C_3F_7 \xrightarrow{H_2NC_4H_9} C_2F_5CF{=}CHC_3F_7 \xrightarrow{H_2NC_4H_9} C_2F_5{-}\underset{HNC_4H_9}{CFCH_2C_3F_7} \xrightarrow{H_2NC_4H_9}$$

$$C_2F_5{-}\underset{NC_4H_9}{\overset{\|}{C}CH_2C_3F_7} \quad + \quad C_2F_5\underset{HNC_4H_9}{C}{=}CHC_3F_7$$

Scheme 3 Reaction between butylamine and a model compound [22]

In addition to difunctional curing agents, strong basic primary, secondary and tertiary amines also create crosslinking of Viton-A, even if they require a rather high press temperature to obtain successful cures when used alone [19, 111]. Indeed, those basic components can help the dehydrofluorination of the polymer backbone. Further, mono tertiary amines are potential cocuring agents for all diamines. Tertiary amines show good efficiency as a cocuring agent in combination with dithiols [111]. Indeed, dithiols do not crosslink Viton when used alone, but in combination with tertiary amines, well-cured vulcanizates can be formulated by their use [19].

At higher temperature or time (12 days at 25 °C with 72% of amine) a primary monoamine such as butylamine can crosslink a poly(VDF-*co*-HFP) copolymer or a poly(VDF-*co*-CTFE) copolymer [20]. The mechanism of crosslinking of the butylamine onto a poly(VDF-*co*-CTFE) copolymer [20] is given in Scheme 4. In a first step, the amine dehydrochlorinates the VDF/CTFE diad. Then, due to a Michael addition, the amine adds onto the CF = CH double bond, creating a secondary amine. Finally, in the last step of the mechanism, the secondary grafted amine can add again onto an unsaturation creating a bridge between two polymeric chains.

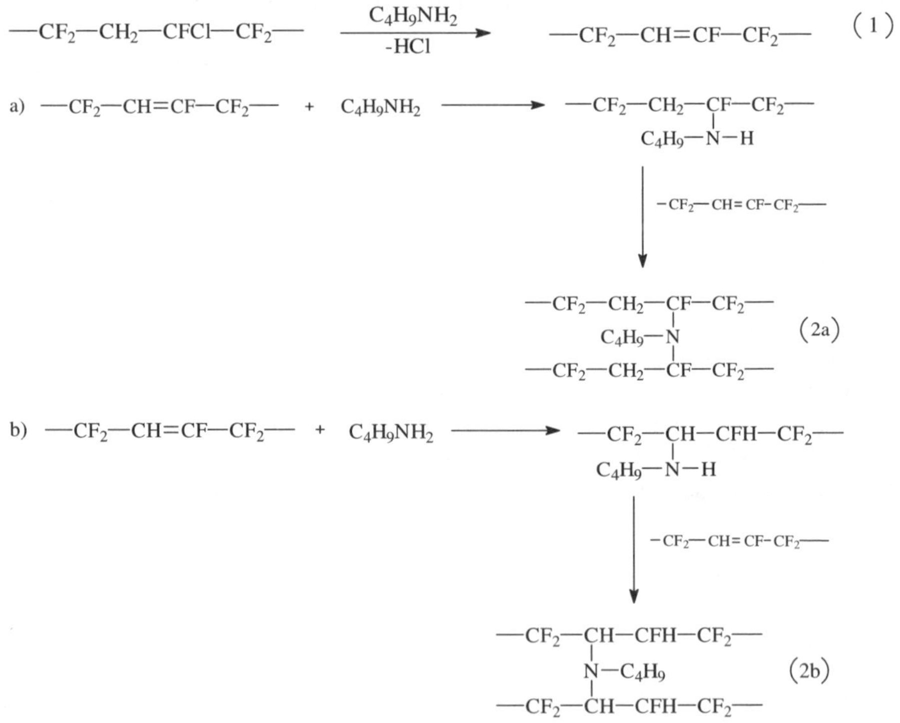

$$-CF_2-CH_2-CFCl-CF_2- \xrightarrow[-HCl]{C_4H_9NH_2} -CF_2-CH=CF-CF_2- \quad (1)$$

a) $-CF_2-CH=CF-CF_2- + C_4H_9NH_2 \longrightarrow -CF_2-CH_2-\underset{\underset{H}{|}}{\underset{C_4H_9-N}{|}}CF-CF_2-$

$\downarrow -CF_2-CH=CF-CF_2-$

$$-CF_2-CH_2-\underset{\underset{-CF_2-CH_2-CF-CF_2-}{|}}{\underset{C_4H_9-N}{|}}CF-CF_2- \quad (2a)$$

b) $-CF_2-CH=CF-CF_2- + C_4H_9NH_2 \longrightarrow -CF_2-\underset{\underset{H}{|}}{\underset{C_4H_9-N}{|}}CH-CFH-CF_2-$

$\downarrow -CF_2-CH=CF-CF_2-$

$$-CF_2-\underset{\underset{-CF_2-CH-CFH-CF_2-}{|}}{\underset{N-C_4H_9}{|}}CH-CFH-CF_2- \quad (2b)$$

Scheme 4 Reaction mechanism between butylamine and a poly(VDF-*co*-CTFE) copolymer [20]

The addition of the butylamine can occur either at the carbon atom bearing a hydrogen as postulated in sequence (b), or at the fluorine-bearing carbon atom as postulated in sequence (a), although a controverse was also found in the literature [17, 121].

3.1.2.2
Mechanism with Diamines

The mechanism of crosslinking with diamine is similar to that involving monoamines. The mechanism of crosslinking with hexamethylenediamine onto a poly(VDF-co-HFP) copolymer is given in Scheme 5 with $R = (CH_2)_6$ [19, 21]. This mechanism occurs in the course of the press-cure treatment of the polymer (150–170 °C, 30 min). As above, in a first step, the diamine dehydrofluorinates the VDF/HFP diad, creating a double bond. Then, by Michael addition the diamine adds onto two $CF = CH$ unsaturated backbones, creating bridges between polymeric chains. The CF – NH bonds are sensitive to the oxygen atmosphere and heating, and submit to a further dehydrofluorination leading to a $C = N$ bond that can degrade into a $C = O$ bond.

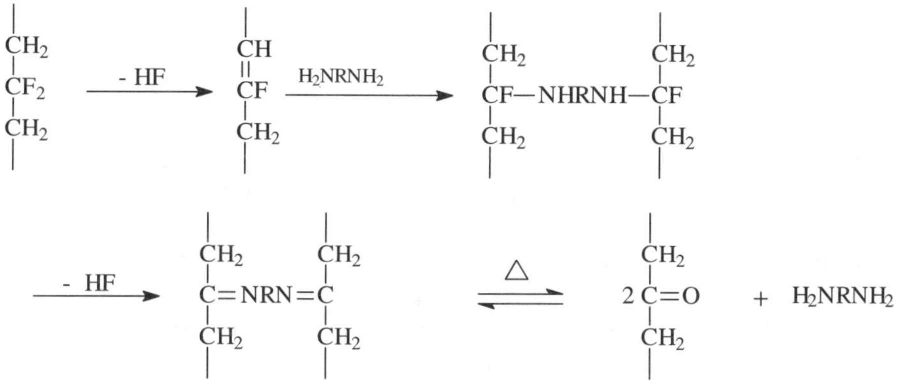

$$\begin{array}{c}|\\CH_2\\|\\CF_2\\|\\CH_2\\|\end{array} \xrightarrow{-HF} \begin{array}{c}|\\CH\\\|\\CF\\|\\CH_2\\|\end{array} \xrightarrow{H_2NRNH_2} \begin{array}{cc}|&|\\CH_2&CH_2\\|&|\\CF-NHRNH-CF\\|&|\\CH_2&CH_2\\|&|\end{array}$$

$$\xrightarrow{-HF} \begin{array}{cc}|&|\\CH_2&CH_2\\|&|\\C=NRN=C\\|&|\\CH_2&CH_2\\|&|\end{array} \xrightarrow{\Delta} \begin{array}{c}|\\CH_2\\|\\2\,C=O\\|\\CH_2\\|\end{array} + H_2NRNH_2$$

Scheme 5 Mechanism of crosslinking with diamine, in three main steps

3.1.3
The Different Amines and Diamines

Although the crosslinking mechanism of amines and diamines onto VDF-based fluoropolymers proceed in three main steps and are similar, the reaction conditions (temperature and time) and the physical properties of aliphatic and aromatic containing mono or diamines are different.

3.1.3.1
Reaction with Aliphatic and Cycloaliphatic Monoamines

As mentioned above, Paciorek et al. [22] studied the addition of butylamine, dibutylamine and triethylamine on model fluoro-compounds. They also studied the reaction between a Viton-A poly(VDF-co-HFP) copolymer and a Kel-F poly(VDF-co-CTFE) copolymer with monoamines [20], in solution of diglyme, at different times and temperatures, and for different amounts of amines:

- butylamine $C_4H_9NH_2$
- dibutylamine $C_4H_9NHC_4H_9$
- piperidine

Structure 1

- triethylamine $(C_2H_5)_3N$
- diethylcyclohexylamine

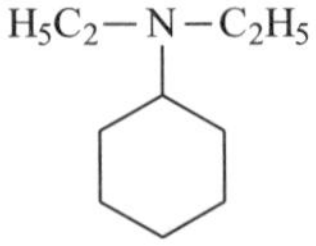

Structure 2

Only one model of addition of amines on partially fluorinated molecules was studied. Indeed, the study of addition of equimolar quantities of monoamines (butylamine, dibutylamine and triethylamine) onto 4,4-dihydro-3-iodoperfluoroheptane as a model molecule, in diethylether at room temperature [22] afforded 80, 94 and 81% of the amine hydroiodides, respectively. But, by determining the time required for a given reaction mixture to reach a pH value of 6, it is concluded that the reaction with butylamine is faster than that using dibutylamine. This latter system is faster than that involving triethylamine.

The reaction between several monoamines and VDF-based fluoropolymers like Viton-A and Kel-F [20] evidences a crosslinking mechanism.

Tables 5 and 6 [20] exhibit the weight percentage of added amine, the temperature of reaction, the formation of a gel, the reaction time, the initial and final pH, and finally the color of the solution.

The presence of a gel from butylamine, dibutylamine, piperidine and diethylcyclohexylamine-cured Kel-F polymer, and the piperidine-cured Viton-A evidences a reaction of crosslinking between the polymer backbone and those monoamines [20].

Table 5 Reaction of a Kel-F poly(VDF-co-CTFE)copolymer with several amines [20]

Amine	%[1]	T (°C)	Crystalline precipitate	Gel	Anal. of solution	Conversion of amine to hydrohalide[2] (%)	Reaction time (days)	Initial pH	Final pH	Final color
Butylamine	72	25	Yes	Yes	Cl^-, F^-	–	21	9+	5-6	Orange
	54						12	9	5	Yellow
Dibutylamine	75	25	Yes	No	Cl^-, F^-	75	39	9+	6-7	Orange
	55	25	Yes	No	Cl^-, F^-	66	15	9	5	–
	11	170	–	Yes	–	–	1	–	–	–
	11	50	Yes	Yes	Cl^-, F^{-4}	–	1	8-	–	Yellow
	60	25	Yes	No	Cl^-	89	27	9	5	Orange
Piperidine	54	25	Yes	No	Cl^-, F^-	86 (5)	35	9	5	Orange
	11	50	Yes	Yes	Cl^-	–	1	8	–	Pale yellow
	11	25	Yes	No	Cl^{-4}	95	2	8	5	Almost colorless
Triethylamine	460	25	Yes	No	Cl^{-4}, F^-	$79^3(0.1)$	61	–	9-10	Brown
	54	25	Yes	No[4]	Cl^-, F^-	65	25	9	7+	–
	11	185	Yes	Yes	Cl^-, F^-	–	1	–	–	Brown
	11	150	Yes	No	Cl^-, F^-	85(19)	1	9	–	Red
Diethylcyclo-hexylamine	53	25	Yes	No	Cl^-, F^-	21(3)	15	9	8	Yellow
	15	172	No	Yes	Cl^-, F^-	–	1	–	–	Brown
	11	150	No	No	Cl^-, F^-	90(22)	1	–	–	Red

[1] Based on the chlorine content of the polymer and equivalent weight of the amine
[2] Hydrochlorides shown; hydrochlorides in parentheses
[3] Based on chlorine available
[4] Trace

Table 6 Reaction of a poly(VDF-*co*-HFP)copolymer Viton-A with monoamines [20]

Amine	%[1]	T (°C)	Crystalline precipitate	Gel	Conversion of amine to hydrohalide (%)	Reaction time (days)	Initial pH	Final pH	Final color
Butylamine	18	25	Few	No	57[2]	6	–	5	Yellow
	89		Yes		28[2]	18	9	5	Orange
Dibutylamine	89	25	Yes	No	45[3]	20	9	7+	Dark red
	18	190	No	Little	–	1	8	7	Dark brown
Piperidine	89	25	No	No	47[2]	8	9	5	Dark red
	18	190	No	Yes	–	1	8	7	Dark brown
Triethylamine	88	25	No	No	73[3]	17	9	8	Dark yellow
	19	190	No	No	–	1	8–	–	Dark brown
Diethylcyclo-	88	25	No	No	53[3]	20	8+	8	Yellow
hexylamine	18	100	No	Little	–	1	8–	8	Dark brown

[1] Based on the tertiary fluorine content of the polymer and equivalent weight of the amine
[2] Based on the formation of amine dihydrofluoride
[3] Based on the formation of amine monohydrofluoride

Thus, the following mechanism can be postulated as a crosslinking from primary monoamines. This is explained in Scheme 4.

Poly(VDF-co-CTFE) copolymers are crosslinked more easily in the presence of monoamines than poly(VDF-co-HFP) copolymers are, because HF elimination should proceed more readily with tertiary fluorine than with the difluoromethylene group [20]. Hence, dehydrofluorination proceeds at a much lower rate with Viton-A than with Kel-F elastomer.

3.1.3.2
The Aliphatic Diamines and Diimines

Several diamines are used as curing systems for fluorocarbon elastomers: acyclic diamines, cyclic diamines, and some times aromatic diamines [28].

Diamines give effective vulcanizates, but those leading to the best properties are very scorchy [19, 21]. Derivatives of diamines designed to stiffen the amine function, so as to reduce the scorching tendency, are by far the most widely used curing agents [19, 21]. The most common examples are the carbamate and bis-cinnamylidene derivatives of the hexamethylene diamine (HMDA):

- bis-cinnamylidene hexamethylene diamine [19, 108]

$$\langle \bigcirc \rangle - CH=CH-CH=N-(CH_2)_6-N=CH-CH=CH-\langle \bigcirc \rangle$$

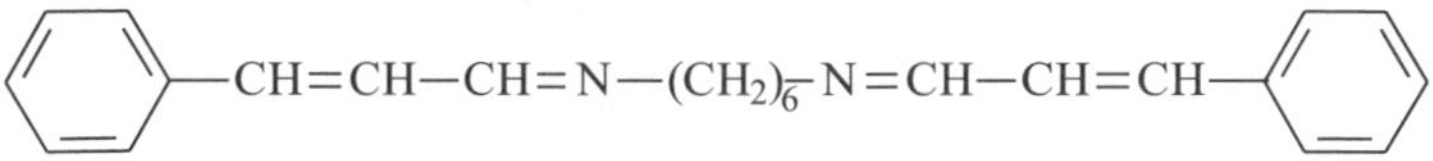

Structure 3

- hexamethylene diamine carbamate (HMDA-C) [108]

$$^+H_3N-(CH_2)_6-N\begin{matrix}COO^-\\ \diagdown\\ H\end{matrix}$$

Structure 4

The bis-cinnamylidene hexamethylene diamine [19,108]:
A Viton-HV poly(VDF-co-HFP) copolymer with a molar percentage of HFP of 28.5%, in the presence of 5 parts of MgO per hundred parts of rubber (phr), was vulcanized with different phr of Diak No3 (N,N'-dicinnamylidene-1,6-hexanediamine) [27].

Table 7 [27] gives the C_1 constant in the Mooney–Rivlin equation [122–124]:

$$2C_1/gRT = 6.1 \times 10^{-5} D \tag{1}$$

where g is the fraction of gel rubber, R is the gas constant, and D the amount of curing agent expressed in phr.

Table 7 also supplies the gel fraction (g), and crosslinking density of the different samples (A-1 to A-8) called v_e and v_e^*, where $v_e = v_e^* \times g$. The higher the amount of Diak No3, the greater the gel fraction, and the higher the crosslinking density. Indeed, for example in sample A-8 (8 phr of curing agent), all the polymer is in gel fraction (no soluble fraction), so it is completely crosslinked, and evidences the good efficiency of Diak No3 as a crosslinking agent for poly(VDF-co-HFP) copolymer.

Moreover, the study of crosslinking with Diak No3 shows an increase of the glass transition temperature (T_g) of the vulcanizate with the amount of crosslinking agent [27], (Table 8).

Table 7 Evolution of the gel fraction and crosslinking density (v_e) of different cured poly(VDF-co-HFP) copolymers with increasing curing agent (N,N'-dicinnamylidene-1,6-hexanediamine) [27]

Viton A-HV	Curing agent, phr	273 ($2C_1/T$), psi	Gel fraction	$v_e^* \times 10^5$, mole/cm^3	$v_e \times 10^5$, mole/cm^3
A-1	0.2	1.5	0.510	0.90	0.46
A-2	0.4	5.5	0.700	2.39	1.67
A-3	0.6	9.4	0.820	3.50	2.87
A-4	0.8	16.0	0.900	5.41	4.87
A-5	1.0	20.2	0.925	6.65	6.15
A-6	2.0	38.8	0.964	12.20	11.80
A-7	4.0	–	≈ 1	24.40	24.40
A-8	8.0	–	≈ 1	48.80	48.80

v_e^* and v_e are the crosslinking densities, and $v_e = v_e^* \times g$, where g is the fraction of gel rubber

Table 8 Evolution of the glass transition temperature (T_g, °C) of samples A-1 to A-7 (indicated in Table 7) with the amounts of N,N'-dicinnamylidene-1,6-hexanediamine [27]

Viton A-HV	Log $(t_B/a_T)_{max}$, min[1]	T(°C) at which $t_{B_{max}} = 1$ min	T_g, °C	$(T - T_g)_{max}$, °C
A-1	– 2.5	43	– 29	72
A-2	– 2.5	42	– 29	71
A-3	– 2.0	47	– 29	76
A-4	– 2.0	47	– 28	75
A-5	– 3.0	47	– 28	75
A-6	– 4.5	17	– 26	43
A-7	– 7.7	0	– 23.5	23.5

[1] for $a_T = 1$ at 90 °C

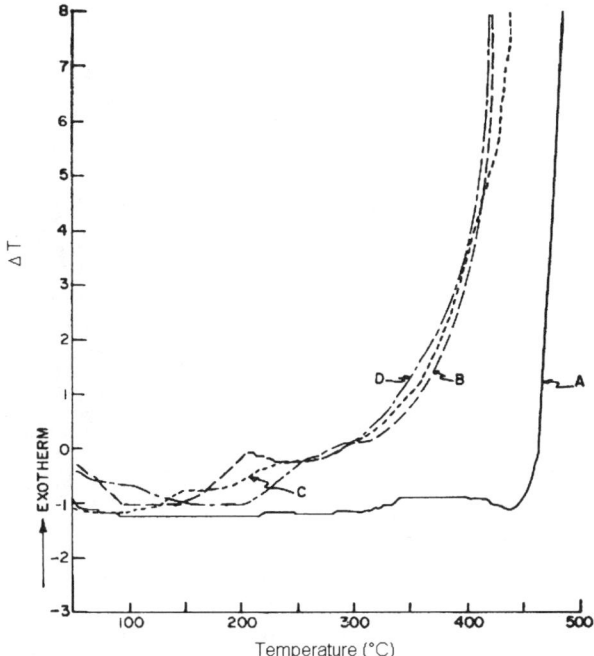

Fig. 7 Differential Thermal Analysis of Viton AHV cured with Diak No3 after various heat treatment: (A) Viton AHV; (B) Viton AHV + 15 phrMaglite + 4 phr Diak No3 (mill mixed); (C) Viton AHV + 15 phrMaglite + 4 phr Diak No3 (press cure); (D) Viton AHV + 15 phrMaglite + 4 phr Diak No3 (oven cure) (Reprinted with permission of Wiley) [23]

Table 8 [27] shows that the T_g increases from sample A-1 (0.2 phr of Diak No3, and T_g =– 29 °C) to sample A-7 (4.0 phr and T_g =– 23.5 °C). This increase of the glass transition temperature value is attributed to the increase of the crosslinking density.

Finally, crosslinking and decomposition temperatures of a Viton A-HV cured by Diak No3 were studied by differential thermal analysis (DTA) [23].

Figure 7 [23] represents the variations of temperature versus temperature for four different samples. Curve (B) (Viton A–HV + MgO + Diak No3, mill-mixed) exhibits an exotherm centered at 200 °C, which is due to the crosslinking reaction occurring in the course of the press cure. Indeed, curves (C) and (D) that correspond to the same samples as curve (B), press cure and post cure, respectively, do not exhibit such an exotherm.

Paciorek et al. [23] also studied the evolution of decomposition temperature of different post-cured systems by DTA. Thermograms of three Viton A-HV vulcanizates crosslinked by press cure (30 min at 150 °C) and post cured for 24 h at 200 °C are shown in Fig. 8 [23].

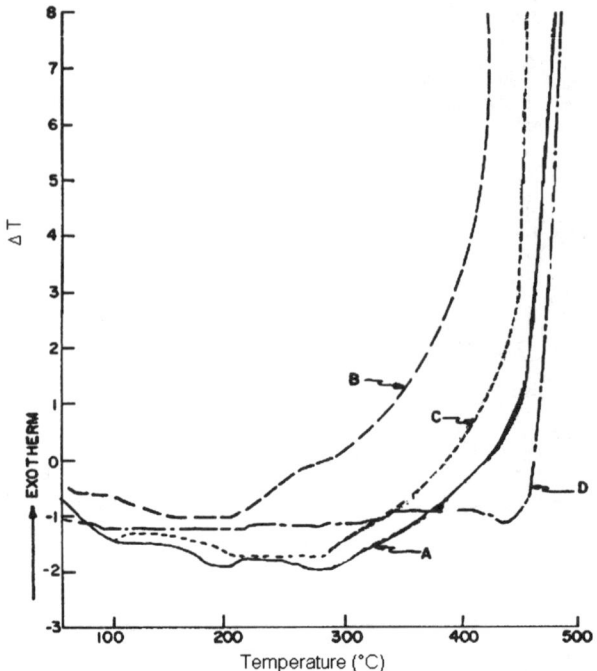

Fig. 8 Differential Thermal Analysis of Viton AHV cured with Diak No3 after oven cure: (A) Viton AHV + 15 phr Maglite; (B) Viton AHV + 15 phrMaglite + 4 phr Diak No3; (C) Viton AHV + 15 phrMaglite + 4 phr benzoyl peroxide; (D) Viton AHV (Reprinted with permission of Wiley) [23]

Both (A) and (C) curves indicate a final exothermic reaction initiated in the vicinity of 285 °C, whereas the elastomer cured by Diak No3 shows some reaction as low as 200 °C. In contrast, the untreated Viton A-HV starts to decompose from 430 °C. Hence, vulcanization results in a decrease of the thermal stability of fluorinated elastomer.

Reaction with HMDA and ethylenediamine:
HMDA-C is the ionic form of HMDA, which is unreactive at room temperature, but decomposes rapidly in the range of 130 to 170 °C, to produce the free reactive diamine [18, 35], as shown in the following scheme:

$$^+H_3N\text{--}(CH_2)_6\text{--}N\begin{matrix}COO^-\\H\end{matrix} \xrightarrow{\Delta} NH_2\text{--}(CH_2)_6\text{--}NH_2 + CO_2$$

Structure 5

A solution of Viton poly(VDF-co-HFP) copolymer in tetrahydrofurane, mixed with HMDA, at room temperature for one day produces gel forma-

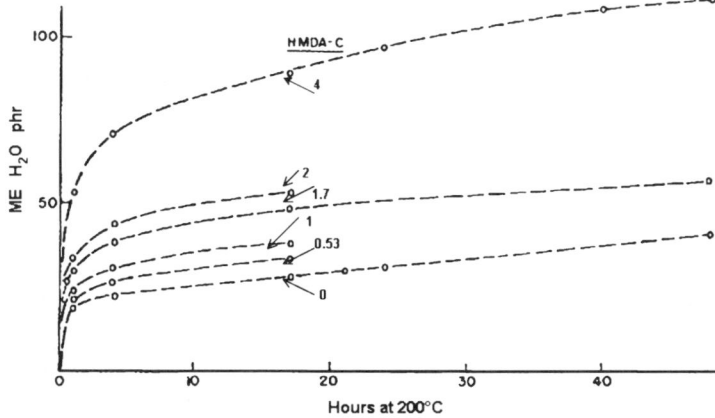

Fig. 9 Effect of the concentration (phr) of curing agent (HMDA-C) on the rate of elimination of H_2O from a VitonA [21]

Table 9 Mechanical properties of HMDA-C, EDA-C or N,N'-biso-(hydroxybenzylidene)-1,2-propylenediamine cured poly(VDF-co-HFP) copolymer of Viton A-HV [18]

Compound	Sample A	Sample B	Sample C
Viton A-HV	100	100	100
MgO[1]	15	15	15
MT Carbon black	20	20	20
HMDA-C[2]	1	–	1
EDA-C[3]	–	0.85	–
N,N'-bis-(o-hydroxybenzylidene)-1,2-propylenediamine[4]	–	–	1.3
Mooney scorch at 250 °F Minutes to 10-point rise	7	36	30
CURING CONDITIONS: press cure at 300 °F for 30 min; oven cure at 400 °F for 24 h			
Stress-strain properties:			
Modulus at 100% (psi)	390	350	580
Modulus at 200% (psi)	1130	1190	2075
Tensile strength (psi)	2500	2875	2525
Elongation at break (%)	320	350	240
Hardness, Shore A	67	69	69
Compression set, 70 h at 250 °F	35	34	16

[1] Darlington 601, Darlington Chemical Co., Philadelphia, Pa
[2] Diak No. 1, Du Pont
[3] Diak No. 2, Du Pont
[4] Active ingredient of Copper Inhibitor 65, Du Pont

tion [20], proving that the HMDA adds onto unsaturations created by dehydrofluorination, and creating crosslinks [26, 29, 125].

Different amounts of HMDA-C are added to a poly(VDF-*co*-HFP) copolymer, and Fig. 9 [21, 114] shows the amount of water evolved in a given time for different amounts of curing agent. This water could have arisen from an HF elimination from the polymeric chain, that reacts with MgO (coagent), according to the following reaction:

$$MgO + 2HF \rightarrow MgF_2 + H_2O.$$

Hence, the higher the amounts of HMDA-C, the higher the HF elimination, thus the higher the amount of addition [21, 114].

Ethylenediamine carbamate of (EDA-C) is particularly advantageous. Results of stress-strain tests and compression set of a poly(VDF-*co*-HFP) copolymer cured by EDA-C show that 0.85 part of EDA-C produces a state of cure equal to that obtained with one part of HMDA-C (Table 9 [111]).

The crosslinking mechanism of the HMDA-C is the same as that with HMDA [Scheme 5, with $R = (CH_2)_6$] [21, 114].

Bis-peroxycarbamates [hexamethylene-*N,N'*bis(*tert*-butylperoxycarbamate) or HBTBP, and methylene bis-4-cyclohexyl-*N,N'*(*tert*-butylperoxycarbamate), or MBTBP] [30]

These carbamates have the following formula, respectively:

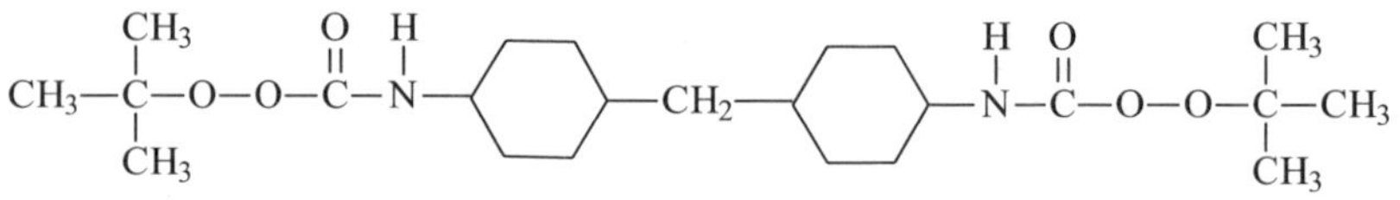

$$CH_3-\underset{\underset{CH_3}{|}}{\overset{\overset{CH_3}{|}}{C}}-O-O-\underset{}{\overset{\overset{O}{\|}}{C}}-\underset{}{\overset{H}{\underset{|}{N}}}-\!\!\!\bigcirc\!\!\!-CH_2-\!\!\!\bigcirc\!\!\!-\underset{}{\overset{H}{\underset{|}{N}}}-\underset{}{\overset{\overset{O}{\|}}{C}}-O-O-\underset{\underset{CH_3}{|}}{\overset{\overset{CH_3}{|}}{C}}-CH_3$$

HBTBP

Structure 6

$$CH_3-\underset{\underset{CH_3}{|}}{\overset{\overset{CH_3}{|}}{C}}-O-O-\underset{}{\overset{\overset{O}{\|}}{C}}-\underset{}{\overset{H}{\underset{|}{N}}}-(CH_2)_6-\underset{}{\overset{H}{\underset{|}{N}}}-\underset{}{\overset{\overset{O}{\|}}{C}}-O-O-\underset{\underset{CH_3}{|}}{\overset{\overset{CH_3}{|}}{C}}-CH_3$$

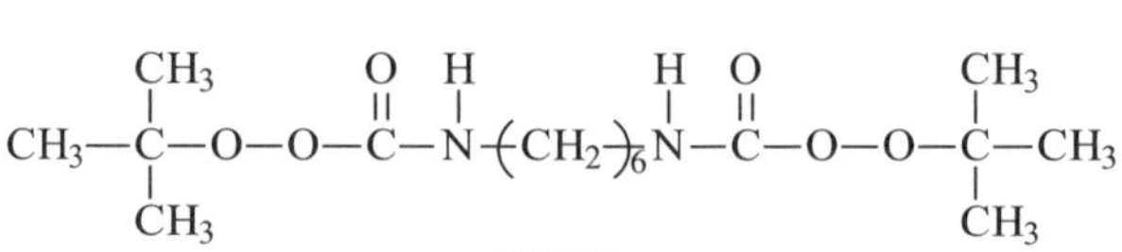

MBTBP

Structure 7

There are two possible crosslinking mechanisms which occur in the presence of those peroxycarbamates (Schemes 6 and 7) [30]. One is different from the usual mechanism of diamine crosslinking, while the second one is the classic one where the diamine adds onto the polymeric backbone by nucleophilic Michael addition. Indeed, those peroxycarbamates can undergo a thermal decomposition creating radicals (Scheme 6), and so the crosslinking mechanism can be a nucleophilic addition.

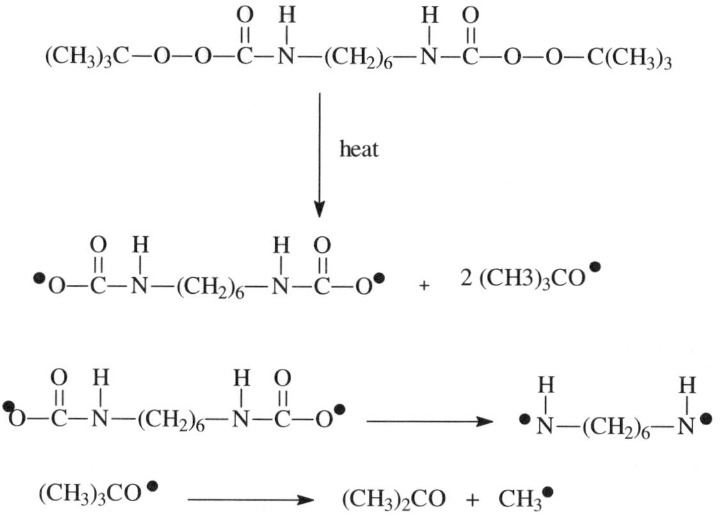

$$(CH_3)_3C-O-O-\overset{\overset{O}{\|}}{C}-\overset{\overset{H}{|}}{N}-(CH_2)_6-\overset{\overset{H}{|}}{N}-\overset{\overset{O}{\|}}{C}-O-O-C(CH_3)_3$$

$$\downarrow \text{heat}$$

$$^\bullet O-\overset{\overset{O}{\|}}{C}-\overset{\overset{H}{|}}{N}-(CH_2)_6-\overset{\overset{H}{|}}{N}-\overset{\overset{O}{\|}}{C}-O^\bullet \quad + \quad 2\,(CH_3)_3CO^\bullet$$

$$^\bullet O-\overset{\overset{O}{\|}}{C}-\overset{\overset{H}{|}}{N}-(CH_2)_6-\overset{\overset{H}{|}}{N}-\overset{\overset{O}{\|}}{C}-O^\bullet \quad \longrightarrow \quad ^\bullet\overset{\overset{H}{|}}{N}-(CH_2)_6-\overset{\overset{H}{|}}{N}^\bullet$$

$$(CH_3)_3CO^\bullet \quad \longrightarrow \quad (CH_3)_2CO \;+\; CH_3^\bullet$$

Scheme 6 The homolytic thermal decomposition of the hexamethylene-N,N'bis(tert-butyl) peroxycarbamate [30]

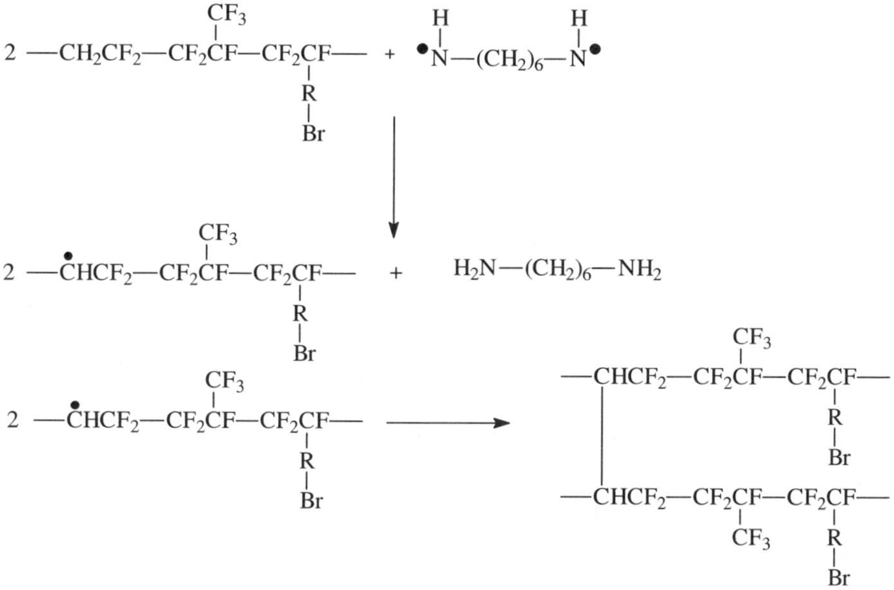

$$2\;-CH_2CF_2-CF_2\underset{\underset{Br}{|}}{\overset{\overset{CF_3}{|}}{C}F}-CF_2CF- \;+\; ^\bullet\overset{\overset{H}{|}}{N}-(CH_2)_6-\overset{\overset{H}{|}}{N}^\bullet$$

$$\downarrow$$

$$2\;-\overset{\bullet}{C}HCF_2-CF_2\underset{\underset{Br}{|}}{\overset{\overset{CF_3}{|}}{C}F}-CF_2CF- \;+\; H_2N-(CH_2)_6-NH_2$$

$$2\;-\overset{\bullet}{C}HCF_2-CF_2\underset{\underset{Br}{|}}{\overset{\overset{CF_3}{|}}{C}F}-CF_2CF- \quad\longrightarrow\quad \begin{array}{c} -CHCF_2-CF_2\overset{\overset{CF_3}{|}}{C}F-CF_2\underset{\underset{Br}{|}}{\overset{}{C}}F- \\ | \\ -CHCF_2-CF_2\underset{\underset{CF_3}{|}}{\overset{}{C}}F-CF_2\underset{\underset{Br}{|}}{\overset{}{C}}F- \end{array}$$

Scheme 7 Crosslinking formation in terpolymer obtained through hydrogen abstraction [30]

HBTBP and MBTBP were mixed with a poly(VDF-ter-HFP-ter-$CF_2 = CF - R - Br$) terpolymer (Viton GF or FKM-G)—where R is a fluorinated methylene spacer—in the presence of MgO. The typical formulation is

Table 10 Typical formulation of fluoroelastomers cured by bis-peroxycarbamate [30]

Materials	Parts per hundred rubber (phr)
Fluoroelastomer	100
Medium thermal carbon Black	25
Calcium oxide	4
Calcium hydroxide	6
Curing agent	1 to 5

given in Table 10 [30]. The samples were then press cured (15 min at 170 °C) and post cured (24 h at 250 °C).

Figure 10 [30] represents a comparison between the ODR cure traces of both samples. HBTBP (curve c) is noted to possess a higher state of cure than that of MBTBP (curve e). Although there is a significant structural similarity between both crosslinking agents, they have different cure responses with FKM-G copolymer.

Figure 11 [30] exhibits the crosslinking density versus the amount of HBTBP (in phr), and the extrapolation of this plot shows that a minimum quantity of HBTBP is required before any formation of crosslinks can occur.

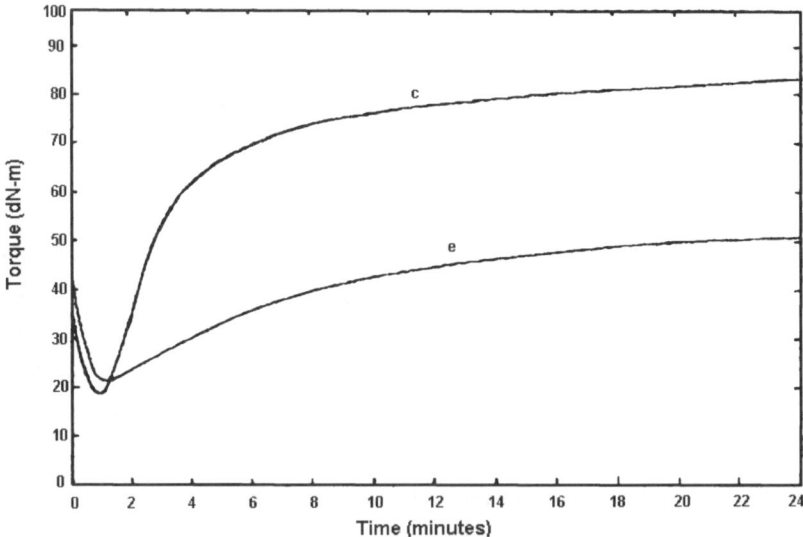

Fig. 10 Comparison of ODR cure traces of a poly(VDF-ter-HFP-ter-$CF_2 = CF$-RBr) terpolymer crosslinked with HBTBP (*curve c*), and MBTBP (*curve e*) at 170 °C. Formulation: terpolymer (100), MT Black (25), CaO (4), Ca(OH)$_2$ (6), and HBTBP (4) for curve c; terpolymer (100), MT Black (25), CaO (4), Ca(OH)$_2$ (6), and MBTBP (4) for curve e (Reprinted with permission of ACS) [30]

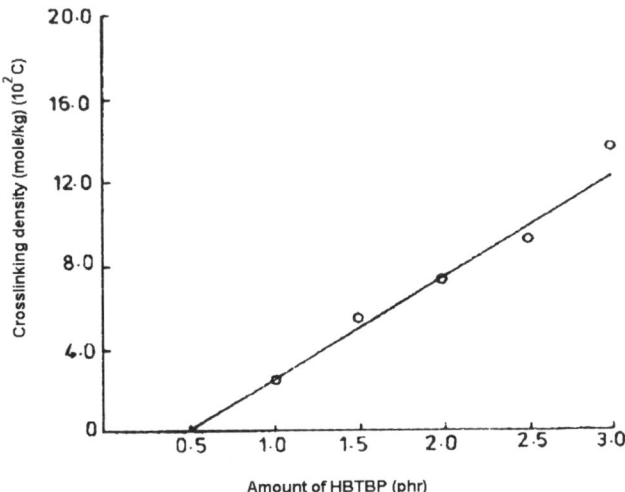

Fig. 11 Effect of HBTBP concentration (in phr) on the crosslinking density (Reprinted with permission of ACS) [30]

In the crosslinking formation by HBTBP, there is a thermal decomposition of the peroxycarbamates to give free radicals in a suitable medium. This thermal decomposition is represented in Scheme 6 [30, 126–129].

Two possible routes of crosslink formation have also been proposed. Scheme 7 [30] leads to the first one [30]. The hexamethylene diradical abstracts the hydrogen atoms of two polymeric chains creating two macroradicals, and regenerating the hexamethylene diamine. Both macroradicals form carbon–carbon crosslinks.

Hexamethylene diamine (intermediate products of the mechanism given in Scheme 7) can then form additional crosslinks. This mechanism is given in Scheme 8. It is the classical Michael addition of the HMDA onto the fluoropolymer.

From those crosslinking mechanisms, it can be understood that the radical of HBTBP is more reactive than that of MBTBP. That is why HBTBP-cured system possesses a higher state of cure.

Piperazine, triethylene diamine, tetramethylethyldiamine, and diethylene triamine:

Different diamines (piperazine, triethylene diamine, and tetramethylethyldiamine), given in Scheme 9 were also used as crosslinking agents for two copolymers: poly(VDF-co-HFP) copolymer (Viton A) and poly(VDF-co-CTFE) copolymer (Kel-F) [19, 20, 111, 130]. Those curing agents were mixed with both copolymers in a solution of diglyme, at different temperatures and for different reaction times. Results are summarized in Tables 11 and 12 [20]. Kel-F exhibits a gel formation when vulcanized with piperazine at 57 °C

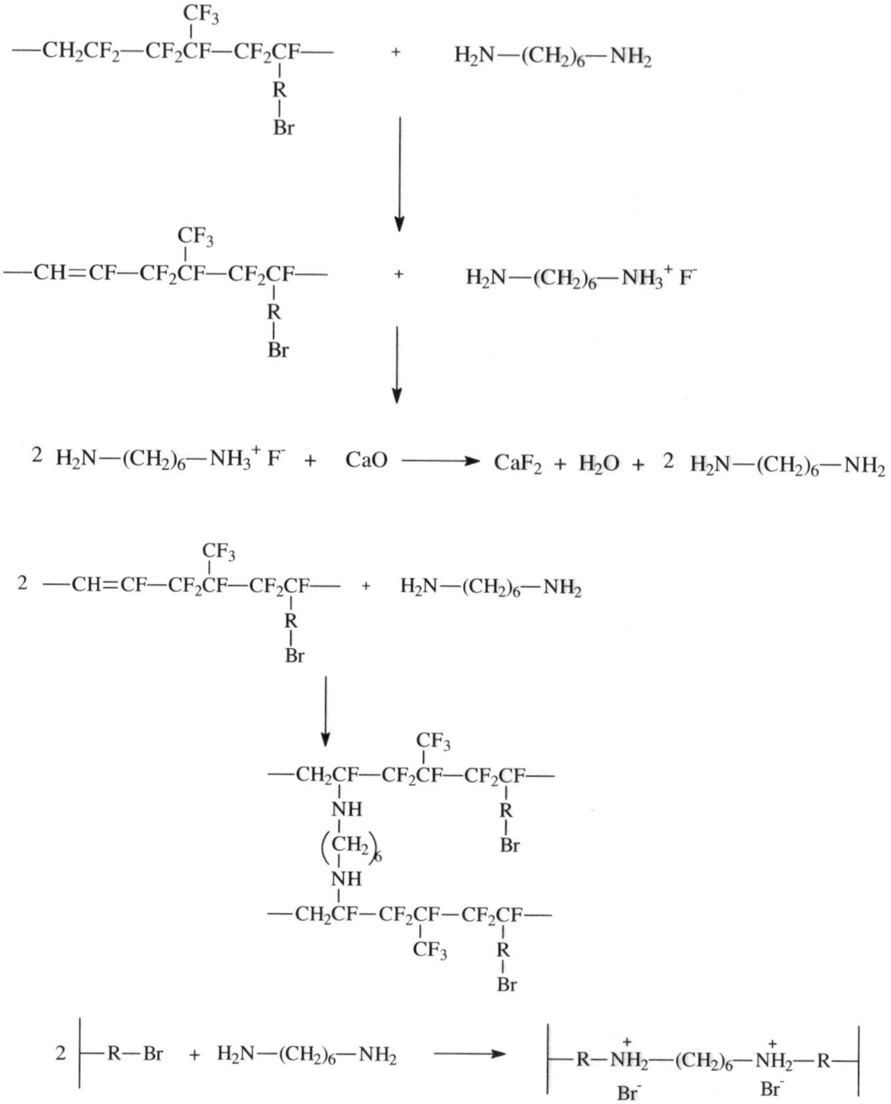

Scheme 8 Further interaction of the reactive intermediate (hexamethylene diamine) with a fluorinated polymer, bearing a bromine side atom [21, 30, 114].

for one day, while the same behavior occurred when it was crosslinked with triethylene diamine and tetramethylethyldiamine at 97 °C for one day. Poly(VDF-*co*-HFP) copolymer exhibits gel formation only with piperazine at 55 °C for 1 day, or at 25 °C for 20 days.

Table 11 Conditions of reaction of diamines with a Kel-F poly(VDF-co-CTFE)copolymer [20]

Diamine	%[1]	T (°C)	Crystalline precipitate	Gel	Anal. of solution	Conversion of amine to hydrohalide[2] (%)	Reaction time (days)	Initial pH	Final pH	Final color
Hexamethylene-diamine	55	25	Yes	Yes	Cl$^-$, F$^-$	–	1	9	–	Light yellow
Piperazine	47	25	Yes	No	Cl^{-4}	80[5]	6	9	5	Light yellow
	18	57	Yes	Yes	–	–	1	–	–	Colorless
	10	25	Yes	No	–	–	6	–	–	Colorless
	55	25	Yes	No	Cl$^-$	92[5]	13	8+	5	Light yellow
Triethylene-diamine	57	25	Yes	No	Cl^{-4}	68[5]	9	8–9	5	Colorless
	11	97	Yes	Yes	Cl$^-$, F$^-$	–	1	–	–	Colorless
Tetramethyl-ethylenediamine	53	25	Yes	No	Cl$^-$	62	15	9	7	Light yellow
	10	97	Yes	Yes	Cl$^-$, F^{-4}	–	1	–	–	Light yellow

[1] Based on the chlorine content of the polymer and equivalent weight of the amine
[2] Hydrochlorides shown
[3] Based on chlorine available
[4] Trace
[5] Based on the formation of amine monohydrochloride

Table 12 Conditions of reaction of diamine onto a Viton-A poly(VDF-co-HFP)copolymer [20]

Amine	%[1]	T (°C)	Crystalline precipitate	Gel	Conversion of amine to hydrohalide[2] (%)	Reaction time (days)	Initial pH	Final pH	Final color
Hexamethylene-diamine	18	25	No	Yes	72[2]	1	–	–	Yellow
	20	25	Few	No	73[2]	6	9	5	Yellow
Piperazine	18	55	Yes	No	–	1	9	–	Orange Brown
	19	190	No	Little	–	1	–	–	
	92	25	Yes	No	76[2]	20	8+	5	Orange
Triethylene-diamine	18	190	No	No	–	1	–	–	Brown
	89	25	No	No	78[2]	34	8	6	Orange
Tetramethyl-ethyldiamine	88	25	Trace	No	72[2]	12	8	8–	Yellow
	18	190	Trace	No	–	1	7+	7	Red brown

[1] Based on the tertiary fluorine content of the polymer and equivalent weight of the amine
[2] Based on the formation of amine dihydrofluoride

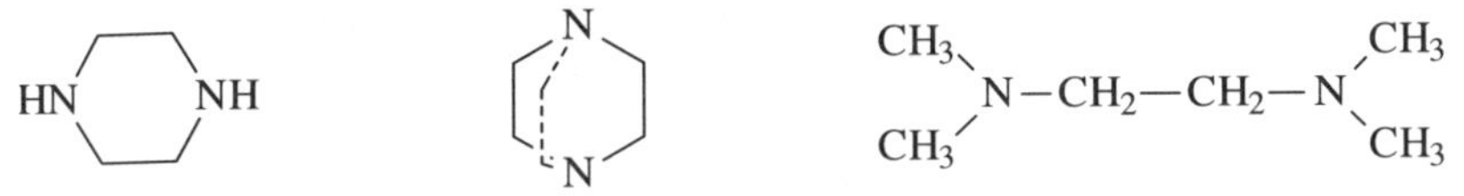

Scheme 9 Piperazine, triethylene diamine, and tetramethylethyldiamine [20]

So, piperazine is less efficient as a curing agent for poly(VDF-co-HFP) copolymer, whereas 2,5-dimethyl piperazine [18] produces well-cured vulcanizates from stocks which are less toxic than those containing HMDA-C.

Compression set resistance at 200 °C was compared between a TecnoflonT poly(VDF-ter-HFP-ter-1-hydropentafluoropropene) terpolymer vulcanized with HMDA-C, and TecnoflonT vulcanized with piperazine carbamate [131]. The results are included in Fig. 12 [131]. Curves 3 and 4 are both poly(VDF-co-HFP) copolymer vulcanizates with piperazine carbamate in the presence of trimethylenediamine carbamate and MgO for curve 3, and bisphenol in the presence of MgO for curve 4. The compression set resistance is better for curve 4, than curve 2, which is better than curve 3, and finally the worst compression set resistance is obtained for Tecnoflon T vulcanized with HMDA-C.

The reaction between PVDF and diethylene triamine [DETA $H_2N-(CH_2)_2-NH-(CH_2)_2-NH_2$] at various temperatures (25–80 °C) was monitored by infrared spectroscopy [25, 31].

Figure 13 (a and b) [31] represents two infrared spectra. The first one (a) illustrates the spectrum of PVDF and DETA when no reaction occurred, whereas spectrum (b) represents PVDF film after heating in DETA for 16 h at 70 °C. For spectrum (a), the absorption bands at 3350 cm^{-1} and 1590 cm^{-1}

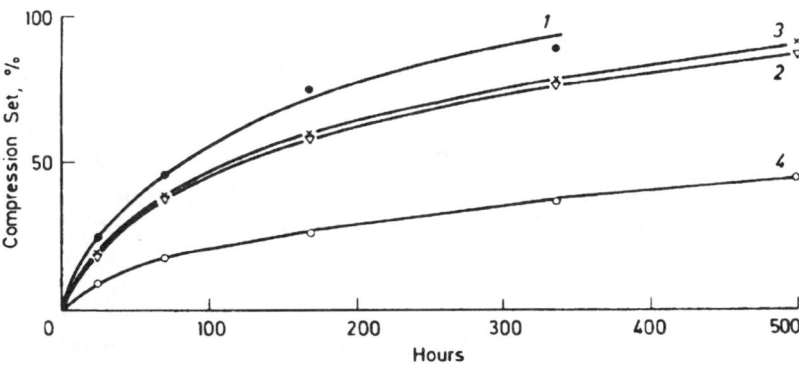

Fig. 12 Compression set versus ageing time at 200 °C for several vulcanizates: TecnoflonT(poly(VDF-ter-HFP-ter-TFE) terpolymer) vulcanized with HMDA-C (*curve 1*); TecnoflonT vulcanized with piperazine carbamate (*curve 2*); Tecnoflon (poly(VDF-co-HFP) copolymer) vulcanized with piperazine carbamate and trimethyldiamine carbamate (*curve 3*); Tecnoflon cured with bisphenolAF (*curve 4*) (Reprinted with permission of Hüthig Fach Verlag) [131]

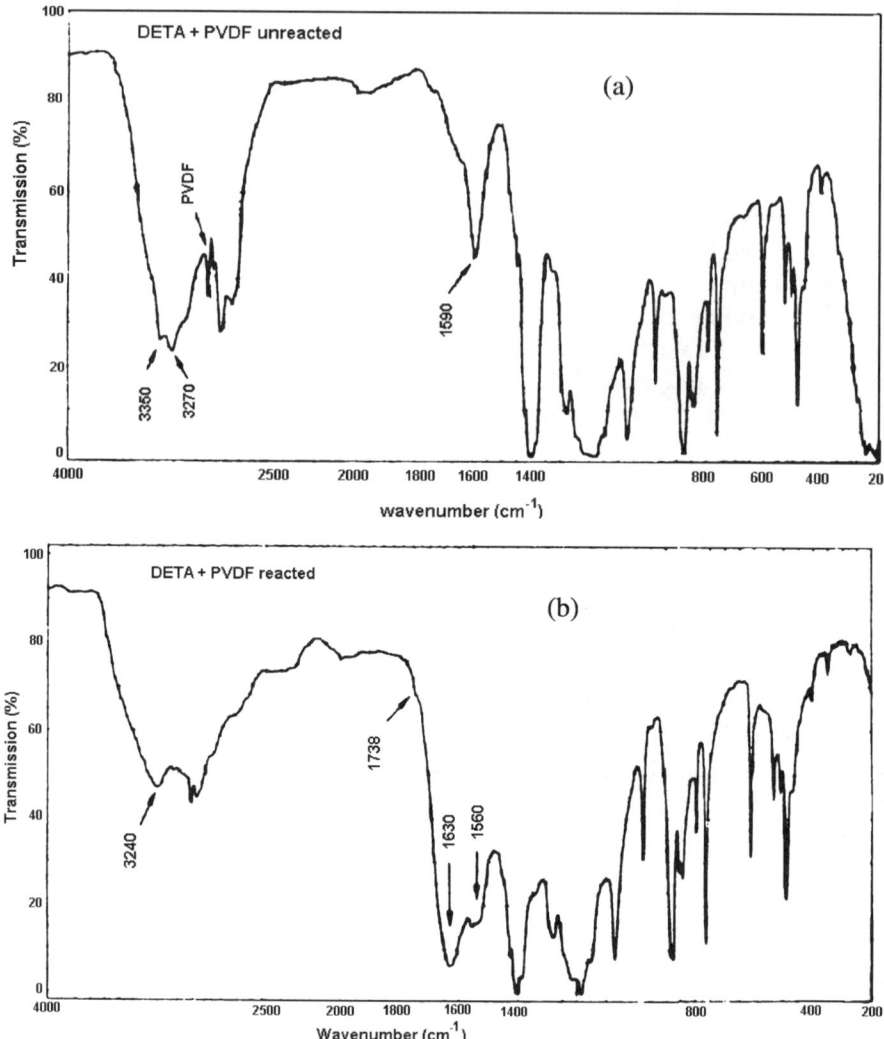

Fig. 13 Transmission infrared spectra of DETA on PVDF unreacted (**a**), and DETA – PVDF composite reacted for 16 h at 70 °C (**b**) (Reprinted with permission of VSP) [31]

are assigned to the NH stretching and to NH_2 deformation in the primary amine DETA, and the additional absorptions are from PVDF. In spectrum (b), the bands of primary amine at 3350 and 3270 cm^{-1} are replaced by that of a secondary one at 3240 cm^{-1}. Two strong bands also appear at 1630 and 1560 cm^{-1} which are also attributed to secondary amide. A shoulder at 1738 cm^{-1} is characteristic of C=N stretching. Thanks to infrared spectroscopy, a crosslinking mechanism was proposed [31] and is likely the same as that of Scheme 5.

Reaction between PVDF and diamine curing agent produces alterations in the polymer's surface, which enables the formation of strong adhesive joints without prior surface modification [31].

3.1.3.3
Aromatic and Aromatic Containing Amines and Diamines

Aromatic containing diamines have been explored at various times, because it was assumed that the aromatic ring structure built into the polymeric network, during the curing process, would be more stable to high temperature oxidation than the aliphatic hydrocarbon crosslinks derived from aliphatic diamines. However, little success has been achieved [17, 28, 132].

Recently, the grafting of aromatic containing amines (benzylamine and phenylpropylamine) onto poly(VDF-co-HFP) copolymers was investigated [133]. First, the ^{19}F NMR spectrum of a benzylamine grafted onto poly(VDF-co-HFP) copolymer was studied, and is represented in Fig. 14. Figure 14 shows the evolution of the ^{19}F NMR signals centered at − 108.5, − 110.4, − 112.3, − 113.6, − 115.2, − 115.9, and − 117.9 to − 118.5 ppm, assigned

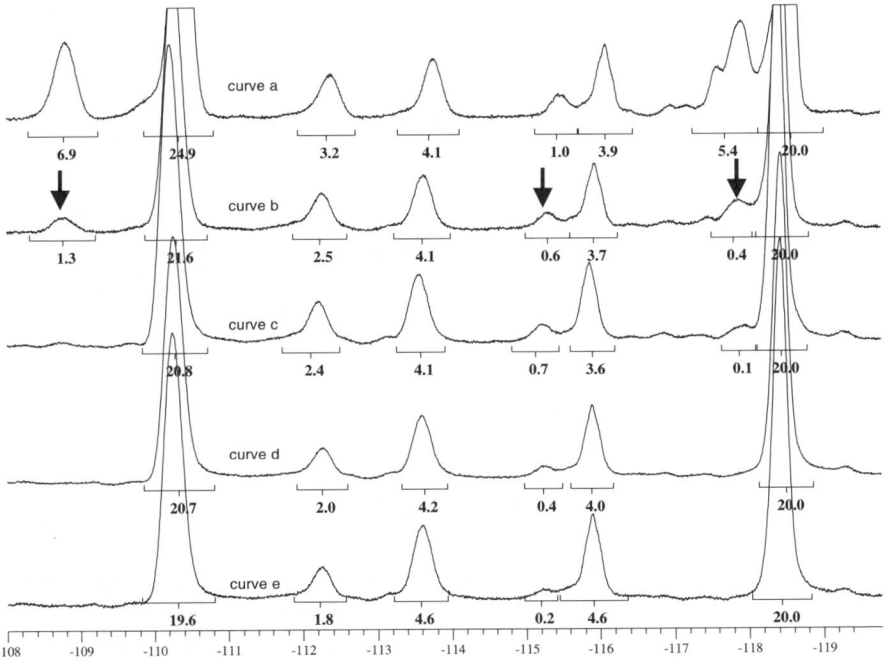

Fig. 14 ^{19}F NMR spectra (zone from − 108 to − 120 ppm) of a poly (VDF-co-HFP) copolymer 20 mole % HFP grafted with 200 mole % of benzylamine at 80 °C: the copolymer (curve a); after 115 min of reaction (curve b); after 475 min (curve c); after 2845 min (curve d); after 6835 min (curve e) [133]

Table 13 Characteristic peaks in ^{19}F NMR between -108 and -120 ppm of fluorinated groups in VDF or HFP units of grafted poly(VDF-co-HFP) copolymer

Chemical shifts in ^{19}F NMR (ppm)	Characteristic group
-108.5	$CF_2-CF(CF_3)-CH_2-\underline{CF_2}-CF_2-CF(CF_3)$
-110.4	$CH_2CF_2-CH_2\underline{CF_2}-CF_2CF(CF_3)$
-112.3	$CF_2CF(CF_3)-CH_2\underline{CF_2}-CF_2CH_2$ $CF_2CH_2-CH_2\underline{CF_2}-CF_2CF(CF_3)$
-113.6	$CH_2CF_2-\underline{CF_2}CH_2-CF_2CF(CF_3)$ $CH_2CF_2-CH_2\underline{CF_2}-CF_2CH_2$
-115.2	$CH_2CF_2-\underline{CF_2}CF(CF_3)\ CH_2CF_2$
-115.9	$CH_2CF_2-\underline{CF_2}CH_2-CH_2CF_2$
-117.9 to -118.5	$CH_2CF_2-\underline{CF_2}CF(CF_3)-CF_2CH_2$

to fluoride groups (given in Table 13) of a benzylamine-grafted copolymer (containing 20% in mol of HFP). This spectrum shows the disappearance of the peaks at -108, -115.2 and 117.9 ppm of the copolymer. This result confirms Schmiegel's conclusions [33, 67] that stated that the addition of the crosslinking agent first occurs onto VDF between two HFP units.

Then, the evolution of the molar percentage of grafted benzylamine as a function of time was studied at different temperature, with different molar percentages of HFP (ranging from 10 to 20% in mol) in the copolymer. Figure 15 plots the molar percentages of grafted benzylamine at 80 °C for a 10% in mol of HFP-containing copolymer. It proves that in the first 300 min, all the

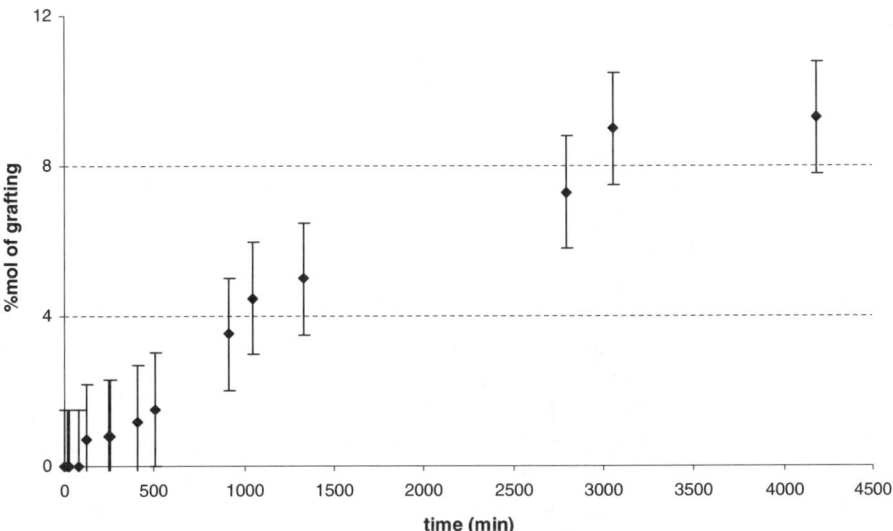

Fig. 15 Kinetics of grafting of the benzylamine (200%) onto a poly(VDF-*co*-HFP) copolymer Kynar containing 10 mole % of HFP. The grafting reaction is carried out in MEK at 80 °C [133]

HFP/VDF/HFP sites were crosslinked. Finally, phenyl propylamine seems to add faster onto copolymer than benzylamine does, that latter amine containing one spacer only.

So, aromatic containing amine can graft onto poly(VDF-*co*-HFP) copolymer if it exhibits a spacer between the amino group and the aromatic ring. As expected, the higher the number of methylene groups in the spacer, the faster the kinetics of grafting. As in the cases above, the grafting is carried out first onto VDF between two HFP units.

Finally, it was deduced that the crosslinking mechanism for aliphatic and aromatic containing amine and diamine was slightly the same and took place in three steps. Aliphatic diamines are more reactive than aliphatic amines; and the most used diamine is HMDA-C and its derivatives (N,N'-dicinnamylidene-1,6-hexanediamine and bis-peroxycarbamates). During the post-cure treatment, two different mechanisms of bond making and bond breaking were evidenced, as explained in Sect. 3.1.4.

3.1.4
Formation of Two Networks During Post Cure

The study of the heat ageing of bisphenol-, peroxide-, and diamine-cured systems shows the formation of different networks more or less stable.

First, a Size Exclusion Chromatography (SEC) study of untreated and unheated FKM gum showed an average molecular weight of 250 000 g/mol. The gum heated at 325–375 °C for 3 h gave a molecular weight of 96 000 g/mol, suggesting chain scissions. Seven to ten percent of this sample could not be dissolved in tetrahydrofurane (THF), showing that crosslinking took place [118].

Second, the infrared spectrum of heated FKM gum (3.1.1.2) shows the presence of HC = O groups at 1745 cm^{-1} [24]. Those aldehyde functions were identified in the last step of the crosslinking mechanism (Scheme 5), and result from the oxidative scissions of the bridges created by the diamine. These scissions induce the decrease in molecular weight.

The compression set percentage of a cured poly(VDF-ter-HFP-ter-TFE) terpolymer, press cured and post cured under nitrogen or air (Table 3) was studied [35]. The authors noted first that post cure drastically improves the compression set resistance of the press-cure sample, and second a nitrogen atmosphere leads to better post cure results than air atmosphere. It can be concluded that post cure improves compression set resistance all the more if it is carried out under nitrogen in order to avoid oxidative scissions. Consequently, the scissions in or at the crosslink are due to oxygen [35].

Kalfayan [118] and Thomas [24] studied the stress relaxation of poly(VDF-ter-HFP-ter-TFE) terpolymers crosslinked with N,N'-dicinnamylidene-1,6-hexanediamine, and of a poly(VDF-co-HFP) copolymer crosslinked with HMDA-C, respectively.

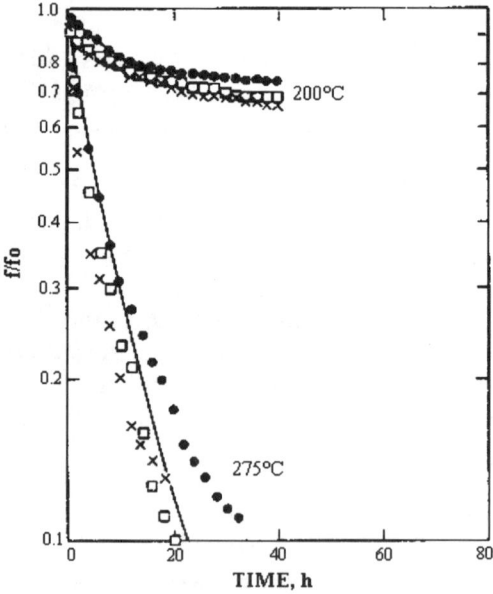

Fig. 16 Stress relaxation of FKM (poly(VDF-co-HFP) copolymer) of various crosslink densities at 200 and 275 °C in air. ● FKM + 1.5 phr of Diak No3, □ FKM + 3.0 phr of Diak No3, × FKM + 4.5 phr of Diak No3 (Reprinted with permission of ACS) [118]

Figures 16 [118] and 17 [24] exhibit the evolution of f/f_0 versus time, where f and f_o represent the tensile forces at t and t_o times, respectively.

In Fig. 16 [118] stress relaxation is measured for three samples that contain different amounts of diamine at 200 and 250 °C in air. In Fig. 17 [24], stress relaxation is measured at 250 °C in air for different amounts of diamine. Those figures show that stress relaxation is independent of both the amount of diamine and of the crosslinking density.

The most readily oxidizable groups in both structures are the methylene groups in α position to the C = N bond.

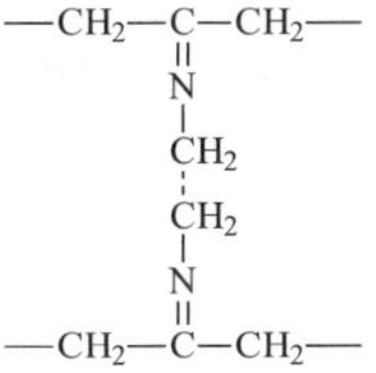

$$-CH_2-\underset{\underset{\underset{\underset{\underset{-CH_2-\overset{\|}{C}-CH_2-}{\|}}{N}}{CH_2}}{CH_2}}{\overset{\|}{C}}-CH_2-$$

Structure 8

It suggests that scissions occur at the crosslinking group (i.e., the grafting group) rather than in the polymeric chain [24, 118].

Hence, in order to avoid these oxidative chain scissions at the crosslink, the sample must be post cured under nitrogen [24].

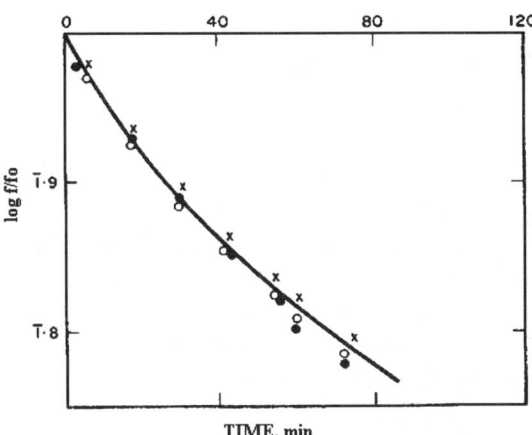

Fig. 17 Continuous stress relaxation of poly(VDF-co-HFP) copolymer vulcanizates at 250 °C in air, with: ● 2.0% of HMDA-C, × 1.5% of HMDA-C, ○ 1.0% of HMDA-C (Reprinted with permission of Wiley) [24]

Post cure contributes to a thermally induced bond-breaking and bond-making process that results in a thermally and mechanically more stable network [35, 58, 134].

3.2
Crosslinking with Bisphenols

Bisphenols are presently the predominant crosslinking agents for curing fluorocarbon-based elastomers. Bisphenols curing was developed in the late 1960s and started replacing the diamine cure in the early 1970s [109, 135–138]. Because of processing and property advantages, the most commonly used compound is bisphenol AF [2,2-bis(4-hydroxyphenyl) hexafluoropropane]. Others, like substituted hydroquinone, and 4,4'-disubstituted bisphenols also work well and are used commercially to a lesser degree [35, 36, 38, 58]. As in the cases above, the crosslinking reaction was evidenced by ^{19}F NMR.

3.2.1
Crosslinking Mechanism

The crosslinking mechanism takes place in three steps: elimination of HF creating double bonds, then reorganization of the double bonds, such as in Scheme 2, and finally substitution of the bisphenol onto the double bond.

Crosslinking agents require accelerators to make the reaction more efficient. For example, to enable the dehydrofluorination of a VDF/HFP diad, bisphenols need to react with a metal oxide to give the phenolate ion, which in turn reacts with the phosphonium or tetraalkylammonium ion to give intermediates I and II.

$R_4P^{+-}OArOH$ $R_4N^{+-}OArOH$
I   II

These intermediates are strong bases [35]. The crosslinking mechanism proposed by Schmiegel is shown in Scheme 10 [33, 35, 63, 139].

In the first step (the dehydrofluorination) Viton copolymer is attacked by the intermediate described below, creating diene. Then, the bisphenol-derived phenolate ($^-$OArOH) attacks the intermediate diene and finally leads to the dienic phenyl ether crosslinks [32, 33]. This reaction is a substitution. The resulting product surprisingly shows good properties, particularly with regard to oxidative and hydrophilic stability.

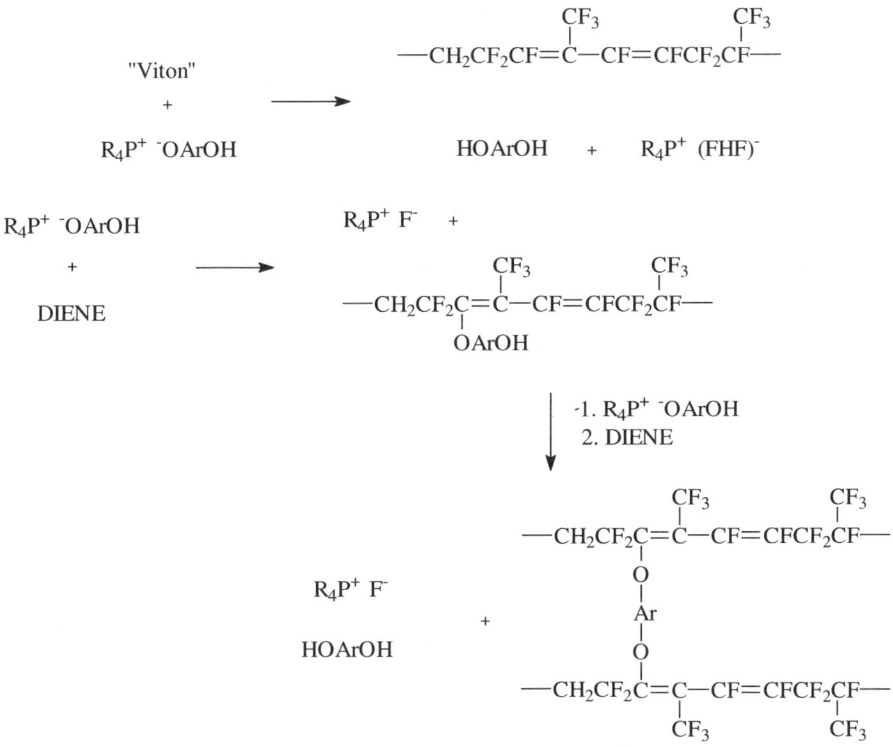

```
                                    CF₃         CF₃
                                    |           |
      "Viton"              —CH₂CF₂CF=C—CF=CFCF₂CF—
          +         ⟶
      R₄P⁺ ⁻OArOH              HOArOH    +    R₄P⁺ (FHF)⁻


  R₄P⁺ ⁻OArOH              R₄P⁺ F⁻   +
                                      CF₃       CF₃
        +             ⟶                |         |
                                 —CH₂CF₂C=C—CF=CFCF₂CF—
      DIENE                            |
                                     OArOH

                               │  1. R₄P⁺ ⁻OArOH
                               │  2. DIENE
                               ▼
                                      CF₃       CF₃
                                      |         |
                                —CH₂CF₂C=C—CF=CFCF₂CF—
                                      |
                                      O
      R₄P⁺ F⁻                         |
                        +             Ar
      HOArOH                          |
                                      O
                                      |
                                —CH₂CF₂C=C—CF=CFCF₂CF—
                                      |         |
                                     CF₃       CF₃
```

Scheme 10 Crosslinking mechanism with the bisphenol [33, 35]

3.2.2
¹⁹F NMR Study

The mechanism of crosslinking was evidenced by ^{19}F NMR, and Schmiegel et al. [3, 32, 33, 67] studied the ^{19}F NMR spectra of poly(VDF-co-HFP) copolymer treated in dimethylacetamide (DMAC), with DBU (1,8-diazabicyclo[5-4-0]-undec-7-ene) and bisphenol AF. DBU is strong enough to enable a dehydrofluorination of the polymer and to ionize the phenol, but is sterically hindered for being an efficient competitor of phenoxide for the fluoroolefin [33].

Figure 18 [33, 34] represents the ^{19}F NMR spectra of a base (DBU)-treated soluble polymer (upper spectrum), and the gel produced by the base in the presence of the bisphenol-treated polymer (lower spectrum). Both spectra exhibit two new peaks at − 55 and − 62 ppm, assigned to − C = C(CF3) − C-isomeric structure of the CF$_3$. So, poly(VDF-co-HFP) copolymers treated with DBU and with a DBU/bisphenol system undergo at least dehydrofluorination and rearrangement, such as in Scheme 2. However, to prove the crosslinking of bisphenol-AF onto the copolymer, a ^{19}F NMR spectrum of the same sample after precipitation must be recorded.

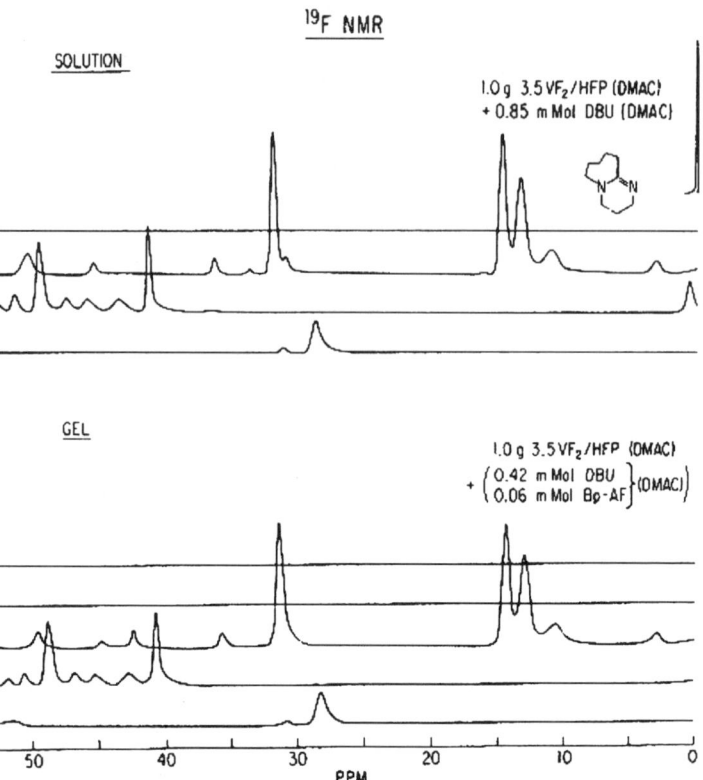

Fig. 18 ^{19}F NMR spectra of a poly(VDF-*co*-HFP) copolymer treated with DBU in a solution of DMAC (*top*), and the gel which results from this reaction in the presence of Bp-AF (*bottom*) (Reprinted with permission of Verlag Chemie) [33]

Figure 19 [33, 34] shows two ^{19}F NMR spectra of poly(VDF-*co*-HFP) copolymers. The first one (top) deals with the spectrum of poly(VDF-*co*-HFP) copolymer treated with DBU and the bisphenol AF in DMAC, while the other one (bottom) represents the same sample but precipitated twice from an appropriate solvent for free phenol or any unreacted phenolate (acetonitrile). The ^{19}F NMR spectra of the washed polymer (bottom) clearly shows the presence of the geminal trifluoromethyl groups. So, after precipitation in acetonitrile of all the phenol and phenolate that did not react with the copolymer, the peak at – 55 ppm was still noted. It proves that a part of the bisphenol-AF enabled the crosslinking of the copolymer. Under those conditions, about 40% of the phenolate were incorporated based on the internal *p*-fluoroanisole standard.

Hence, ^{19}F NMR results allowed us to evidence that crosslinking was achieved.

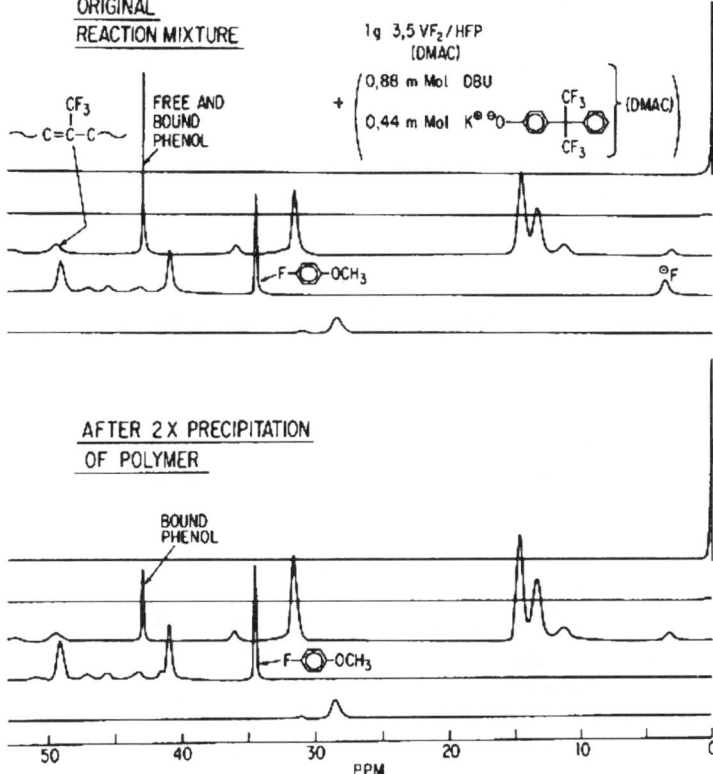

Fig. 19 ^{19}F NMR spectra of a poly(VDF-*co*-HFP) copolymer treated with DBU and Bp-AF in a solution of DMAC (*top*), and the same sample after several purification (*bottom*) (Reprinted with permission of Verlag Chemie) [33]

3.2.3
Oscillating Disc Rheometer (ODR) Response

Bisphenol-cured fluoropolymers are usually analyzed by ODR. Reaction time and crosslinking density can be deduced from the ODR curve.

This equipment can plot the evolution of the torque (in N m) as a function of time (in min), at a given temperature, for a crosslinkable mixing (copolymer, crosslinking agent, accelerators, coagent ...). Usually, the torque starts to decrease (during an induction period), and when the crosslinking reaction occurs it increases rapidly, reaching a maximum when the reaction is finished.

Bisphenol curing systems are usually used for O-ring applications [26]. Indeed, they exhibit a high resistance to high temperature compression set. Figure 20 [3, 33, 35] depicts the evolution of a 177 °C cure response by ODR, of a bisphenol AF (Bp-AF) curing poly(VDF-*co*-HFP) copolymer.

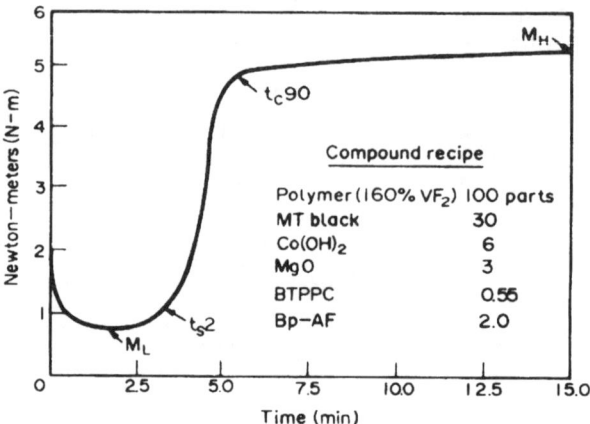

Fig. 20 Cure response by Oscillating disc rheometer at 177 °C of a VDF-based polymer cured with Bisphenol-AF (Reprinted with permission of Elsevier) [35]

The ODR response is characterized by an induction period, which depends on the amount of the accelerator (benzyltriphenylphosphonium chloride or BTPPC), or amount of bisphenol. High Bp-AF amounts increase the length of the induction period and lead to high cure states. The maximum cure state is the initial slope of the curve; t_s2, the time to initiation; t_c90, the time to 90% completion of cure; M_L, the minimum torque; M_H, the maximum torque;

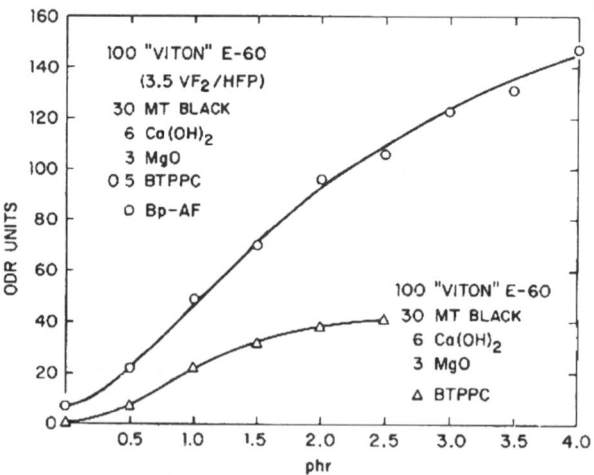

Fig. 21 Evolution of the ODR units (crosslink density) at 177 °C: (○) with variation of bisphenol-AF concentration (in phr), in the presence of BTPPC (benzyltriphenyl phosphonium chloride), (△) with BTPPC concentration, in the absence of Bisphenol-AF (Reprinted with permission of Verlag Chemie) [33]

and $M_H - M_L$, the degree of state of cure [3, 33, 35]. In Fig. 20, at 177 °C, and after a 2.5 min induction period, the reaction of crosslinking is practically complete after 5 min. Only a 2% increase in cure state occurs between 13 and 60 min. The final state of cure does not change with increasing temperature [3, 33, 35]. When BTPPC is omitted from the standard recipe, no cure occurs within 1 h at 177 °C.

Figure 21 [3, 33] shows the dependence of the ODR cure state versus Bp-AF concentration, in the presence of a standard concentration of $Ca(OH)_2$, MgO, BTPPC and carbon black. It is noted that the greater the concentration in bisphenols, the higher the ODR cure state, so the higher the crosslinking density. The lower line shows that the accelerator BTPPC in the absence of the bisphenol can also lead to a substantial cure state, although only at very high concentrations [3, 33].

3.2.4
Limitations of the Bisphenol-Cured Fluoroelastomers

Bisphenol-cure is a very rapid crosslinking system, as shown by ODR, but this system also presents some limitations.

The crosslinking mechanism between poly(VDF-ter-PMVE-ter-TFE) terpolymer and bisphenols generates elimination of a trifluoromethoxide and a fluoride ion, giving $CF=CF$ double bonds. The trifluoromethoxide reacts with hydrogen, giving trifluoromethanol that is further degraded in air to hydrogen fluoride and carbon dioxide, which results in the formation of a large amount of volatiles [3, 33, 35]. The cured system, therefore, shows excessive porosity and poor vulcanizate properties due to volatiles produced during the curing process. For this reason, it is advisable that VDF-based polymers containing perfluoroalkyl vinyl ethers have a special cure site with curing chemistry different from nucleophilic attack on the backbone. Such a chemistry is the peroxide-induced crosslinking which was specially developed to bypass these kinds of problems.

3.3
Crosslinking with Peroxides

Another technique to crosslink VDF-containing fluoropolymers requires peroxides.

The first peroxide cure agents were used in 1929. But the vulcanizates obtained had poor physical properties, and poor resistance to heat ageing when compared to sulfur-cured vulcanizates.

Braden and Fletcher [140] described the vulcanization of natural rubber with dicumyl peroxide using different compounding ingredients and comparing it with sulfur-cured compounds.

Since the 1950s, peroxide/triallylisocyanurate systems, which enable crosslinking of fluoropolymers through a free radical mechanism, have been established as the best-known non-sulfurated crosslinking agent.

3.3.1
Reaction Conditions

That kind of crosslinking is more easily achieved when the polymer bears a specific group. This group or atom can be introduced into the polymer from the direct terpolymerization of VDF and fluoroalkene.

A fluorinated monomer susceptible to copolymerizing or terpolymerizing vinylidene fluoride is needed to undergo free-radical attack to render peroxide curable elastomeric co- or terpolymers of VDF [35]. So, this monomer must be functionalized or halogenated to ensure a free-radical crosslinking. The monomers mainly used are bromine-containing fluoroolefins such as [42, 43]:

- Bromotrifluoroethylene, $BrCF=CF_2$ [42, 141, 142];
- 1-bromo-2,2-difluoroethylene, $BrCH=CF_2$ [143–145];
- 4-bromo-3,3,4,4-tetrafluorobutene-1, $CH_2=CHCF_2CF_2Br$ [146–148];
- 3-bromoperfluoropropylene, $BrCF_2CF=CF_2$ [149];
- Fluorobutylene $BrCF_2CF_2CF=CF_2$, $BrCF_2CF_2CH=CF_2$, $F_2C=CFOC_2F_4Br$ [150–152] or 1,1,2-trifluoro-4-bromobutene, $F_2C=CFC_2H_4Br$ [87].

The VDF-based polymer containing the brominated monomer gives free radical intermediates on its polymeric backbone upon attack by peroxides [3, 35, 40, 41, 62–65, 134].

Fluoroelastomers containing iodine or bromine atoms can be cured with peroxides. Indeed, modifications of fluorocarbon elastomers with perfluoroalkyl iodides allow us to introduce iodine end groups on the polymeric chain [35, 153–158]. These polymers also lead to free radical intermediates upon attack by peroxides, which in turn crosslink into a network in the presence of a radical trap. Thus, the peroxide needs a coagent to trap the polymeric radicals.

Aromatic as well as aliphatic peroxides can be used. Diacyl peroxides give low crosslinking efficiency and usually require 10 phr for adequate curing. Some dialkyl peroxides and peresters give high crosslinking efficiencies. However, mainly di-tertiary butyl peroxide and dicumyl peroxide are able to cure compounds containing reinforcing carbon black fillers [16].

The peroxides mainly used are:

- dibenzoyl peroxide, $t_{1/2} = 1$ h at 92 °C [16]:

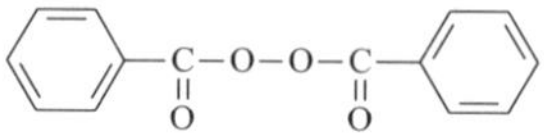

Structure 9

- di-t-butyl peroxide, $t_{1/2} = 1$ h at 141 °C [16]:

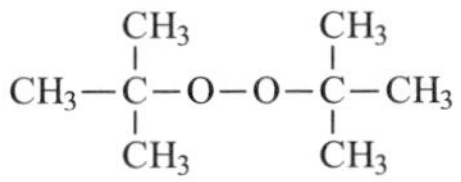

$$CH_3-\underset{\underset{CH_3}{|}}{\overset{\overset{CH_3}{|}}{C}}-O-O-\underset{\underset{CH_3}{|}}{\overset{\overset{CH_3}{|}}{C}}-CH_3$$

Structure 10

- dicumyl peroxide, $t_{1/2} = 1$ h at 132 °C [16, 159]:

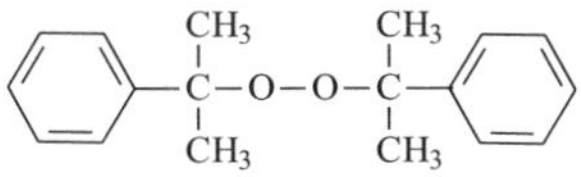

$$Ph-\underset{\underset{CH_3}{|}}{\overset{\overset{CH_3}{|}}{C}}-O-O-\underset{\underset{CH_3}{|}}{\overset{\overset{CH_3}{|}}{C}}-Ph$$

Structure 11

- 1,1-bis(*tert*-butylperoxy)-3,3,5-trimethylcyclohexane, $t_{1/2} = 1$ h at 105 °C [159]:

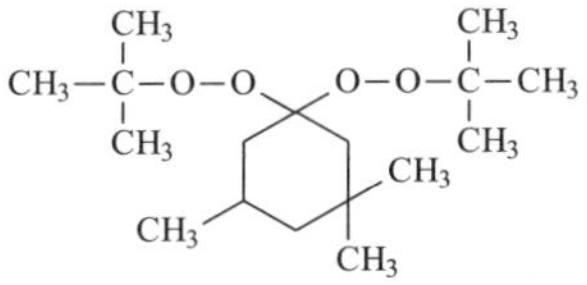

Structure 12

- 2,5-bis-(t-butylperoxy)-2,5-dimethylhexane, $t_{1/2} = 1$ h at 134 °C [40, 159]:

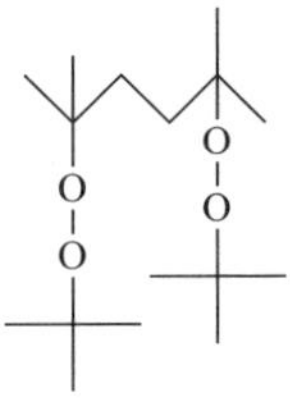

Structure 13

- 2,5-bis-(t-butylperoxy)-2,5-dimethylhexyne, $t_{1/2} = 1$ h at 141 °C [40, 159]:

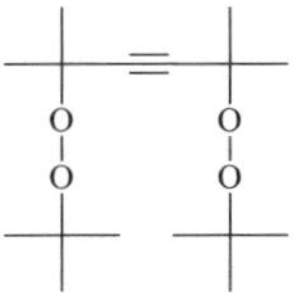

Structure 14

- α,α'-bis(t-butylperoxy)diisopropylbenzene, $t_{1/2} = 1$ h at 134 °C [39, 159]:

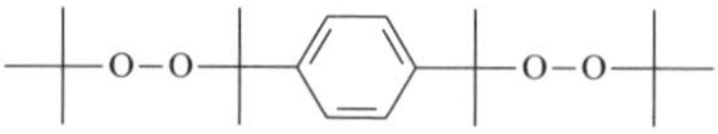

$$+\!\!-\!\!O\!-\!\!O\!-\!\!+\!\!\!\bigcirc\!\!\!+\!\!-\!\!O\!-\!\!O\!-\!\!+$$

Structure 15

The coagents are used to enhance the crosslinking efficiency of peroxide-cured compounds. They are generally di- and trifunctional vinyl compounds, such as:

- 1,2-polybutadiene [16, 159]:

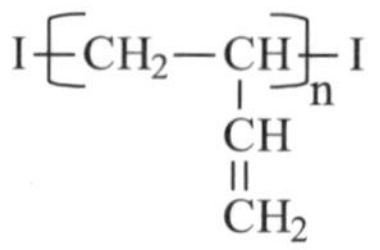

$$I\!-\!\!\!\left[\!CH_2\!-\!\!\underset{\underset{CH_2}{\overset{\|}{CH}}}{CH}\!\right]_n\!\!\!-\!I$$

Structure 16

- ethylene glycol dimethacrylate [16, 159]:

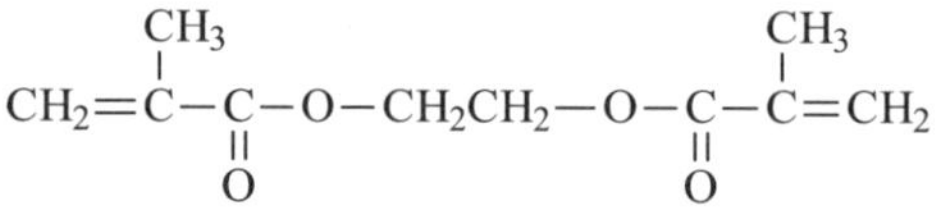

$$CH_2\!=\!\underset{\underset{O}{\overset{\|}{C}}}{\overset{CH_3}{C}}\!-\!C\!-\!O\!-\!CH_2CH_2\!-\!O\!-\!\underset{\underset{O}{\overset{\|}{C}}}{C}\!-\!\overset{CH_3}{\underset{}{C}}\!=\!CH_2$$

Structure 17

- triallyl phosphate [16]:

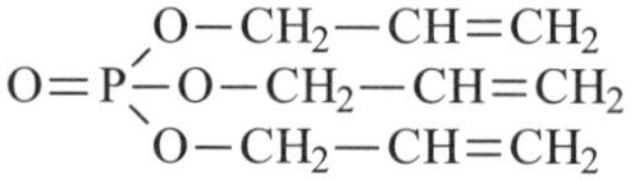

$$O\!=\!P\!\!\begin{array}{l}\diagup O\!-\!CH_2\!-\!CH\!=\!CH_2\\ -\!O\!-\!CH_2\!-\!CH\!=\!CH_2\\ \diagdown O\!-\!CH_2\!-\!CH\!=\!CH_2\end{array}$$

Structure 18

- the triallylisocyanurate (TAIC) or triallyl-1,2,5-triazine-2,4,6-(1H,3H,5H)-trione [35, 40, 95, 97, 160]:

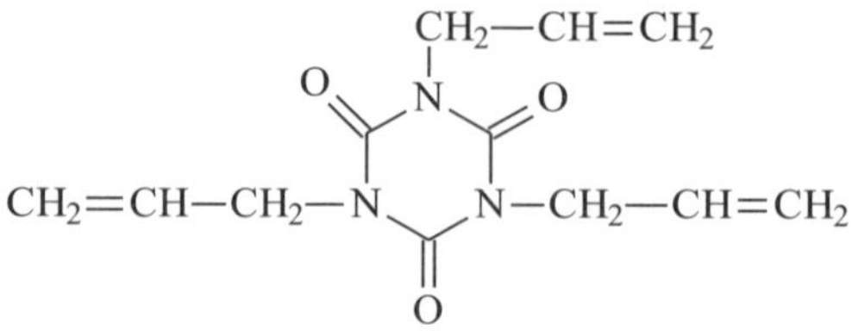

Structure 19

- the triallylcyanurate (TAC), or 2,4,6-triallyloxy-1,2,5-triazine [16]:

$$CH_2=CH-CH_2O-\underset{\underset{\underset{\underset{CH_2}{\overset{\|}{CH}}}{\overset{|}{CH_2}}}{\overset{|}{O}}}{\underset{N\diagdown\diagup N}{\overset{N}{\diagup\diagdown}}}-OCH_2-CH=CH_2$$

Structure 20

The triazine ring is chemically and thermally stable. So, it reinforces the crosslinking network. But the best coagent is TAIC.

The crosslinking reaction also needs metal oxides such as $Ca(OH)_2$, CaO, MgO, ZnO, and PbO to absorb traces of HF generated during the curing process [35], MgO, being the most efficient one, as shown in a case above (see Sect. 3.1.1.4).

3.3.2
Importance of the Coagent

The coagent, the most efficient one being TAIC, is essential for the peroxide-cure mechanism. Indeed, it permits the reaction of crosslinking and improves the compression set resistance.

A poly(TFE-alt-P) copolymer is mixed with the α,α'-bis(t-butylperoxy)diisopropylbenzene (5 phr), different coagents, such as divinylbenzene, N,N'-m-phenylenedimaleimide, 1,2-polybutadiene, trimethylolpropane trimethylmethacrylate, diallylmelamine, TAC, TAIC (3 phr), MgO as the acid acceptor (10 phr), and carbon black (35 phr), to investigate the influence of the different coagents on the gel fraction and the compression set resistance [39]. Each sample is press cured at 160 °C for 30 min, and oven cured at 200 °C for 2 h.

Table 14 [39] exhibits the effects of the coagent on peroxide vulcanization. By considering the compression set percentage, TAIC is found to exhibit the lowest compression set, so it is the most efficient coagent. Basic metal also contributes to improve the compression set resistance. Indeed, the percentage of compression decreases from 75 to 62% thanks to calcium carbonate in the presence of TAIC.

Figure 22 [39, 40] shows the gel formation as a function of the peroxide level. The gel fraction gradually increases with the peroxide level, when co-agent is not present. The cure-promoting effect of the TAIC is remarkable, yielding a gel fraction of nearly 90% at low peroxide dose. So, the crosslinking

Table 14 Influence of the coagent on the gel fraction and the compression set resistance of a peroxide-cured poly(VDF-co-HFP) copolymer ($M_n = 100\,000$ g/mol) [39]

Coagent	Gel fraction	Compression set (%)
None	0.44	100
Divinylbenzene	0.50	100
N,N'-m-phenylenedimaleimide	0.70	98
1,2-polybutadiene ($M_w = 2000$)	0.66	100
Trimethylolpropane trimethacrylate	0.77	94
Diallylmelamine	0.76	95
Triallyl cyanurate (TAC)	0.80	85
Triallyl isocyanurate (TAIC)	0.84	75
Triallyl isocyanurate + $CaCO_3$	0.89	62

Receipe in phr: polymer, 100; α,α'-bis-(t-butylperoxy)-p-diisopropylbenzene, 5; MgO, 10; MT carbon black, 35. Vulcanisation conditions: press cure at 160 °C for 30 min, and oven cure at 200 °C for 2 h

density is not really influenced by the level of peroxide when coagent (TAIC) is introduced.

The same result is obtained in the presence of the 2,5-bis-(t-butylperoxy)-2,5-dimethylhexane [40]. A poly(VDF-co-HFP) copolymer was crosslinked by

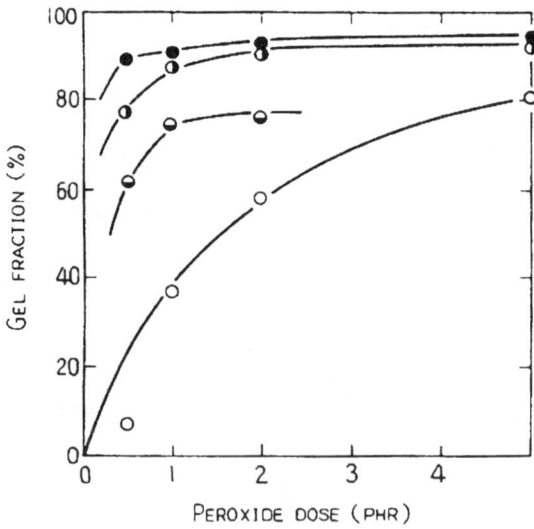

Fig. 22 Evolution of the gel fraction as a function of peroxide dose (phr) for a peroxide-cured poly(TFE-alt-P) copolymer without any coagent (○); with 3 phr of TAIC (●); with 2.4 phr of divinylbenzene (◡); with 3 phr of TAC (▸) (Reprinted with permission of ACS) [39]

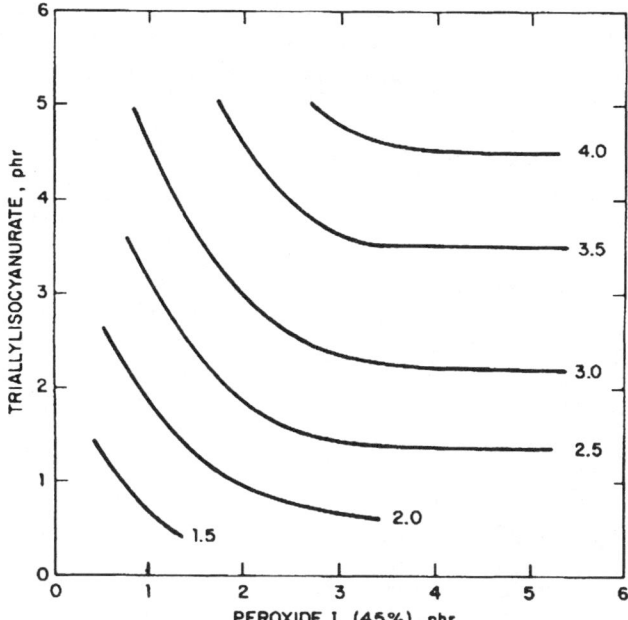

Fig. 23 Interactions between peroxide and coagent, in the evolution of initial cure state (N m) of a poly(VDF-co-HFP) copolymer, measured by ODR at 177 °C for 30 min (Reprinted with permission of Sage) [40]

this peroxide in the presence of TAIC, by oscillating disk rheometer (ODR) at 177 °C for 30 min. $M_H - M_L$ represents the measured cure state (or crosslinking density in N m).

Figure 23 [40] shows that cure state tends to be more drastically influenced by coagent concentration than peroxide concentration. Indeed, Fig. 23 shows that with an unchanged amount of 3 phr for the coagent, and by increasing the quantity of peroxide, the cure state remains constant, whereas with a constant amount of 3 phr of the peroxide, and by increasing the amount of the coagent, the cure state increases. However, cure rate is influenced by both TAIC concentration and peroxide concentration.

3.3.3
Influence of the Nature and the Amount of the Peroxide

Lots of peroxides enable the curing of VDF-based fluoropolymers, but the nature of the peroxides, and the molar amount can influence many different factors such as the curing temperature and the gel fraction.

Identical cure systems are crosslinked, with either 2,5-bis-(t-butylperoxy)-2,5-dimethylhexane, or 2,5-bis-(t-butylperoxy)-2,5-dimethylhexyne [35, 40]. Table 15 [35] shows the different cure states obtained when changing

Table 15 Crosslinking of two peroxides (2,5-bis(t-butylperoxy)-2,5-dimethylhexane, and 2,5-bis(t-butylperoxy)-2,5-dimethylhexyne-3) onto poly(VDF-co-HFP) copolymer, and the influence of their cure temperature on ODR values (t_s2, t_s90, and $M_H - M_L$) [35]

Cure temperature (°C)	Peroxide half-life (min)	ODR values		
		t_s2 (min)	t_c90 (min)	Cure state, $M_H - M_L$ (N-m)
2,5-bis(t-butylperoxy)-2,5-dimethylhexane				
160	4.80	4.0	24.0	8.0
177	0.80	1.6	8.3	8.6
190	0.24	1.4	4.6	8.2
204	0.07	0.8	2.8	7.9
2,5-bis(t-butylperoxy)-2,5-dimethylhexyne-3				
160	18.7	7.1	41.0	–
177	3.4	3.4	14.0	7.7
190	1.0	2.1	7.5	7.7
204	0.3	1.2	4.2	7.5

Table 16 Half life and gel fraction values of poly(TFE-*co*-P) copolymers (M_n = 100 000 g/mol) cured with acyl-, alkyl- or hydroperoxides [39]

Peroxides	Type	Peroxide group	$t_{1/2}$ (min)	Gel fraction
Ph-C(O)-O-O-C(CH₃)₂-CH₂-CH₂-C(CH₃)₂-O-O-C(O)-Ph	Acyl	2	1.2	0.05
‒(O‒O)‒C₆H₄‒(O‒O)‒	Alkyl	2	4.0	0.44
CH₃-C(CH₃)₂-O-O-C(CH₃)₂-CH₃	Alkyloxy or methyl	1	12.0	0.10
CH₃-C(CH₃)₂-O-O-H	Hydroxy	1	6.0	0.00

The samples are press cured at 160 °C for 30 min, and post cured at 200 °C for 2 h

the peroxide and the temperature. For both peroxides, the cure state exhibits a maximum at a fixed temperature. For 2,5-bis-(t-butylperoxy)-2,5-dimethylhexane, the cure state is maximum at 177 °C, whereas with 2,5-

bis-(t-butylperoxy)-2,5-dimethylhexyne, it is maximum at 182 °C. Moreover, 2,5-bis-(t-butylperoxy)-2,5-dimethylhexane is also more efficient than 2,5-bis-(t-butylperoxy)-2,5-dimethylhexyne. Indeed, the first peroxide reaches a cure state of 8.6 N m, whereas it is 7.7 N m for the second one.

Gel fractions were measured from different poly(TFE-alt-P) copolymers cured by peroxide. In Table 16 [39], 30 eq mol^{-1} polymer of peroxide (acyl-, alkyl- or hydro-) are added to a poly(TFE-alt-P) copolymer and vulcanized in a mold at 160 °C for 30 min, and post cure at 200 °C for 2 h. Gel fraction results indicate that these peroxides achieve vulcanization, except for the hydroperoxide which tends to decompose ionically. The best result was obtained with α, α'-bis(t-butylperoxy)diisopropylbenzene, but even in this case, the gel fraction was only 44% because of the absence of a coagent.

A rheometric study [159] was carried out by different tested poly(VDF-ter-HFP-ter-TFE) terpolymers cured by peroxides [dicumyl peroxide (40%) for P-1; 1,3-bis(*tert*-butylperoxisopropyl)-benzene (40%) for P-2; 1,1-bis(*tert*-butylperoxy)-3,3,5-trimethylcyclohexane (40%) for P-3; 2,5-bis-(*tert*-butylperoxy)-2,5-dimethylhexane (45%) for P-4; 2,5-bis-(t-butylperoxy)-2,5-dimethylhexyne (45%) for P-5] with the same coagent (TAIC). Table 17 [159] gives the composition of the compounds used in the study. Compounds (FP-1 to FP-5) differ only from the type and the amount of the peroxide.

Figure 24 [159] is the cure response of an oscillating disc rheometer of the compounds mentioned in Table 17. Figure 24 demonstrates that the rate of crosslinking varies drastically for the industrial peroxides. Efficiencies of P-4

Table 17 Composition of different Viton GF compounds cured with peroxide [159]

Component	Compound				
	FP-1	FP-2	FP-3	FP-4	FP-5
Viton GF[1]	100	100	100	100	100
Carbon black MT	30	30	30	30	30
PbO	3	3	3	3	3
TAIC	3	3	3	3	3
P-1[2]	3				
P-2[3]		4.8			
P-3[4]			3.4		
P-4[5]				3	
P-5[6]					3

[1] poly(VDF-ter-HFP-ter-TFE) terpolymer
[2] P-1 = dicumyl peroxide (Struct 11)
[3] P-2 = 1,3-bis(tert-butylperoxisopropyl)-benzene (Struct 15)
[4] P-3 = 1,1-bis(tert-butylperoxy)-3,3,5-trimethylcyclohexane (Struct 12)
[5] P-4 = 2,5-bis-(t-butylperoxy)-2,5-dimethylhexane (Struct 13)
[6] P-5 = 2,5-bis-(t-butylperoxy)-2,5-dimethylhexyne-3 (Struct 14)

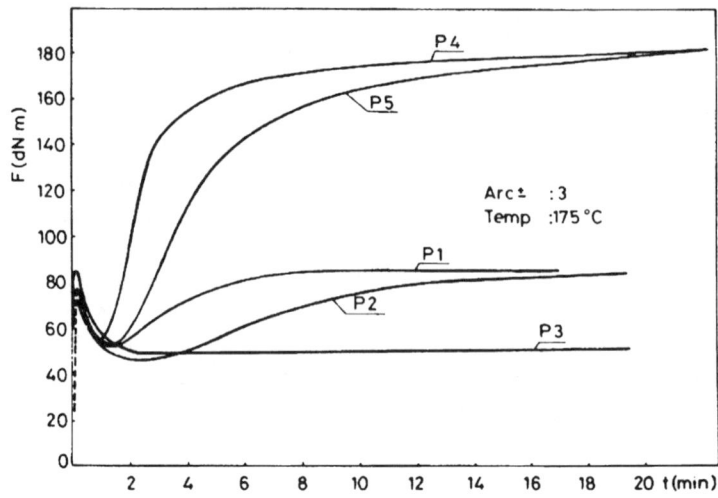

Fig. 24 Rheometer curve of the compounds containing different peroxides (Reprinted with permission of Ferenc Wettl) [159]

and P-5 are outstanding as compared to the three other peroxides, but in the presence of P-3, no crosslinking could be detected by rheometric curve.

So, the following decreasing activity series of the peroxides can be suggested: P-5 ≈ P-4 ≫ P-1 ≈ P-2.

On the other hand for low temperature properties, Dyneon LTFE 6400X, prepared by crosslinking a poly(VDF-tetra-TFE-tetra-PAAVE-tetra-CSM) tetrapolymer (where PAAVE stands for perfluoralkoxyalkylvinylether) in the presence of peroxide/triallylisocyanurate demonstrates superior thermal resistance, high thermostability and outstanding low temperature flexibility [105]. For example its TR-10 performance (or T_g) is – 40 °C and its fuel resistance is the best among the FKM fluorelastomers (fluorine content of 70 wt %).

3.3.4
Mechanism of Crosslinking

The crosslinking mechanism with the peroxide/TAIC system is slightly different from that of diamine or bisphenol ones.

Scheme 11 [3, 16, 35] shows the most probable reaction taking place in a fully compounded stock. An initial process is the thermally induced homolytic cleavage of a peroxide molecule to yield two oxy radicals. The primary decomposition of the 2,5-bis-(t-butylperoxy)-2,5-dimethylhexane leads to the formation of a t-butoxy radical, which may, in a minor reaction, abstract a hydrogen atom to give a t-butanol, and in a major reaction (usually from 120 °C), decomposes into acetone and methyl radical. The methyl rad-

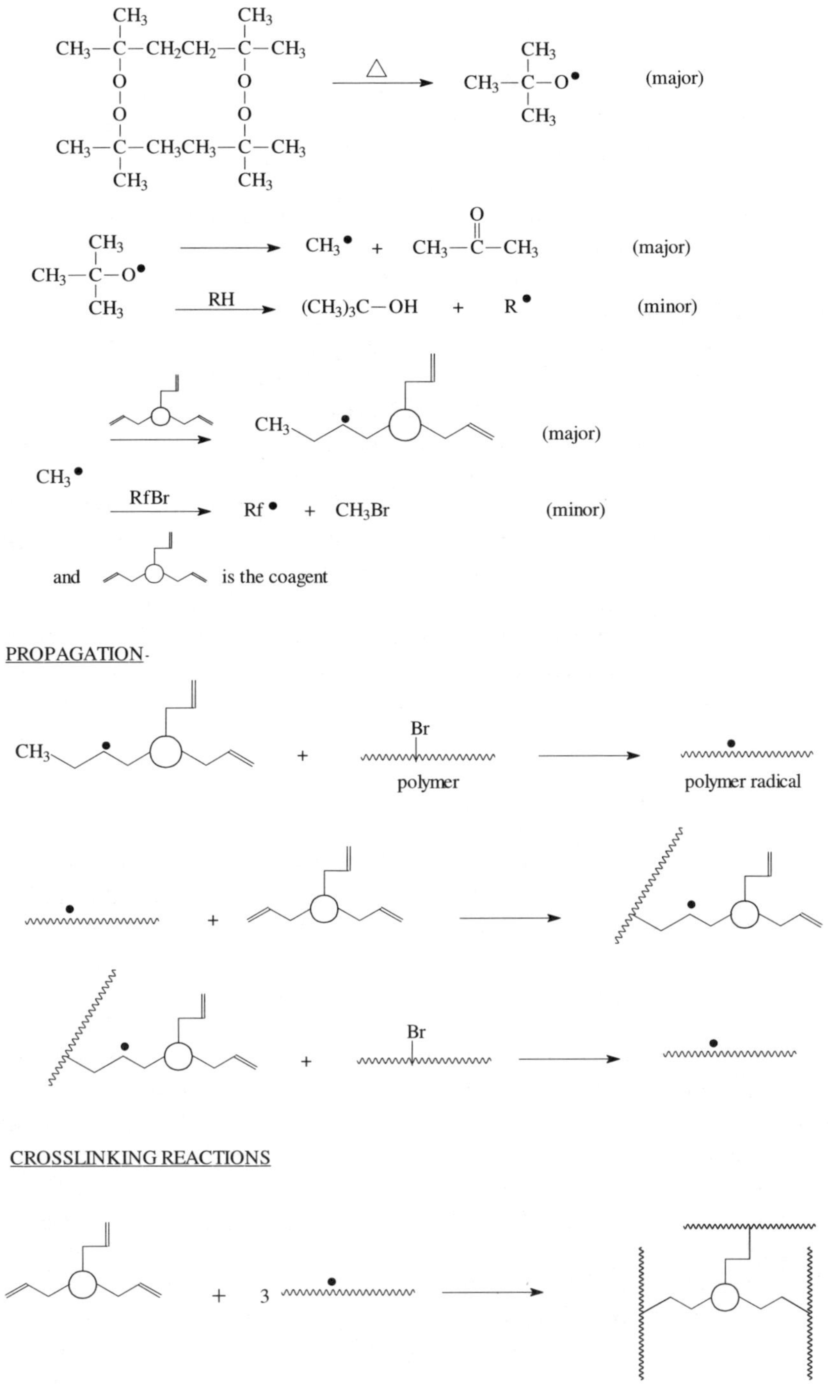

Crosslinking of Vinylidene Fluoride-Containing Fluoropolymers

Scheme 11 Crosslinking mechanism with peroxide [3, 35]

ical, in turn, can abstract a bromine atom from the polymer, to give methyl bromine, or add to the triallyl(iso)cyanurate to give a more stable radical intermediate. These intermediate radicals abstract bromine from the polymer to generate polymeric radicals. The driving force for the chain reaction during propagation is the transfer of a bromine atom from the electron-poor fluoropolymer to an electron-rich hydrocarbon radical on the coagent. Crosslinking takes place when the polymeric radicals add to allylic bonds of the trifunctional coagent. The coagent, therefore becomes the crosslinker.

The cure temperature chosen in peroxide formulations depends on the stability, and the half life of the peroxide. Peroxides with acid groups decompose at lower temperatures than those involving dialkyl or diaryl peroxide. Thus, dicumyl peroxide offers better processing safety than dibenzoyl peroxide. According to Bristow [161], to obtain peroxide-cured natural vulcanizates endowed with the best properties, cure times should not be less than 1 h at 160 °C, 3 h at 150 °C, and 8 h at 140 °C.

Peroxide and bisphenol or diamine-cured systems are differentiated by the type of reaction of crosslinking. Bisphenols or diamines exhibit a dehydrofluorination of the fluoropolymer backbone, followed by a Michael addition for the diamine, and a substitution of the fluorine atom for the bisphenols. The peroxide-cure reaction is a free radical attack, and so this system needs special cure site monomers. Another important crosslinking system that involves a technique requiring radicals is the radiation crosslinking.

3.4
Radiation Crosslinking

The last important way to crosslink VDF-based fluoroelastomers occurs by high-energy radiation. This method only needs the use of specific radiation (without any agent or coagent), but can lead to the formation of other undesirable reactions.

VDF-based copolymers containing hydrogen can be crosslinked with different degrees of efficiency by high-energy radiation [50]. In 1957, Dixon et al. [96] were the first to disclose that poly(VDF-co-HFP) copolymers could be cured successfully with high-energy radiation. Then, Florin and Wall [162] performed further studies of VDF-based copolymers, and Yoshida et al. [163] studied the stress relaxation of irradiated fluorocarbon elastomers. Moreover, many reviews and articles reported the radiation crosslinking of fluoroelastomers, published by Lyons [164–166], Logothetis [167], and Forsythe [168, 169], respectively.

The mechanism of crosslinking by irradiation is the same as that of grafting. That is why the grafting process is briefly explained.

Different types of high-energy radiation are available to be used for the grafting process [170–176]. This radiation may be either electromagnetic radiation such as X-rays and γ-rays (^{60}Co, ^{137}Cs) or charged particles, such as β particles and electrons.

The purpose of the use of γ-rays or an electron beam is to generate radicals in the grafting system. Three different methods may be used to generate radiation [164, 166, 170–178]:

(a) simultaneous radiation grafting;
(b) pre-irradiation in air (hydroperoxide method);
(c) pre-irradiation in vacuum (trapped-radicals method).

In simultaneous radiation grafting, the polymer and the monomer are exposed to radiation at the same time. A chemical reaction of the monomer with the polymer backbone radical initiates the grafting reaction [177]. Alternatively, a two-step grafting procedure may be adapted. In the first step, the polymer is exposed to radiation which leads to the formation of radicals on the macromolecular chain. If the irradiation is carried out in air, radicals react with oxygen, leading to the formation of peroxides and hydroperoxides (hydroperoxide method). When in contact with a monomer, the irradiated polymer initiates grafting by thermal decomposition of the hydroperoxides.

In the absence of air, these macromolecular radicals remain trapped in the polymer matrix and initiate the grafting in the presence of a monomer (trapped radical method).

Simultaneous radiation grafting is, therefore, a single step process while the pre-irradiation method involves two steps [170, 177]. The lifetime of irradiated PVDF at room temperature is about one year [179].

The free generated radicals can undergo several reactions, such as combination to form crosslinks [164, 169], chain scission and recombination or disproportionation, and elimination with the formation of double bonds [170]. In the presence of vinyl monomers, the free radical centers can initiate graft copolymerization [177].

3.4.1
Crosslinking Mechanism by Electron Beam Radiation

When irradiated by electron beam, a VDF-based fluoropolymer can undergo many radical reactions that are going to be detailed in this section. The crosslinking reaction is a particular one.

A PVDF polymer (0.08 mm of thickness, $1.76\,\text{mg/cm}^3$ of density) was electron-beam irradiated at different doses (from 0 to 1200 kGy) [51]. Scheme 12 [45, 51] proposes the radiation induced reaction in this PVDF.

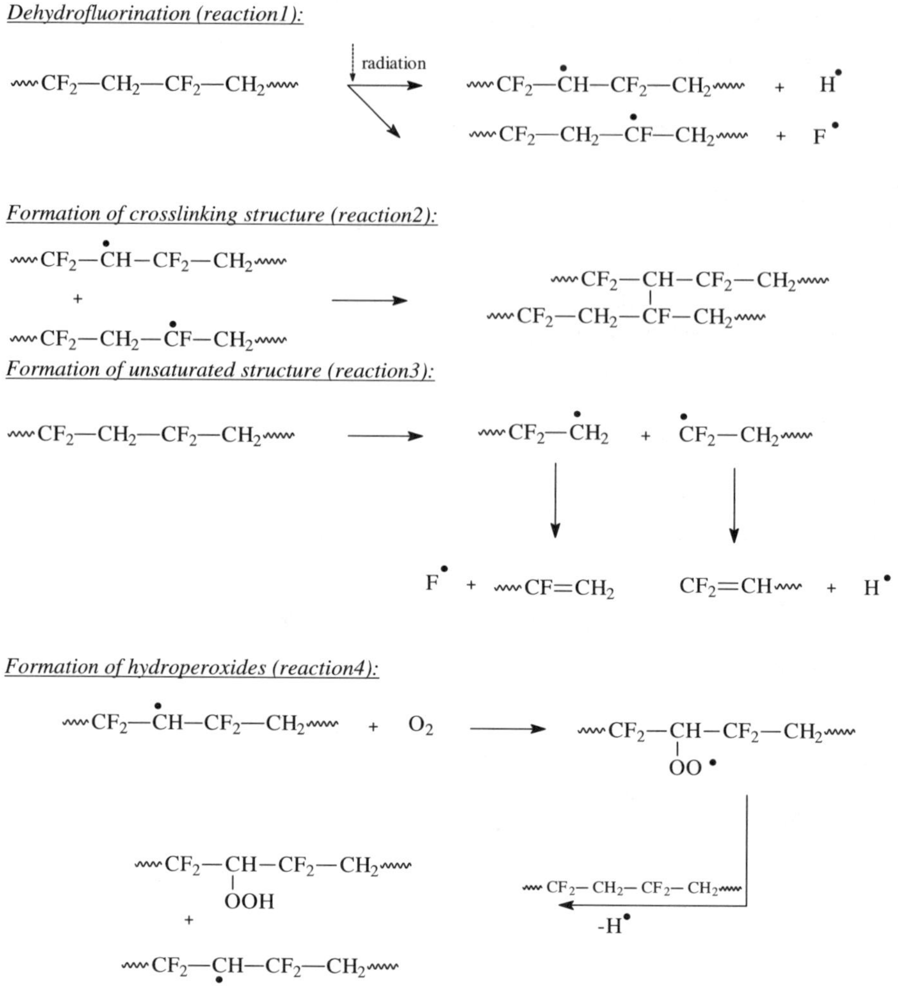

Dehydrofluorination (reaction1):

$$\text{\textasciitilde\textasciitilde CF}_2-\text{CH}_2-\text{CF}_2-\text{CH}_2\text{\textasciitilde\textasciitilde} \xrightarrow{\text{radiation}} \begin{cases} \text{\textasciitilde\textasciitilde CF}_2-\overset{\bullet}{\text{CH}}-\text{CF}_2-\text{CH}_2\text{\textasciitilde\textasciitilde} + \text{H}^{\bullet} \\ \text{\textasciitilde\textasciitilde CF}_2-\text{CH}_2-\overset{\bullet}{\text{CF}}-\text{CH}_2\text{\textasciitilde\textasciitilde} + \text{F}^{\bullet} \end{cases}$$

Formation of crosslinking structure (reaction2):

$$\text{\textasciitilde\textasciitilde CF}_2-\overset{\bullet}{\text{CH}}-\text{CF}_2-\text{CH}_2\text{\textasciitilde\textasciitilde} + \text{\textasciitilde\textasciitilde CF}_2-\text{CH}_2-\overset{\bullet}{\text{CF}}-\text{CH}_2\text{\textasciitilde\textasciitilde} \longrightarrow \begin{matrix} \text{\textasciitilde\textasciitilde CF}_2-\text{CH}-\text{CF}_2-\text{CH}_2\text{\textasciitilde\textasciitilde} \\ | \\ \text{\textasciitilde\textasciitilde CF}_2-\text{CH}_2-\text{CF}-\text{CH}_2\text{\textasciitilde\textasciitilde} \end{matrix}$$

Formation of unsaturated structure (reaction3):

$$\text{\textasciitilde\textasciitilde CF}_2-\text{CH}_2-\text{CF}_2-\text{CH}_2\text{\textasciitilde\textasciitilde} \longrightarrow \text{\textasciitilde\textasciitilde CF}_2-\overset{\bullet}{\text{CH}}_2 + \overset{\bullet}{\text{CF}}_2-\text{CH}_2\text{\textasciitilde\textasciitilde}$$

$$\downarrow \qquad\qquad \downarrow$$

$$\text{F}^{\bullet} + \text{\textasciitilde\textasciitilde CF}=\text{CH}_2 \qquad \text{CF}_2=\text{CH}\text{\textasciitilde\textasciitilde} + \text{H}^{\bullet}$$

Formation of hydroperoxides (reaction4):

$$\text{\textasciitilde\textasciitilde CF}_2-\overset{\bullet}{\text{CH}}-\text{CF}_2-\text{CH}_2\text{\textasciitilde\textasciitilde} + \text{O}_2 \longrightarrow \begin{matrix} \text{\textasciitilde\textasciitilde CF}_2-\text{CH}-\text{CF}_2-\text{CH}_2\text{\textasciitilde\textasciitilde} \\ | \\ \text{OO}^{\bullet} \end{matrix}$$

$$\begin{matrix} \text{\textasciitilde\textasciitilde CF}_2-\text{CH}-\text{CF}_2-\text{CH}_2\text{\textasciitilde\textasciitilde} \\ | \\ \text{OOH} \end{matrix} + \text{\textasciitilde\textasciitilde CF}_2-\overset{\bullet}{\text{CH}}-\text{CF}_2-\text{CH}_2\text{\textasciitilde\textasciitilde} \xleftarrow{-\text{H}^{\bullet}} \text{\textasciitilde\textasciitilde CF}_2-\text{CH}_2-\text{CF}_2-\text{CH}_2\text{\textasciitilde\textasciitilde}$$

Scheme 12 Schematic representation of the mechanism of electron radiation-induced reactions taking place in PVDF films [51]

It is shown that three different reactions are possible after irradiation. Both macroradicals can react together leading to a crosslinking structure as shown in reaction (2) [45, 51, 180]. The macroradicals can be rearranged leading to a disproportionation end-mechanism, which is given in reaction (3) of Scheme 12. Finally, as shown in reaction (4), the macroradical can react with the oxygen atom leading to the formation of hydroperoxide. Those two last reactions [reactions (3) and (4)] do not lead to the formation of a crosslinked structure. These mechanisms are identified by FTIR.

Figure 25 [51] represents FTIR spectra of irradiated PVDF films with their corresponding unirradiated samples. The spectra of irradiated films do not

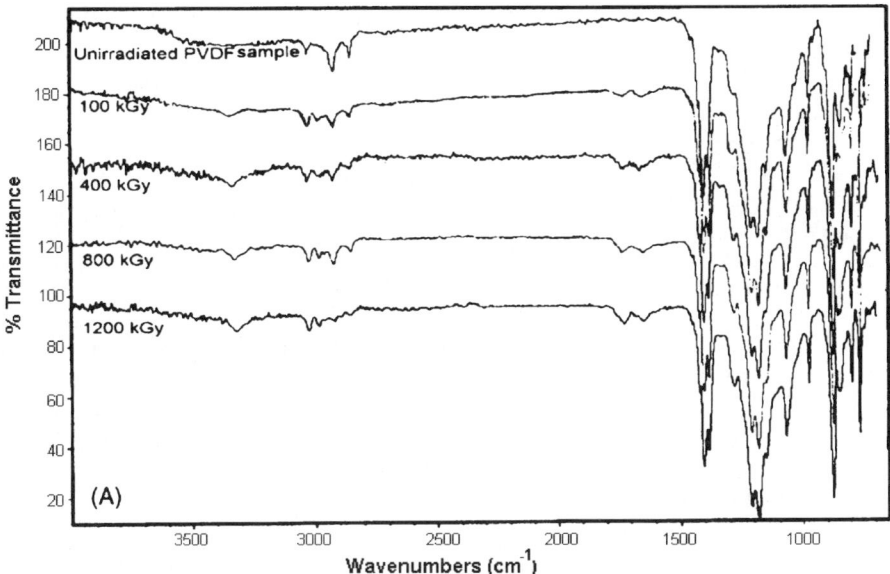

Fig. 25 FTIR spectra of irradiated PVDF films from 0 to 1200 kGy (Reprinted with permission of Elsevier) [51]

show major changes in the main absorption band compared to the unirradiated film. However, two small adjacent bands, in the range of 1650–1750 cm^{-1} appear in the spectra of the irradiated films. The peak at 1654 cm^{-1} is assigned to the C=C bond resulting from the dehydrofluorination reaction in PVDF. Whereas the bands at 1739 and 3300 cm^{-1} represent the C=O group of hydroperoxide initiated by irradiation in air, and the –OH group of hydroperoxides, respectively.

Activated PVDF or poly(VDF-co-HFP) copolymer prepared by γ pre-irradiation (at 25 kGy/h with ^{60}Co source), underwent a grafting by styrene [14, 52]. Indeed, styrene-grafted copolymers based on VDF find a wide range of applications. The degree of grafting (d.o.g.) is determined after assessing the weight of the membrane before irradiation (w_1) and after grafting (w_2):

$$\text{d.o.g.}(\%) = \frac{w_1 - w_2}{w_2} 100$$

Table 18 [52] reports the d.o.g. (%) of different membranes under the same experimental conditions. The difference in d.o.g. between the homopolymer and the dense-membrane copolymer is evident. A 100% grafting yield is detected in the case of the poly(VDF-co-HFP) copolymer, whereas 6% is the d.o.g. obtained in the case of the PVDF membrane. This difference can be due to the different compatibility of the two polymers with styrene, and the kinetics of grafting which is faster in the case of the copolymer. Hence, irra-

Table 18 d.o.g. (degree of grafting) of starting membranes of PVDF and poly(VDF-co-HFP)copolymer [52]

Membrane	d.o.g. (%)
Compact PVDF	6
Compact P(VDF-5 mol % HFP)	100
Porous PVDF (100 nm)	74
Porous PVDF (10 μm)	53

diated VDF-based copolymers can give higher grafting densities than PVDF homopolymers can.

3.4.2
Influence of Irradiation Parameters on the Properties of Crosslinked Fluoropolymers

The method of irradiation and the dose can influence several properties such as thermal properties (glass transition temperature T_g, decomposition temperature T_{dec}, crystallization temperature T_c, and melting temperature T_m) and mechanical properties (tensile strength, modulus at 50% elongation, gel fraction ...).

3.4.2.1
Thermal Properties

A PVDF film was irradiated with doses of electron irradiation (ranging from 0 to 1200 kGy). Figures 26 and 27 [51] represent the evolution of the DSC melting thermograms, and cooling thermograms, respectively of irradiated PVDF films. The heat of melting which is obtained from the area under the peaks of curves noted in Fig. 26, is found to increase for the lower melting peak, whereas that under the higher melting peak decreases with the increase in the irradiation dose up to 200 kGy. Table 19 shows that the heat of melting under both peaks together increases with the increase in irradiation dose. Moreover, both melting temperature, T_m, (Fig. 26) and the crystallization temperature, T_c, (Fig. 27) decrease with the increasing irradiation dose [180, 181]. Table 19 [51] summarizes those results.

This behavior indicates that crosslinking and the main chain scission reaction play an important role during irradiation of PVDF. Furthermore, the decreasing of T_c with the increasing irradiation dose indicates that crosslinking is the predominant reaction.

The evolution of the glass transition temperature (T_g) of a FKM [poly(VDF-ter-HFP-ter-TFE) terpolymer] crosslinked by electron beam irradiation in the presence of MgO, and HMDA-C is measured as a function of radiation dose.

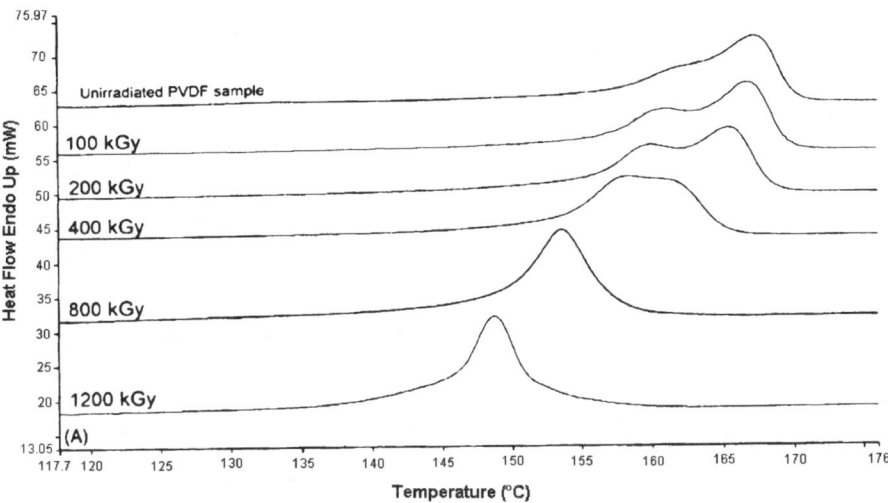

Fig. 26 DSC melting thermograms of irradiated PVDF films from 0 to 1200 kGy (Reprinted with permission of Elsevier) [51]

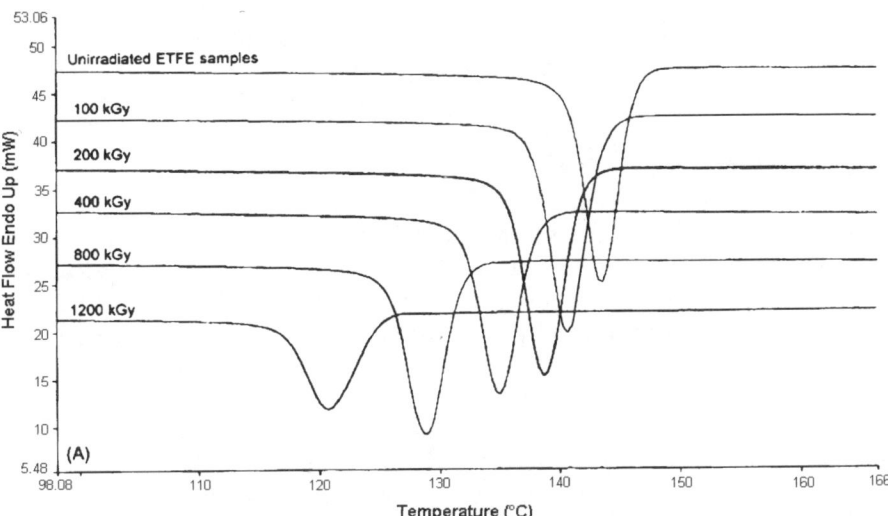

Fig. 27 DSC cooling thermograms of an irradiated PVDF film from 0 to 1200 kGy (Reprinted with permission of Elsevier) [51]

Table 20 [48] represents the evolution of the T_g. The higher the radiation dose, the higher the T_g. Indeed, crosslinking hinders the macromolecular rotation, thus requiring a higher temperature for the inception of rotation. An increase in glass transition temperature at higher radiation doses is due to an increase in the extent of crosslinking.

Table 19 Variation of heat of melting, melting temperature and crystallization temperature of PVDF films with the irradiation dose [51]

Irradiation dose (kGy)	Crystallisation temperature (°C)	Heat of melting (J/g)	Melting temperature (°C)	
			Higher peak	Lower peak
0	143	34.3	167.3	162.1
100	141	38.2	166.8	160.7
200	139	40.6	165.4	159.8
400	135	39.1	158.5	158.2
800	129	39.9	–	153.5
1200	121	40.2	–	148.7

Table 20 Influence of electron beam irradiation on the glass transition temperature of a poly(VDF-ter-HFP-ter-TFE)terpolymer [48]

Radiation dose (kGy)	T_g (°C)
0	4
1000	27
1500	38

3.4.2.2
Mechanical Properties

The interaction of high energy radiation with partially fluorinated polymer causes various changes in its thermal properties, but also in its physical and mechanical properties, depending on the irradiation conditions.

Polyvinylidene fluoride and poly(VDF-ter-HFP-ter-TFE) terpolymer were irradiated in the presence or the absence of a crosslinking agent, and several mechanical properties were measured.

A PVDF film was exposed to high energy electron radiation in air with a 1 MeV electron beam to different dose levels up to 10^6 Gy [46]. Gel fraction analysis was carried out on the virgin and irradiated films. The samples were extracted in dimethylacetamide (DMAC) at 75 °C for 1 h. The residue contents were then dried out in a vacuum oven at 80 °C for 16 h. According to Charlesby [182], the gel fraction analysis is a measurement of the degree of crosslinking, and is defined as the ratio of weight of the insoluble residue, to the weight of the original sample. The evolution of the gel fraction versus total absorbed dose is given in Fig. 28 [46]. The curve increases gradually to about 20% with an increase in total absorbed dose up to 10^4 Gy, and exhibits a larger increase to about 82% with a further increase in the radiation dose to 10^6 Gy. The presence of a gel is a clear evidence of extensive crosslinking.

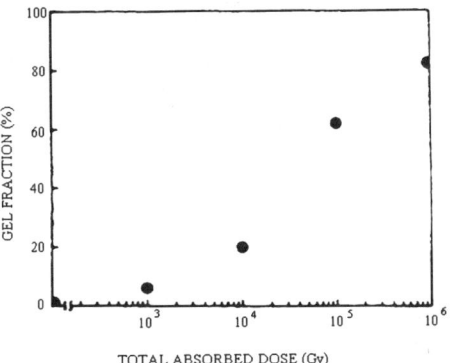

Fig. 28 Gel fraction analysis in DMAC at 75 °C of virgin and irradiated PVDF films (Reprinted with permission of Kluwer) [46]

A PVDF and a poly(VDF-co-HFP) copolymer were irradiated by Kr ions at 6.2 MeV/amu energy [47]. The evolution of the insoluble F fraction versus the absorbed dose D, and the fluence is represented in Fig. 29 [47]. In both cases, an increase of the gel fraction F as the absorbed dose increases is observed.

A poly(VDF-ter-HFP-ter-TFE) terpolymer was irradiated with an electron beam accelerator at different doses (0–200 kGy) in air, and cured afterwards with 1 phr of HMDA-C. The gel fraction was carried out using the solvent methylethylketone at 25 °C [48, 49]. Figure 30 [49] shows the gel fraction and the crosslinking density of the irradiated pure or vulcanized poly(VDF-ter-HFP-ter-TFE) terpolymer at different doses. It is observed an increase in crosslinking density (- □ -) on irradiating rubber vulcanizates with 50 kGy dose and above. Both the control (pure rubber) and the rubber vulcanizate

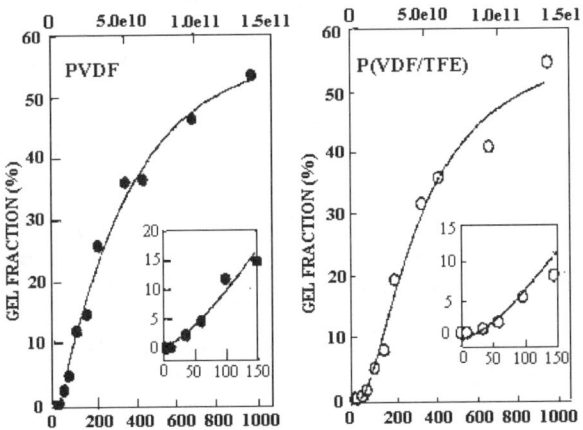

Fig. 29 Evolution of the insoluble fraction versus the absorbed dose and the fluence for a PVDF and a poly(VDF-co-TrFE) copolymer (Reprinted with permission of Elsevier) [47]

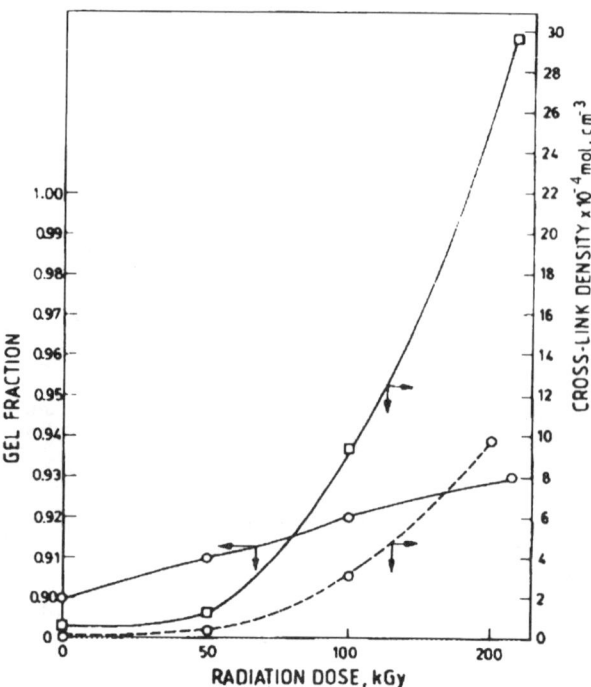

Fig. 30 Variation of the gel fraction (full line o) and the crosslink density (□) of irradiated poly(VDF-ter-HFP-ter-TFE) terpolymer vulcanized with HMDA-C, and variation of the crosslink density of pure terpolymer rubber (broken line) versus radiation dose (Reprinted with permission of Kluwer) [49]

behave similarly. The increase in crosslink density is about 10–25 times, and the change in crosslink density is more important for the rubber vulcanizate.

Nasef and Dahlan [51] investigated the irradiation of PVDF using an electron beam accelerator of 2.0 MeV. The effect of irradiation dose on the mechanical properties (tensile strength (a), elongation at break (b)) is given in Figs. 31 and 32 [51]. The tensile strength (Fig. 31) of PVDF increases until a dose of 800 kGy, then it slightly decreases as the irradiation dose goes higher.

The elongation at break (Fig. 32) decreases gradually with the increase in the irradiation dose, due to crosslinking. This leads to an increase in molecular weight and to the formation of an insoluble three dimensional network [46, 51]. The crosslinking progressively immobilizes the insoluble oriented molecules and prevents them from moving laterally without breaking the bonds.

These radiation induced changes in the mechanical properties of a PVDF are permanent since crosslinking causes damage to the crystal structure which is unalterable [46]. A poly (VDF-ter-HFP-ter-TFE) terpolymer was ir-

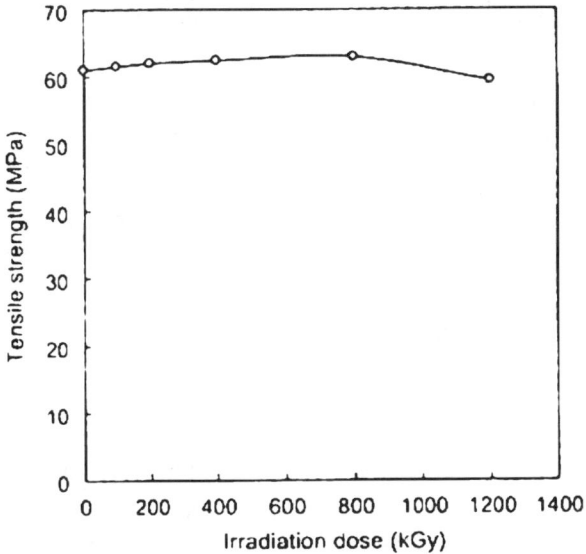

Fig. 31 Variation of the tensile strength of PVDF films versus irradiation dose (Reprinted with permission of Elsevier) [51]

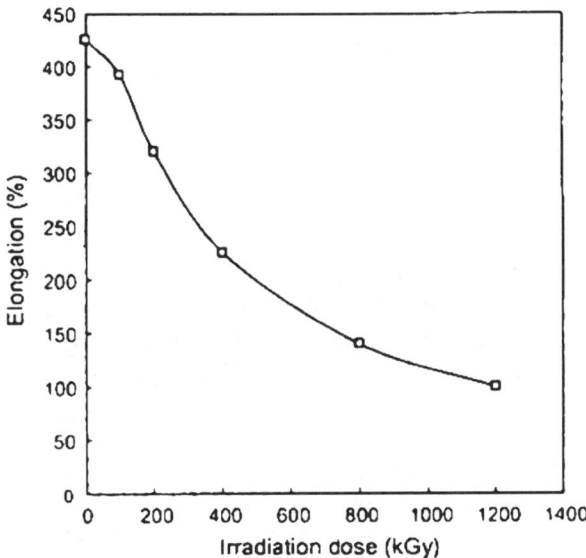

Fig. 32 Variation of the elongation percent of PVDF films versus the irradiation dose. (Reprinted with permission of Elsevier) [51]

radiated by an electron accelerator (at energy of 2 MeV) in the presence of MgO (5 phr) and HMDA-C (1 phr). Mechanical properties (tensile strength, elongation at break and modulus at 50% elongation) were investigated and

Table 21 Influence of electron beam irradiation on the mechanical properties of a terpolymer [48]

Radiation dose (kGy)	Tensile strength (MPa)	Elongation at break (%)	Modulus at 50% elongation	Gel fraction
0	3.6	513	0.73	0.90
250	3.2	108	1.26	0.92
500	2.3	81	1.32	0.94
750	1.8	50	1.60	0.95
1000	3.3	38	–	0.96
1500	4.3	23	–	0.97

are presented in Table 21 [48, 49]. The modulus increases and the elongation at break decreases with an increase in irradiation dose. The tensile strength, however, does not change much when increasing irradiation dose.

Radiation curing is the last main method of crosslinking for VDF-based fluoroelastomers. It is more efficient for copolymers and terpolymers than for PVDF.

3.5
Crosslinking with a Thiol-ene System

There are a few other methods to crosslink VDF-based fluoroelastomers, like the thiol-ene system, but these are very rarely used.

Although thiol-ene systems easily lead to the vulcanization of hydrogenated elastomers, this process has scarcely been used to cure fluoroelastomers because of the limited availability of mercapto side groups. However, we have recently used an original trifluorovinyl ω-thioacetate monomer able to copolymerize with VDF [44, 50] and the resulting copolymer was hydrolyzed to generate elastomers bearing mercapto lateral groups. Curing was performed in the presence of non-conjugated dienes under radical conditions as in Scheme 13 [44].

Other systems of crosslinking exist, but they are more adapted to TFE-containing copolymers, although crosslinking in the presence of water from PVDF-bearing trialkoxysilanes was efficient [94]. A nitrile-containing cure-site monomer with the same reactivity as perfluoromethylvinylether was copolymerized with TFE. Crosslinking was brought about by the catalytic interaction of tetraphenyltin or silver oxide on the pendant nitrile groups. The structure of the crosslinks is assumed to be mainly triazine [35, 183].

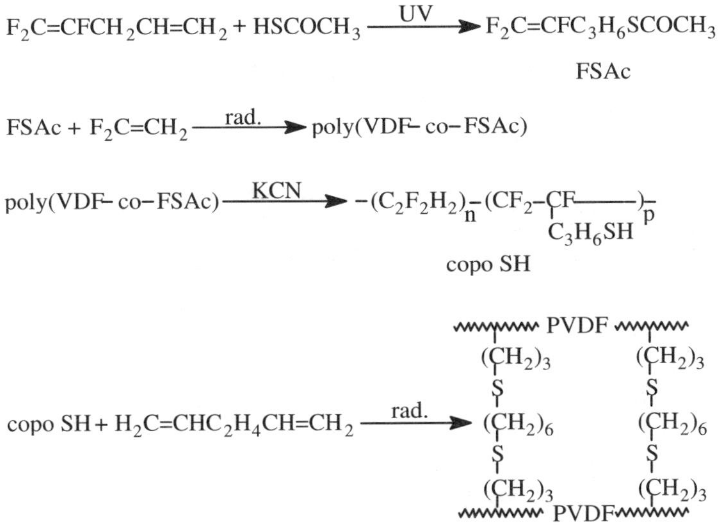

$F_2C=CFCH_2CH=CH_2 + HSCOCH_3 \xrightarrow{UV} F_2C=CFC_3H_6SCOCH_3$
$\phantom{F_2C=CFCH_2CH=CH_2 + HSCOCH_3 \xrightarrow{UV} F_2C=CFC_3H_6SCOC}$FSAc

$FSAc + F_2C=CH_2 \xrightarrow{rad.} poly(VDF\text{-}co\text{-}FSAc)$

$poly(VDF\text{-}co\text{-}FSAc) \xrightarrow{KCN} -(C_2F_2H_2)_n\text{-}(CF_2\text{-}\underset{C_3H_6SH}{CF}\text{---})_p$
copo SH

copo SH + $H_2C=CHC_2H_4CH=CH_2 \xrightarrow{rad.}$ [crosslinked structure with PVDF chains connected via $(CH_2)_3$–S–$(CH_2)_6$–S–$(CH_2)_3$ bridges]

Scheme 13 Crosslinking of VDF-containing copolymer by thiol ene systems [44]

4
Comparison of Physical and Mechanical Properties

All the different cured systems mentioned above differ in crosslinking agents and mechanism, but also in the crosslinking density, and the mechanical properties, such as compression set resistance, elongation at break, modulus at 100 and 200%, tensile strength, hardness of the cured material, etc. ... The comparison of the crosslinking density and the different mechanical properties for the main cured systems (diamines, bisphenols and peroxides) is the subject of this part of the article.

Regarding resistance to acids (H_2SO_4, HNO_3), to bases (NaOH, NaClO), and to water, by measuring the volume increase after immersion, it is clear that peroxide-cured elastomers are more resistant to acids, to bases and to water than those from the diamine-cured elastomers. Indeed, this last system decomposes when immersed in a strong base or an acid solution.

Nevertheless, diamine-cured systems have a lower percentage of volume increase after immersion in oil and fuel oil than peroxide ones.

Figure 33 [3, 35, 134] compares the crosslinking density of post-cure diamine-, bisphenol- or peroxide-cured systems at 204 °C under nitrogen. The vulcanizate crosslinking density of bisphenol- and peroxide-cured systems, before and after post curing remains the same, whereas that of the diamine vulcanizate increases substantially. This implies that diamine is a better crosslinking agent than bisphenol or peroxide regarding crosslinking densities.

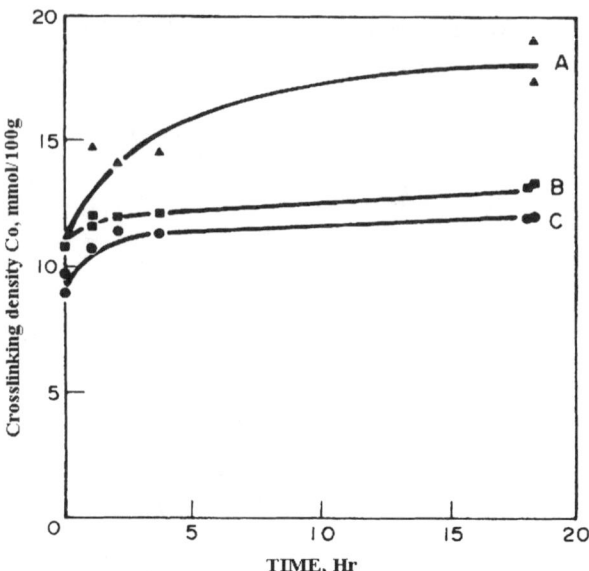

Fig. 33 Crosslinking density of cured gum stocks versus post cure time at 204 °C, under nitrogen: (A) diamine cured system; (B) peroxide cured system; (C) bisphenol cured (Reprinted with permission of ACS) [134]

A Viton A-HV (poly(VDF-co-HFP) copolymer) is crosslinked with a same amount of a diamine (biscinnamylidene hexamethylene diamine or LD-214) and a peroxide (benzoyl peroxide), in the presence of Maglite D (MgO) [23]. Different mechanical properties (tensile strength, elongation at break, and hardness) are evaluated for both press-cure and post-cure systems. Tables 22 and 23 [23] show that the peroxide-cure system leads to materials which exhibit higher tensile strengths, whereas diamine improves the elongation and hardness of the resulting crosslinked macromolecules. Table 23 also shows that post-cure step improves readily each mechanical property (tensile strength, elongation and hardness). So, this step is essential in the crosslinking mechanism.

Other mechanical properties and resistance to bases and to acids are studied for a poly(TFE-alt-P) elastomer cured with a peroxide (α, α'-bis-(t-butylperoxy)-p-diisopropylbenzene), and a poly(VDF-ter-HFP-ter-TFE) terpolymer cured with a diamine (N,N' dicinnamylidene-1,6-hexanediamine) (Table 24) [39].

As in Table 23, peroxide-cured systems exhibit a better tensile strength, and a better compression set resistance, whereas diamines exhibit higher elongation at break and hardness.

Table 2 [35] displays different mechanical properties (modulus at 100%, tensile strength, elongation at break and compression set for O-rings and pellets) for bisphenol-cured and peroxide-cured poly(VDF-ter-HFP-ter-

Table 22 Composition of vulcanizates I and II [23]

Components, parts	Vulcanisate I	Vulcanisate II
Viton A-HV[1]	100	100
Maglite D[2]	15	15
LD-214[3]	4	–
Benzoyl peroxide, 95%	–	4

[1] Viton A-HV = poly(VDF-co-HFP)copolymer
[2] Maglite D = MgO
[3] LD-214 = biscinnamylidene hexamethylene diamine

Table 23 Mechanical properties of vulcanizates I (diamine cure) and II (peroxide cure) after press cure at 150 °C for 30 min, and post cure from 120 to 200 °C at a heating rate of 25 °C/h, then heat at 200 °C for 24 h [23]

Properties	Vulcanisate I		Vulcanisate II	
	Press cure	Post cure	Press cure	Post cure
Tensile strength (psi)	2230	3250	3550	3650
Elongation (%)	460	310	460	420
Hardness, shore A	64	68	60	63

TFE)terpolymer, after press cure at 177 °C for 10 min and after post cure at 232 °C for 24 h. As in Table 23, Table 2 shows an improvement of the mechanical properties after post cure. Moreover, the bisphenol-cured system has a better compression set resistance than the peroxide-cured system, whereas the peroxide-cured system has better modulus at 100%, better tensile strength, better elongation than those resulting from bisphenol crosslinking.

A study of several mechanical properties (100% Modulus, tensile strength, elongation at break, and compression set resistance) of bisphenol AF and peroxide (2,5-dimethyl-2,5-di-t-butyl-peroxyhexane)-cured poly(VDF-co-HFP) copolymers is supplied in Table 25 [40]. First, the post-cure step improves all mechanical properties. Then, bisphenol post-cure systems exhibit a better compression set resistance than the peroxide one, whereas peroxide-cured polymer exhibits better modulus at 100%, better tensile strength, and better elongation at break than bisphenol-cured polymer.

Flisi [131] studied the evolution of elongation at break and tensile strength for a poly(VDF-co-HFP) Tecnoflon N copolymer cured with a bisphenol AF and diamines (melting of piperazine carbamate and trimethylene diamine carbamate). Figures 34 and 35 [131] represent elongation at break and tensile strength versus time for those samples. This author showed that elongation at break (Fig. 34) decreased continuously because the network chains became shorter. Moreover, crosslinking with bisphenol yielded materials with

Table 24 Mechanical and chemical properties of a peroxide-cured poly(TFE-alt-P) elastomer, and a diamine-cured poly(VDF-ter-HFP-ter-TFE) elastomer [39]

Properties	Peroxide[1] – cured TFE/P elastomer	Diamine[2]-cured poly(VDF-ter-HFP-ter-TFE) terpolymer
Physical properties:		
Specific gravity (g/cm^3)	1.60	1.93
Tensile strength (MPa)	∼ 20.0–25.0	∼ 14.0–17.0
Elongation at break (%)	∼ 200–400	∼ 400–500
Tensile modulus at 100%, MPa	∼ 2.5–3.5	∼ 4.0–5.0
Hardness, shore A	70	83
Compression set[3] (%)	35	49
Brittle point (°C)	– 40	– 45
Retraction temperature (°C)	3	– 20
Chemical resistance:		
Volume increase after immersion (%)		
96% H$_2$SO$_4$, 100 °C, 3 d	4.4	45
60% HNO$_3$, 70 °C, 3 d	10.0	Decomposed
50% NaOH, 100 °C, 3 d	1.1	Decomposed
10% NaClO, 100 °C, 7 d	1.0	Decomposed
H$_2$O, 100 °C, 3 d	1.1	5.9
Steam, 160 °C, 7 d	4.6	12.8
Oil #1, 150 °C, 3 d	2.0	0.5
Oil #3, 150 °C, 3 d	10.0	2.0
Fuel oil, 25 °C, 7 d	38	3.0

[1] a 180 000 g/mol poly(TFE-alt-P) elastomer is cured with a peroxide (α,α'-bis-(t-butylperoxy)-*p*-diisopropylbenzene) and TAIC
[2] diamine = N,N' dicinnamylidene-1,6-hexanediamine
[3] at 200 °C for 22 h

a higher elongation than those achieved from diamine. Tensile strength (Fig. 35) decreased slowly in bisphenol vulcanizate during the whole period of 32 days, while the curve of the diamine vulcanizate presents an irregular trend. Indeed, the curve first decreased, then increased, and finally decreased again.

Flisi [131] also study the compression set resistance at 200 °C versus ageing time for different diamine and bisphenol Tecnoflon T [poly(VDF-ter-HFP-ter-TFE) terpolymer] vulcanizates (Fig. 12).

First, the best compression set resistance was obtained for curve 4 (bisphenol-AF cured poly(VDF-*co*-HFP) copolymer). Second, little improvement was obtained in Tecnoflon T by changing the curing agent from HMDA-C (curve 1) to piperazine carbamate (curve 2). Finally, no practical difference

Table 25 Mechanical properties for press cure and post cure poly(VDF-co-HFP)copolymer cured with bisphenol AF or peroxide (2,5-dimethyl-2,5-di-t-butyl-peroxyhexane) [40]

Compounds	Poly(VDF-co-HFP) copolymer cured by Bisphenol AF[1]		Poly(VDF-co-HFP) copolymer cured by Peroxide (2,5-dimethyl-2,5-di-t-butyl-peroxyhexane)[1]	
Post cure[2]	No	Yes	No	Yes
Stress-strain, 25 °C				
100% Modulus (Mpa)	4.5	6.3	5.3	11.4
Tensile strength (Mpa)	11.7	14.5	10.4	15.9
Elongation at break (%)	250	185	200	140
Compression set[3] (%)				
Pellets	85	15	45	18
O-ring	40	16	44	25

[1] both cured samples are press cured at 177 °C for 30 min
[2] post cured at 232 °C for 24 h
[3] at 200 °C for 70 h

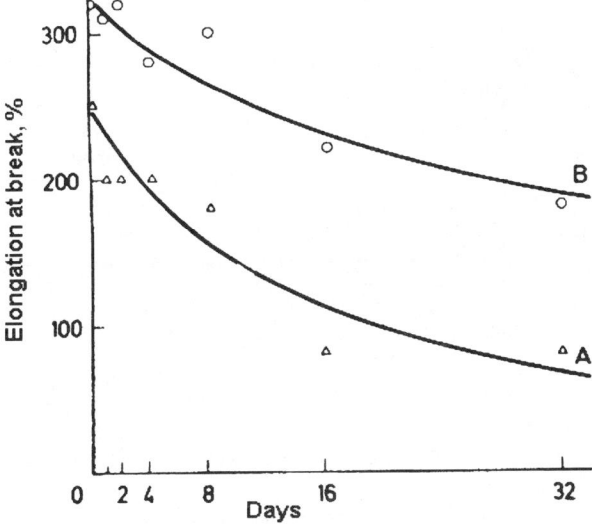

Fig. 34 Dependence of elongation at break versus aging time of a TecnoflonN (poly(VDF-co-HFP) copolymer) crosslinked with piperazine carbamate and trimethyl diamine carbamate (*curve A*), and a TecnoflonN cured with a bisphenol AF (*curve B*) (Reprinted with permission of Hüthig Fach Verlag) [131]

was observed by changing the polymer with the same formulation, since curves 2 and 3 have the same behavior.

Table 26 summarizes the efficiency of bisphenol, peroxide and diamine cure systems regarding each mechanical property. Bisphenol and peroxide-

Table 26 Comparison of the three main cure systems regarding mechanical properties

	Bisphenol	Peroxide	Diamine
Tensile strength	++	+++	+
Elongation at break	+	+++	++
Chemical resistance to oil		+	++
Chemical resistance to acid and base		+++	--
Hardness	++	++	+++
Compression set resistance	+++	++	+
Thermostability	+ (hydroquinone)	+++	++

```
--  = very bad
+   = more or less good
++  = good
+++ = very good
```

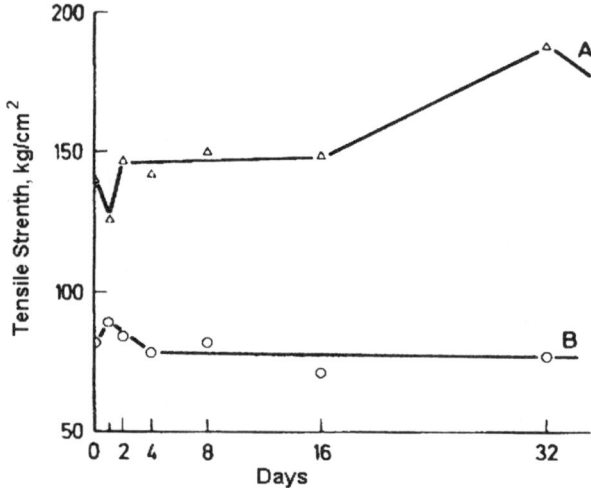

Fig. 35 Dependence of tensile strength versus ageing time of the same sample than Fig. 34 (Reprinted with permission of Hüthig Fach Verlag) [131]

cured fluorinated polymer exhibit better tensile strength, and resistance to bases and acids than diamine-cured systems. However, diamine-cured VDF-based polymers show a higher hardness and resistance to oil than those of peroxide- and bisphenol-cured systems. Indeed bisphenol was shown to be the best crosslinking agent for a high compression set resistance. By regarding the thermostability of each cure system, the diamine (biscinnamylidene hexamethylene diamine) cure copolymer decomposes at 457 °C whereas the peroxide one decomposes at 472 °C [23, 184]. The diamine cure is thermally more stable than the hydroquinone one [29].

5
Applications

The chemical, physical and mechanical properties mainly depend on the crosslinking agent. Those properties are crucial for the applications.

The thermal stability, sealing capability and chemical resistance of fluoroelastomers have led to an increase in their use in a broad variety of industries. Applications for fluorocarbon elastomers in automotive, petroleum (transmission fluids, gear lubricants and automotive engine oils), aerospace, and energy related industries illustrate the potential for those high-performance elastomers [115]. Although the demand for methanol containing fuels has been diminished over the past years, methanol (optionally ethanol) isooctane/toluene blends are still being considered as "flex fuels" and test fuels by design engineers to assess the overall chemical robustness of a fluoroelastomer. With the advent of stricter environmental regulations such as California Air Resource Board's CARB LEV II (globally, the strictest evaporate emission regulation to date), there is a demand for fluoroelastomers that offer low temperature flexibility, low permeation and a low permeation service performance of 15 years of 150 000 miles.

Fluoroelastomers are presently widely used in the industry as O-rings, V-rings, gaskets and other types of static and dynamic seals, as diaphragms, valve seals, hoses, coated clothes, shaft seals quick connector o-rings, dynamic sealing applications, membranes, [185], expansion joints, etc. [186]. They are also used in cars as O-rings for fuel, shaft seals and other components of fuel and transmission systems [35, 58, 62–65].

Moreover, some properties of fluoroelastomers, and especially those of VDF-based elastomers can be improved by crosslinking. These improved properties allow one to use cured fluoroelastomers in new applications as mentioned below.

The elastomeric poly(VDF-co-HFP) copolymers crosslinked with polyamine possess high temperature stability, good resistance to a wide variety of solvents, oils, and fuels [187]. So, these cured elastomers are particularly suitable for use in the manufacture of tubing employed as aircraft hoses, used to carry fuel lubricants, at high temperature and under high pressure [188]. Moreover, poly(VDF-co-HFP) copolymers crosslinked with aminosilane are used in the aircraft construction industry because they are also odorless [189].

Other applications of cured fluoroelastomers are sealings, O-rings [26] and oil seals [26, 115, 131, 190]. It is mentioned above that a cured VDF-based copolymer has a better compression set resistance than a raw rubber. This property is essential for the sealing application.

Peroxide curable VDF-based copolymer and terpolymer offer improved extrusion characteristics. They can be vulcanized at atmospheric pressure and eliminates fissuring in thick sections. They have applications as cords, tubes or irregular-profile items of any dimension [40, 191].

A poly(VDF-co-HFP) copolymer is applied to a metallic substrate, as coating composition and crosslinked with amine, diamine, or ethoxysilane [192]. This cured polymer is used as thick or thin free standing films, or thick or thin films with good adhesion to metallic or other rigid surfaces [192]. Moreover, diamine-cured PVDF can be used as strong adhesive joints without prior surface modification [31].

Crosslinkable fluoropolymers based on TFE, TrFE, HFP, VDF, CTFE, and perfluoro(alkylvinylether) can form corrosion resistant structures [105, 193], especially in the case of crosslinked poly(VDF-tetra-TFE-tetra-PAAVE-tetra-CSM) tetrapolymer (Dyneon LTFE 6400X) [105] highly resistant to a variety of commonly used test fuels, including CE10, CM85 and JP-8+100 (jet engine fuel).

Another application of cured fluoroelastomer is a multi layer insulator system for electrical conductors. This system possesses an extruded crosslinked fluoroelastomer outer layer with the fluoropolymer selected from copolymer or terpolymer of ethylene and TFE.

Irradiated PVDF and poly(VDF-co-TrFE) copolymer possess ferroelectric properties that allow the use of such fluorinated polymer in the domain of captors, sensors, and detectors [47, 194]. Another interesting property of crosslinked poly(VDF-co-HFP) copolymer is their insolubility in organic solvent [195]. Cured fluorinated polymers can be processed as membranes for many electrochemical applications such as fuel cell and batteries [196]. For example, a poly(VDF-co-HFP) copolymer has been crosslinked with various systems such as polyols [197], by irradiation with electron beam or γ-rays [197] or with aliphatic amines [198] in order to elaborate a solid polymer electrolyte for non aqueous lithium battery [197, 198]. This electrolyte is particularly interesting for its ionic conductivity, its adhesion with an electroconductive substrate and also remarkably enhanced heat resistance.

6
Conclusion

A wide variety of copolymers containing VDF are either commercially available or prepared at laboratory scale.

VDF-based fluoropolymers are usually crosslinked by four main agents: diamines, bisphenols, peroxides/coagent systems and by radiation. These methods of crosslinking differ in the polymer used, the mechanism, the conditions of reaction, the searched properties, and the desired applications.

First, it is observed that whatever the agent, the mechanism of crosslinking needs two steps, the press cure and the post cure, in order to activate the reaction, and to improve the physical, mechanical and thermal properties.

The oldest and the easiest processing system concerns the amine and diamine cure system, although the mechanical properties are the worst. The mechanism proceeds in three steps: dehydrofluorination, then addition of the (di)amine onto a HFP/VDF/HFP triad, and finally elimination of HF from the polymer. Primary and secondary monoamines can crosslink poly(VDF-co-CTFE) copolymers. Primary amines are faster than secondary ones. They can be added at lower temperatures. The tertiary amines are potential cocuring agents for all diamines. Aliphatic diamines and diimines can crosslink VDF-based fluoropolymers, whereas diamines containing aromatic group must exhibit nucleophilicity and hence always require a spacer between the aromatic ring and the amino group to react with the poly(VDF-co-HFP) copolymer, and they have a lower reaction rate. The most common diamines are hexamethylene diamine carbamate that crosslinks poly(VDF-co-HFP) copolymer at 200 °C, and bis-cinnamylidene hexamethylene diamine. Bisperoxycarbamates, like hexamethylene-N,N'bis(tert-butyl peroxycarbamate), and methylene bis-4-cyclohexyl-N,N'(tert-butylperoxycarbamate, introduced in a minimum quantity, can add onto VDF-based fluoropolymers, thanks to a radical mechanism. HBTBP is more reactive than MBTBP.

Piperazine can crosslink poly(VDF-co-HFP) and poly(VDF-co-CTFE) copolymers at 57 °C in one day, whereas triethylene diamine and tetramethylethyldiamine can crosslink poly(VDF-co-CTFE) copolymer at 97 °C in one day. Finally, diethylene triamine crosslinks PVDF at 70 °C in 16 h.

Crosslinking with bisphenols in the presence of a metal oxide and phosphonium or tetraalkylammonium ions also proceeds in three main steps. A dehydrofluorination, then a double bond reorganization, and finally the substitution by the bisphenol onto the double bond. This mechanism was identified by ^{19}F NMR. Interestingly, ODR measurements show that 98% of the bisphenol-AF—which is the most used bisphenol—is crosslinked in 5 min at 177 °C. The crosslinking density, measured by ODR, can be improved by increasing the initial concentration in bisphenol-AF. But the bisphenol-cure of a perfluoroalkylvinyl ether exhibits limitations, like porosity or poor vulcanizate properties produced during the curing process. In that case, those VDF-based polymers must be crosslinked by peroxides/coagent systems.

However, peroxide/coagent systems require that the copolymer bears a bromine or iodine content to insure free-radical crosslinking. The monomer precursor must be co- or terpolymerized with VDF. The peroxide cure must be carried out in the presence of a coagent and a metal oxide and the most efficient ones are triallylisocyanurate, and MgO, respectively. 2,5-bis-(t-butylperoxy)-2,5-dimethylhexane is more efficient than 2,5-bis-(t-butylperoxy)-2,5-dimethylhexyne because it reacts at a lower temperature and also leads to a better cure state. Both those peroxides have a higher rate of crosslinking (measured by ODR) than that of α,α'-bis(t-butylperoxy)diisopropylbenzene or dicumyl peroxide. The crosslinking mechanism is slightly different from the two previous ones. Indeed, the decomposition of the peroxide leads to a radical that adds onto the TAIC co-agent. This intermediate radical, by abstracting a bromine from the polymer, becomes the crosslinker.

VDF-based fluoroelastomers can also be cured by different types of high-energy radiation (X-rays, γ-rays, β-particles or electrons). However, the generated free radicals can undergo several reactions, different from the crosslinking reaction (e.g., they can cause chain scission). This last one comes from the recombination between two macroradicals. Several properties vary with the radiation cure. For instance, the glass transition temperature, the gel fraction and the crosslinking density increase with the radiation dose.

Other systems, like thiol-ene systems are also efficient as crosslinking systems for VDF-based fluoroelastomers.

Finally, peroxide-cure fluoroelastomers are more resistant to acids, to bases and to solvents, than diamine-cure systems. But diamines exhibit a better crosslinking density. Regarding the mechanical properties, peroxide-cure leads to a high tensile strength, bisphenols to a high compression set resistance, and diamines are the best agents to exhibit a low elongation and a high hardness.

Fluoropolymers crosslinked with polyamines are particularly suitable for use in the manufacture of tubing employed as aircraft hoses, to carry fuel lubricants at high temperature and under high pressure, and in the aircraft construction industry. They can also be used as strong adhesive joints without prior surface modification.

Bisphenol-cure polymers are exploited for their compression set resistance.

Peroxide curable VDF-based polymers have found applications as cords, tubes or irregular-profile items of any dimension.

Irradiated VDF-based polymers are suitable in the domain of captors, sensors, and detectors.

Another application of cured fluoroelastomers is sealing, O-ring, oil seals, multi layer insulator systems for electrical conductors, and membranes for many electrochemical applications such as fuel cell and batteries.

However, it can be assumed that other curing systems will be found and will attract the attention of many academic and industrial researchers.

References

1. Montermoso JC (1961) Rubber Chem Techn 34:1521
2. Cooper JR (1968) High Polymers 23:273
3. Schmiegel WW, Logothetis AL (1984) ACS Symp Series, No. 260, Polymers for Fibers and Elastomers 260:159

4. Anderson RF, Punserson JO (1979) Organofluorine Chemicals and Their Industrial Applications, Banks RE (ed). Horwood, Chichester
5. Abu-Isa IA, Trexler HE (1985) Rubber Chem Techn 58:326
6. Frapin B (1987) Revue Generale des Caoutchoucs & Plastiques 672:125
7. Wall L (1972) Fluoropolymers. Wiley, New York
8. Banks RE, Smart BE, Tatlow JC (1994) (eds) Organofluorine Chemistry: Principles and Commercial Applications. Wiley, New York
9. Scheirs J (1997) Modern Fluoropolymers. Wiley, New York
10. Ajroldi G (1997) Chimica e l'Industria 79:483
11. Hougham G, Cassidy PE, Johns K, Davidson T (1999) (eds) Fluoropolymers 2: Properties. Kluwer Academic/Plenum Publishers, New York
12. Johns K, Stead G (2000) J Fluorine Chem 104:5
13. Imae T (2003) Current Opinion in Colloid & Interface Science 8:307
14. Ameduri B, Boutevin B (2004) Well-Architectured Fluoropolymers: Synthesis, Properties and Applications. Elsevier, Amsterdam
15. Schmiegel WW (2004) Kaut Gum Kunst 57:313
16. Ogunniyi DS (1999) Prog Rubber Plastics Techn 15:95
17. Pruett RL, Barr JT, Rapp KE, Bahner CT, Gibson JD, Lafferty RH (1950) J Am Chem Soc 72:3646
18. Moran AL, Kane RP, Smith JF (1959) J Ind Eng Chem 51:831
19. Smith JF (1960) Rubber World 142:102
20. Paciorek KL, Mitchell LC, Lenk CT (1960) J Polym Sc 45:405
21. Smith JF, Perkins GT (1961) Rubber and Plastics Age 42:59
22. Paciorek KL, Merkl BA, Lenk CT (1962) J Org Chem 27:266
23. Paciorek KL, Lajiness WG, Lenk CT (1962) J Polym Sc 60:141
24. Thomas DK (1964) J Appl Polym Sc 8:1415
25. Thomas DK (1969) GB Patent 1 175 417
26. Moran AL, Pattison DB (1971) Rubber Age 103:37
27. Smith TL, Chu WH (1972) J Polym Sc, Polym Phys Ed 10:133
28. Arnold RG, Barney AL, Thompson DC (1973) Rubber Chem Techn 46:619
29. Knight GJ, Wright WW (1973) British Polym J 5:395
30. Ogunniyi DS (1988) Rubber Chem Techn 61:735
31. Schonhorn H, Luongo JP (1989) J Adh Sc Techn 3:277
32. Schmiegel WW (1978) Kaut Gum Kunst 31:137
33. Schmiegel WW (1979) Angew Makromol Chem 76/77:39
34. Pianca M, Bonardelli P, Tato M, Cirillo G, Moggi G (1987) Polymer 28:224
35. Logothetis AL (1989) Progress in Polymer Science 14:251
36. Carlson DP, Schmiegel WW (1989) Eur Patent 333062
37. Arcella V, Brinati G, Apostolo M (April 1997) Chem Ind p 490
38. Schmiegel WW (2002) US Patent 2003065132
39. Kojima G, Wachi H (1978) Rubber Chem Techn 51:940
40. Finlay JB, Hallenbeck A, MacLachlan JD (1978) J Elast Plast 10:3
41. Apotheker D, Krusic PJ (1980) US Patent 4214060
42. Ameduri BM, Armand M, Boucher M, Manseri A (2001) PCT WO2001096268
43. Coggio WD, Scott PJ, Hintzer K, Hare ED (2004) US Patent 2004014900
44. Ameduri B, Boutevin B, Kostov GK, Petrova P (1999) Designed Monomers and Polymers 2:267
45. Clark DT, Brennan WJ (1988) J El Spectr Rel Phen 47:93
46. Suther JL, Laghari JR (1991) J Mat Sc Let 10:786

47. Betz N, Petersohn E, Le Moel A (1996) Nuclear Instruments & Methods in Physics Research, Section B: Beam Interactions with Materials and Atoms 116:207
48. Banik I, Bhowmick AK (2000) Rad Phys Chem 58:293
49. Banik I, Bhowmick AK (2000) J Mat Sc 35:3579
50. Ameduri B, Boutevin B, Kostov G (2001) Prog Polym Sc 26:105
51. Nasef MM, Dahlan KZM (2003) Nuclear Instruments & Methods in Physics Research, Section B: Beam Interactions with Materials and Atoms 201:604
52. Soresi B, Quartarone E, Mustarelli P, Magistris A, Chiodelli G (2004) Solid State Ionics 166:383
53. Lee WA, Rutherford RA (1975) The glass transition temperatures of polymers. In: Brandrup J, Immergut EH (eds) Polymer Handbook. Wiley-Interscience, New York
54. Seilers DA (1997) PVDF in the chemical processing industry. In: Scheirs J (ed) Modern Fluoropolymers. Wiley, New York, chap. 25, p 487
55. Smith S (1982) Fluorelastomers. In: Banks RE (ed) Preparation, Properties, and Industrial Applications of Organofluorine Compounds. Ellis Harwood, Chichester, chap. 8, p 235
56. England DC, Uschold RE, Starkweather H, Pariser R (1983) Proc of the Robert A. Welch Foundation Conf on Chemical Research, Houston, Texas, vol. 26, p 192
57. Uschold RE (1985) Polym J 17:253
58. Logothetis AL (1994) Fluoroelastomers. In: Banks RE, Tatlow JC (eds) Organofluorine Chemistry: Principles and Commercial Applications. Wiley, New York, chap. 16, p 373
59. Bowers S (1997) Proc of Fluoroelastomers. In: Scheirs J (ed) Modern Fluoropolymers. Wiley, New York, chap. 5, p 115
60. Tournut C (1994) Macromol Symp 82:99; Tournut C (1997) Thermoplastic copolymers of vinylidene fluoride, Modern Fluoropolymers. In: Scheirs J (ed) Modern Fluoropolymers. Wiley, New York, chap. 31, p 577
61. Lynn MM, Worm AT (1987) Encycl Polym Sci Eng 7:257
62. Cook D, Lynn M (1990) Rapra Review Reports 3:32/1
63. Arcella V, Ferro R (1997) Fluorocarbon elastomers. In: Scheirs J (ed) Modern Fluoropolymers. Wiley, New York, chap. 2, p 71
64. Van Cleeff A (1997) Fluoroelastomers. In: Scheirs J (ed) Modern Fluoropolymers. Wiley, New York, chap. 32, p 597
65. Logothetis AL (1997) Perfluoroelastomers and their Functionalization. Macromolecular Design of Polymeric Materials. M. Dekker Inc., New York, chap. 26, p 447
66. Ameduri B, Boutevin B (2005) J Fluorine Chem 126:221
67. Schmiegel WW (2003) Proceedings of ACS Rubber Technology Conference, Cleveland, USA, Oct 14-17
68. Sorokin AD, Volkova EV, Naberezhnykh RA (1972) Radiat Khim 2:295
69. Baradie B, Shoichet MS (2002) Macromolecules 35:3569
70. Guiot J (2003) PhD Thesis, University of Montpellier
71. Souzy R, Ameduri B, Boutevin B (2004) Macromol Chem Phys 205:476
72. Sianesi D, Caporiccio G (1968) J Polym Sc, Part A1: Polym Chem 6:335
73. Caporiccio G, Sianesi D (1970) Chimica e l'Industria 52:37
74. Ameduri B, Bauduin G (2003) J Polym Sc, Part A: Polym Chem 41:3109
75. Yagi T, Tatemoto M (1979) Polym J 11:429
76. Usmanov KU, Yul'chibaev AA, Mukhamadaliev N, Sarros TK (1975) Izvestiya Vysshikh Uchebnykh Zavedenii, Khimiya i Khimicheskaya Tekhnologiya (Chem. Abstr. 83, 28687) 18:464

77. Otazaghine B, Ameduri B (2000) The 16th Int Symp in Fluorine Chemistry. Durham, United Kingdom and Otazaghine B, Sauguet L, Ameduri B (2005) J Fluorine Chem (in press)
78. Moggi G, Bonardelli P, Bart JCJ (1984) J Polym Sc, Polym Phys Ed 22:357
79. Dohany RE, Dukert AA, Preston SS (1989) Encycl Polym Sci Technol 17:532
80. Bonardelli P, Moggi G, Turturro A (1986) Polymer 27:905
81. Naberezhnykh RA, Sorokin AD, Volkova EV, Fokin AV (1974) Izvestiya Akademii Nauk SSSR, Seriya Khimicheskaya:232
82. Moggi G, Bonardelli P, Russo S (1983) Con Ital Sci Macromol 6th 2:405
83. Gelin MP, Ameduri B J (2005) J Fluorine Chem 126:577
84. Otazaghine B, Sauguet L, Ameduri B (in press) Eur Polym J
85. Ameduri BM, Manseri A, Boucher M (2002) PCT WO2002050142
86. Guiot J, Ameduri B, Boutevin B (2002) J Polym Sc, Part A: Polym Chem 40:3634
87. a) Sauguet L, Guiot J, Neouze MA, Ameduri B, Boutevin B (2005) J Polym Sc Polym Chem 43:917; b) Coggio W. Proceedings of "Fluoropolymers 2004" Conference, Savannyh, Ga, USA, October 7–9
88. Ameduri B, Bauduin G, Boutevin B, Kostov G, Petrova P (1999) Macromolecules 32:4544
89. Souzy R, Guiot J, Ameduri B, Boutevin B, Paleta O (2003) Macromolecules 36:9390
90. Khodzhaev SG, Yusupbekova FZ, Yul'chibaev AA (1981) Sbornik Nauchnykh Trudov. Tashkentskii Gosudarstvennyi Universitet im. V. I. Lenina (Chem. Abstr. 97, 163545) 667:34
91. Guiot J, Ameduri B, Boutevin B, Lannuzel T (2003) Eur Polym J 39:887
92. Souzy R, Ameduri B, Boutevin B (2004) J Polym Sc, Part A: Polym Chem 42:5077
93. Souzy R, Ameduri B, Boutevin B, Capron P, Gebel G (2005) Fuel Cell (in press)
94. Lannuzel T, Ameduri B, Guiot J, Boutevin B (2004) French Patent 20042852316
95. Dittman AL, Passino HJ, Wrightson JM (1954) US Patent 2689241
96. Dixon S, Rexford DR, Rugg JS (1957) J Indus Eng Chem 49:1687
97. Rugg JS, Stevenson AC, Rexford DS (1957) Rubber World 82:102
98. Pailthorp JR, Schroeder HE (1961) US Patent 2968649
99. Rexford DR (1962) US Patent 3051677
100. Conroy ME, Honn FJ, Robb LE, Wolf DR (1955) Rubber Age 76:543
101. Griffis CB, Montermoso JC (1955) Rubber Age 77:559
102. Jackson WW, Hale D (1955) Rubber Age 77:865
103. Ameduri B, Boutevin B, Armand M, Boucher M (2001) WO 049758
104. Worm AT, Vladimirovich N, Volkova MA (2001) US Patent 6294627
105. a) Bach D, Van Gool G, Steffens J (2003) Proceedings of "International Rubber Technology Conference 2003", Cleveland, OH, Oct. 14-17, and b) Dyneon Fluoroelastomer LTFE 6400X (2003) Technical Information brochure
106. Sianesi D, Bernardi C, Diotalleri G (1967) US Patent 3333106
107. Migmierina A, Ceccato G (1969) 4th Int Sun Rubber Symp 2:65
108. Ogunniyi DS (1989) Prog Rubber Plast Techn 5:16
109. Schmiegel WW (2000) US Patent 2000011072
110. Barney AL, Kalb GH, Khan AA (1971) Rubber Chem Techn 44:660
111. Smith JF (1959) Rubber World 140:263
112. Hepburn C, Ogunniyi DS (1985) Proc Int Rubber Conference. Kyoto, Japan
113. Ogunniyi DS, Hepburn C (1986) Plastics and Rubber Processing and Applications 6:3
114. Smith J (1961) J Appl Polym Sc 5:460
115. Albin LD (1982) Rubber Chem Techn 55:902

116. Wright WW (1974) British Polym J 6:147
117. Bryan CJ (1977) Rubber Chem Techn 50:83
118. Kalfayan SH, Silver RH, Liu SS (1976) Rubber Chem Techn 49:1001
119. Ogunniyi DS, Hepburn C (1995) Iranian J Polymer Science & Technology (English Edition) 4:242
120. Barton JM (1978) British Polym J 10:151
121. Bentley FE (1957) PhD Thesis, University of Florida
122. Mullins L (1959) J Appl Polym Sc 2:1
123. Smith TL (1967) J Polym Sc, Polym Symp 841
124. Van der Hoff BME, Buckler EJ (1967) J Macromol Sci, Part A 1:747
125. Moran AL, Kane RP, Smith JF (1959) J Chem Eng Data 4:276
126. O'Brien EL, Beringer FM, Mesrobian RB (1957) J Am Chem Soc 79:6238
127. Pedersen CJ (1958) J Org Chem 23:255
128. Pedersen CJ (1958) J Org Chem 23:252
129. O'Brien EL, Beringer FM, Mesrobian RB (1959) J Am Chem Soc 81:1506
130. Spain RG (1958) Division of Rubber Chemistry, Am Chem Soc Meeting. Cincinnati
131. Flisi U, Giunchi G, Geri S (1976) Kaut Gum Kunst 29:118
132. Schmiegel WW (1985) US Patent 4496682
133. Taguet A, Ameduri B, Boutevin B submitted in J Polym Sci, Part A: Polym Chem
134. Apotheker D, Finlay JB, Krusic PJ, Logothetis AL (1982) Rubber Chem Techn 55:1004
135. Schmiegel WW (1975) US Patent 3872065
136. Schmiegel WW (1975) US Patent 1413837
137. Schmiegel WW (1984) US Patent 127318
138. Hung MH, Schmiegel WW (2001) US Patent 2001081464
139. Arcella V, Albano M, Barchiesi E, Brinati G, Chiodini G (1993) Rubber World 207:18
140. Braden M, Fletcher WP (1955) Transactions, Institution of the Rubber Industry 31:155
141. Udagawa R (2001) Eur Patent 2001081391
142. Staccione A, Albano M (2003) Eur Patent 1347012
143. Davis RA, Tigner RG (1970) US Patent 3505416
144. Kryukova AB, Demidova NM, Khmelevskaya VM, Sankina GA, Dontsov AA, Chulyukina AV, Kosteltsev VV, Zavyalova AD, Savchenkova GL, et al. (1993) Russ Patent 1815268
145. Shimizu T, Enokida T, Naraki A, Tatsu H (2000) Jpn Patent 2000230096
146. Saito M, Kamya H, Miwa T, Hirai H (1994) Jpn Patent 06306245
147. Bowers S, Schmiegel WW (2000) PCT WO2000011050
148. Schmiegel WW (2003) US Patent 2003208003
149. Banks RE, Birchall JM, Haszeldine RN, Nicholson WJ (1982) J Fluorine Chem 20:133
150. Gafurov AK, Isamukhamedov SI, Yul'chibaev AA, Usmanov KU (1978) Uzbekskii Khimicheskii Zhurnal:25
151. Funaki A, Kato K, Takakura T, Myake H (1994) Jpn Patent 06306196
152. Tamura M, Miyake H (1998) Jpn Patent 10158376
153. Tatemoto M, Nagakawa T (1979) US Patent 4158678
154. Tatemoto M (1979) IX Int Symp on Fluorine Chemistry, Avignon, France
155. Tatemoto M, Suzuki T, Tomota M, Furukawa Y, Ueta Y (1981) US patent 4243770
156. Tatemoto M, Morita S (1982) US Patent 4361678
157. Oka M, Tatemoto M (1984) Contemp Top Polym Sc 4:763
158. Ishiwari K, Sakakura A, Yuhara S, Yagi T, Tatemoto M (1985) Int Rubber Conference, Kyoto, Japan

159. Erdos P, Balazs G, Doszlop S, Varga J (1985) Periodica Polytechnica, Chemical Engineering 29:165
160. Ogunniyi DS, Hepburn C (2003) Iranian Polymer J 12:367
161. Bristow GM (1976) Natural Rubber Technology 7 3:61
162. Florin RE, Wall LA (1961) J Res Nat Bur Stand 65A:375
163. Yoshida T, Florin RE, Wall LA (1965) J Polym Sc, Part A: General Papers 3:1685
164. Lyons BJ (March 1984) The Crosslinking of Fluoropolymer with Ionising Radiation. Second Int Conf on Radiation Processing for Plastics and Rubbers, Canterbury, UK
165. Lyons BJ (1994) Radiat Phys Chem 45:158
166. Lyons BJ (1997) The Radiation Crosslinking of Fluoropolymers. In: Scheirs J (ed) Modern Fluoroelastomer. Wiley, New York, chap. 18, p 335
167. Logothetis AL (1999) Polym Int 48:993
168. Forsythe JS, Hill DJT, Whittaker AK, Logothetis AL (1999) Polym Int 48:1004
169. Forsythe JS, Hill DJT (2000) Prog Polym Sc 25:101; Dargaville TR, George GA, Hill DJJ, Whittaker AK (2003) Prog Polym Sci 28:1355
170. Chapiro A (1962) (ed) Radiation Chemistry of Polymeric Systems. Wiley, New York
171. Mandelkern L (1972) Radiat Chem Macromol 1:287
172. Florin RE (1972) Radiation Chemistry of Fluorocarbon Polymers. In: Wall LA (ed) Fluoropolymers. Wiley, New York, p 317
173. Geymer DO (1973) Radiat Chem Macromol 2:3
174. Okamoto J (1987) Rad Phys Chem 29:469
175. Ivanov VS (1992) Radiation Chemistry of Polymers, New Concepts in Polymer Science. Wiley, New York
176. Singh A, Silverman J (1992) (ed) Progress in Polymer Processing, Vol. 3: Radiation Processing of Polymers
177. Gupta B, Scherer GG (1994) Chimia 48:127
178. Uyama Y, Kato K, Ikada Y (1998) Adv Polym Sc 137:1
179. Zhen ZX (1990) Rad Phys Chem 35:194
180. Daudin B, Legrand JF, Macchi F (1991) J Appl Phys 70:4037
181. Macchi F, Daudin B, Legrand JF (1990) Nuclear Instruments & Methods in Physics Research, Section B: Beam Interactions with Materials and Atoms B46:324
182. Charlesby A (1960) Atomic Rad Polym Vol. I
183. Henne AL, Pelley RL (1952) J Am Chem Soc 74:1426
184. Knight GJ, Wright WW (1982) Polym Deg S4:465
185. Wlassics I, Giannetti E (1997) Can Patent 2182328
186. Ogunniyi DS (1990) Elastomerics 122:22
187. Harrell JR, Schmiegel WW (1975) US Patent 3859259
188. Moran AL (1960) US Patent 2951832
189. Allen CM, Hincklieff IR (1982) Eur Patent 53002
190. Fogiel AW (1975) J Polym Sc, Polym Symp 53:333
191. Honn FJ, Sims WM (1960) US Patent 2965619
192. Ehrlich GM, Puglia FJ (2002) US Patent 2002122950
193. Baczek SK, McCain GH, Benezra LL, Covitch MJ (1983) US Patent 4391844
194. Xu Y (1991) Ferroelectric Materials and their Application. Elsevier, Amsterdam
195. Yuan EL (1962) US Patent 3025183
196. Lester PR (1990) Eur Pat 370149
197. Katsurao T, Horie K, Nagai A, Ishikawa Y (2000) US Patent 6372388
198. Coulon M, Silvert PY, Irissin-Mangata J, Ameduri B (2002) WO 082571

Author Index Volumes 101-184

Author Index Volumes 1-100 see Volume 100

de, Abajo, J. and *de la Campa, J. G.*: Processable Aromatic Polyimides. Vol. 140, pp. 23-60.
Abe, A., Furuya, H., Zhou, Z., Hiejima, T. and *Kobayashi, Y.*: Stepwise Phase Transitions of Chain Molecules: Crystallization/Melting via a Nematic Liquid-Crystalline Phase. Vol. 181, pp. 121-152.
Abetz, V. see Förster, S.: Vol. 166, pp. 173-210.
Adolf, D. B. see Ediger, M. D.: Vol. 116, pp. 73-110.
Aharoni, S. M. and *Edwards, S. F.*: Rigid Polymer Networks. Vol. 118, pp. 1-231.
Albertsson, A.-C. and *Varma, I. K.*: Aliphatic Polyesters: Synthesis, Properties and Applications. Vol. 157, pp. 99-138.
Albertsson, A.-C. see Edlund, U.: Vol. 157, pp. 53-98.
Albertsson, A.-C. see Söderqvist Lindblad, M.: Vol. 157, pp. 139-161.
Albertsson, A.-C. see Stridsberg, K. M.: Vol. 157, pp. 27-51.
Albertsson, A.-C. see Al-Malaika, S.: Vol. 169, pp. 177-199.
Al-Malaika, S.: Perspectives in Stabilisation of Polyolefins. Vol. 169, pp. 121-150.
Améduri, B., Boutevin, B. and *Gramain, P.*: Synthesis of Block Copolymers by Radical Polymerization and Telomerization. Vol. 127, pp. 87-142.
Améduri, B. and *Boutevin, B.*: Synthesis and Properties of Fluorinated Telechelic Monodispersed Compounds. Vol. 102, pp. 133-170.
Ameduri, B. see Taguet, A.: Vol. 184, pp. 127-211.
Amselem, S. see Domb, A. J.: Vol. 107, pp. 93-142.
Anantawaraskul, S., Soares, J. B. P. and *Wood-Adams, P. M.*: Fractionation of Semicrystalline Polymers by Crystallization Analysis Fractionation and Temperature Rising Elution Fractionation. Vol. 182, pp. 1-54.
Andrady, A. L.: Wavelenght Sensitivity in Polymer Photodegradation. Vol. 128, pp. 47-94.
Andreis, M. and *Koenig, J. L.*: Application of Nitrogen-15 NMR to Polymers. Vol. 124, pp. 191-238.
Angiolini, L. see Carlini, C.: Vol. 123, pp. 127-214.
Anjum, N. see Gupta, B.: Vol. 162, pp. 37-63.
Anseth, K. S., Newman, S. M. and *Bowman, C. N.*: Polymeric Dental Composites: Properties and Reaction Behavior of Multimethacrylate Dental Restorations. Vol. 122, pp. 177-218.
Antonietti, M. see Cölfen, H.: Vol. 150, pp. 67-187.
Aoki, H. see Ito, S.: Vol. 182, pp. 243-281.
Armitage, B. A. see O'Brien, D. F.: Vol. 126, pp. 53-58.
Arndt, M. see Kaminski, W.: Vol. 127, pp. 143-187.
Arnold Jr., F. E. and *Arnold, F. E.*: Rigid-Rod Polymers and Molecular Composites. Vol. 117, pp. 257-296.
Arora, M. see Kumar, M. N. V. R.: Vol. 160, pp. 45-118.
Arshady, R.: Polymer Synthesis via Activated Esters: A New Dimension of Creativity in Macromolecular Chemistry. Vol. 111, pp. 1-42.

Auer, S. and *Frenkel, D.*: Numerical Simulation of Crystal Nucleation in Colloids. Vol. 173, pp. 149–208.
Auriemma, F., De Rosa, C. and *Corradini, P.*: Solid Mesophases in Semicrystalline Polymers: Structural Analysis by Diffraction Techniques. Vol. 181, pp. 1–74.

Bahar, I., Erman, B. and *Monnerie, L.*: Effect of Molecular Structure on Local Chain Dynamics: Analytical Approaches and Computational Methods. Vol. 116, pp. 145–206.
Ballauff, M. see Dingenouts, N.: Vol. 144, pp. 1–48.
Ballauff, M. see Holm, C.: Vol. 166, pp. 1–27.
Ballauff, M. see Rühe, J.: Vol. 165, pp. 79–150.
Baltá-Calleja, F. J., González Arche, A., Ezquerra, T. A., Santa Cruz, C., Batallón, F., Frick, B. and *López Cabarcos, E.*: Structure and Properties of Ferroelectric Copolymers of Poly(vinylidene) Fluoride. Vol. 108, pp. 1–48.
Baltussen, J. J. M. see Northolt, M. G.: Vol. 178, (in press).
Barnes, M. D. see Otaigbe, J. U.: Vol. 154, pp. 1–86.
Barshtein, G. R. and *Sabsai, O. Y.*: Compositions with Mineralorganic Fillers. Vol. 101, pp. 1–28.
Barton, J. see Hunkeler, D.: Vol. 112, pp. 115–134.
Baschnagel, J., Binder, K., Doruker, P., Gusev, A. A., Hahn, O., Kremer, K., Mattice, W. L., Müller-Plathe, F., Murat, M., Paul, W., Santos, S., Sutter, U. W. and *Tries, V.*: Bridging the Gap Between Atomistic and Coarse-Grained Models of Polymers: Status and Perspectives. Vol. 152, pp. 41–156.
Bassett, D. C.: On the Role of the Hexagonal Phase in the Crystallization of Polyethylene. Vol. 180, pp. 1–16.
Batallán, F. see Baltá-Calleja, F. J.: Vol. 108, pp. 1–48.
Batog, A. E., Pet'ko, I. P. and *Penczek, P.*: Aliphatic-Cycloaliphatic Epoxy Compounds and Polymers. Vol. 144, pp. 49–114.
Baughman, T. W. and *Wagener, K. B.*: Recent Advances in ADMET Polymerization. Vol. 176, pp. 1–42.
Becker, O. and *Simon, G. P.*: Epoxy Layered Silicate Nanocomposites. Vol. 179, pp. 29–82.
Bell, C. L. and *Peppas, N. A.*: Biomedical Membranes from Hydrogels and Interpolymer Complexes. Vol. 122, pp. 125–176.
Bellon-Maurel, A. see Calmon-Decriaud, A.: Vol. 135, pp. 207–226.
Bennett, D. E. see O'Brien, D. F.: Vol. 126, pp. 53–84.
Berry, G. C.: Static and Dynamic Light Scattering on Moderately Concentraded Solutions: Isotropic Solutions of Flexible and Rodlike Chains and Nematic Solutions of Rodlike Chains. Vol. 114, pp. 233–290.
Bershtein, V. A. and *Ryzhov, V. A.*: Far Infrared Spectroscopy of Polymers. Vol. 114, pp. 43–122.
Bhargava, R., Wang, S.-Q. and *Koenig, J. L.*: FTIR Microspectroscopy of Polymeric Systems. Vol. 163, pp. 137–191.
Biesalski, M. see Rühe, J.: Vol. 165, pp. 79–150.
Bigg, D. M.: Thermal Conductivity of Heterophase Polymer Compositions. Vol. 119, pp. 1–30.
Binder, K.: Phase Transitions in Polymer Blends and Block Copolymer Melts: Some Recent Developments. Vol. 112, pp. 115–134.
Binder, K.: Phase Transitions of Polymer Blends and Block Copolymer Melts in Thin Films. Vol. 138, pp. 1–90.
Binder, K. see Baschnagel, J.: Vol. 152, pp. 41–156.
Binder, K., Müller, M., Virnau, P. and *González MacDowell, L.*: Polymer+Solvent Systems: Phase Diagrams, Interface Free Energies, and Nucleation. Vol. 173, pp. 1–104.

Bird, R. B. see Curtiss, C. F.: Vol. 125, pp. 1–102.
Biswas, M. and *Mukherjee, A.*: Synthesis and Evaluation of Metal-Containing Polymers. Vol. 115, pp. 89–124.
Biswas, M. and *Sinha Ray, S.*: Recent Progress in Synthesis and Evaluation of Polymer-Montmorillonite Nanocomposites. Vol. 155, pp. 167–221.
Blankenburg, L. see Klemm, E.: Vol. 177, pp. 53–90.
Blumen, A. see Gurtovenko, A. A.: Vol. 182, pp. 131–242.
Bogdal, D., Penczek, P., Pielichowski, J. and *Prociak, A.*: Microwave Assisted Synthesis, Crosslinking, and Processing of Polymeric Materials. Vol. 163, pp. 193–263.
Bohrisch, J., Eisenbach, C. D., Jaeger, W., Mori, H., Müller, A. H. E., Rehahn, M., Schaller, C., Traser, S. and *Wittmeyer, P.*: New Polyelectrolyte Architectures. Vol. 165, pp. 1–41.
Bolze, J. see Dingenouts, N.: Vol. 144, pp. 1–48.
Bosshard, C.: see Gubler, U.: Vol. 158, pp. 123–190.
Boutevin, B. and *Robin, J. J.*: Synthesis and Properties of Fluorinated Diols. Vol. 102, pp. 105–132.
Boutevin, B. see Améduri, B.: Vol. 102, pp. 133–170.
Boutevin, B. see Améduri, B.: Vol. 127, pp. 87–142.
Boutevin, B. see Guida-Pietrasanta, F.: Vol. 179, pp. 1–27.
Boutevin, B. see Taguet, A.: Vol. 184, pp. 127–211.
Bowman, C. N. see Anseth, K. S.: Vol. 122, pp. 177–218.
Boyd, R. H.: Prediction of Polymer Crystal Structures and Properties. Vol. 116, pp. 1–26.
Bracco, S. see Sozzani, P.: Vol. 181, pp. 153–177.
Briber, R. M. see Hedrick, J. L.: Vol. 141, pp. 1–44.
Bronnikov, S. V., Vettegren, V. I. and *Frenkel, S. Y.*: Kinetics of Deformation and Relaxation in Highly Oriented Polymers. Vol. 125, pp. 103–146.
Brown, H. R. see Creton, C.: Vol. 156, pp. 53–135.
Bruza, K. J. see Kirchhoff, R. A.: Vol. 117, pp. 1–66.
Buchmeiser, M. R.: Regioselective Polymerization of 1-Alkynes and Stereoselective Cyclopolymerization of a, w-Heptadiynes. Vol. 176, pp. 89–119.
Budkowski, A.: Interfacial Phenomena in Thin Polymer Films: Phase Coexistence and Segregation. Vol. 148, pp. 1–112.
Bunz, U. H. F.: Synthesis and Structure of PAEs. Vol. 177, pp. 1–52.
Burban, J. H. see Cussler, E. L.: Vol. 110, pp. 67–80.
Burchard, W.: Solution Properties of Branched Macromolecules. Vol. 143, pp. 113–194.
Butté, A. see Schork, F. J.: Vol. 175, pp. 129–255.

Calmon-Decriaud, A., Bellon-Maurel, V., Silvestre, F.: Standard Methods for Testing the Aerobic Biodegradation of Polymeric Materials. Vol. 135, pp. 207–226.
Cameron, N. R. and *Sherrington, D. C.*: High Internal Phase Emulsions (HIPEs)-Structure, Properties and Use in Polymer Preparation. Vol. 126, pp. 163–214.
de la Campa, J. G. see de Abajo, J.: Vol. 140, pp. 23–60.
Candau, F. see Hunkeler, D.: Vol. 112, pp. 115–134.
Canelas, D. A. and *DeSimone, J. M.*: Polymerizations in Liquid and Supercritical Carbon Dioxide. Vol. 133, pp. 103–140.
Canva, M. and *Stegeman, G. I.*: Quadratic Parametric Interactions in Organic Waveguides. Vol. 158, pp. 87–121.
Capek, I.: Kinetics of the Free-Radical Emulsion Polymerization of Vinyl Chloride. Vol. 120, pp. 135–206.
Capek, I.: Radical Polymerization of Polyoxyethylene Macromonomers in Disperse Systems. Vol. 145, pp. 1–56.

Capek, I. and *Chern, C.-S.*: Radical Polymerization in Direct Mini-Emulsion Systems. Vol. 155, pp. 101–166.
Cappella, B. see Munz, M.: Vol. 164, pp. 87–210.
Carlesso, G. see Prokop, A.: Vol. 160, pp. 119–174.
Carlini, C. and *Angiolini, L.*: Polymers as Free Radical Photoinitiators. Vol. 123, pp. 127–214.
Carter, K. R. see Hedrick, J. L.: Vol. 141, pp. 1–44.
Casas-Vazquez, J. see Jou, D.: Vol. 120, pp. 207–266.
Chandrasekhar, V.: Polymer Solid Electrolytes: Synthesis and Structure. Vol. 135, pp. 139–206.
Chang, J. Y. see Han, M. J.: Vol. 153, pp. 1–36.
Chang, T.: Recent Advances in Liquid Chromatography Analysis of Synthetic Polymers. Vol. 163, pp. 1–60.
Charleux, B. and *Faust, R.*: Synthesis of Branched Polymers by Cationic Polymerization. Vol. 142, pp. 1–70.
Chen, P. see Jaffe, M.: Vol. 117, pp. 297–328.
Chern, C.-S. see Capek, I.: Vol. 155, pp. 101–166.
Chevolot, Y. see Mathieu, H. J.: Vol. 162, pp. 1–35.
Choe, E.-W. see Jaffe, M.: Vol. 117, pp. 297–328.
Chow, P. Y. and *Gan, L. M.*: Microemulsion Polymerizations and Reactions. Vol. 175, pp. 257–298.
Chow, T. S.: Glassy State Relaxation and Deformation in Polymers. Vol. 103, pp. 149–190.
Chujo, Y. see Uemura, T.: Vol. 167, pp. 81–106.
Chung, S.-J. see Lin, T.-C.: Vol. 161, pp. 157–193.
Chung, T.-S. see Jaffe, M.: Vol. 117, pp. 297–328.
Clarke, N.: Effect of Shear Flow on Polymer Blends. Vol. 183, pp. 127–173.
Cölfen, H. and *Antonietti, M.*: Field-Flow Fractionation Techniques for Polymer and Colloid Analysis. Vol. 150, pp. 67–187.
Colmenero, J. see Richter, D.: Vol. 174, (in press).
Comanita, B. see Roovers, J.: Vol. 142, pp. 179–228.
Comotti, A. see Sozzani, P.: Vol. 181, pp. 153–177.
Connell, J. W. see Hergenrother, P. M.: Vol. 117, pp. 67–110.
Corradini, P. see Auriemma, F.: Vol. 181, pp. 1–74.
Creton, C., Kramer, E. J., Brown, H. R. and *Hui, C.-Y.*: Adhesion and Fracture of Interfaces Between Immiscible Polymers: From the Molecular to the Continuum Scale. Vol. 156, pp. 53–135.
Criado-Sancho, M. see Jou, D.: Vol. 120, pp. 207–266.
Curro, J. G. see Schweizer, K. S.: Vol. 116, pp. 319–378.
Curtiss, C. F. and *Bird, R. B.*: Statistical Mechanics of Transport Phenomena: Polymeric Liquid Mixtures. Vol. 125, pp. 1–102.
Cussler, E. L., Wang, K. L. and *Burban, J. H.*: Hydrogels as Separation Agents. Vol. 110, pp. 67–80.
Czub, P. see Penczek, P.: Vol. 184, pp. 1–95.

Dalton, L.: Nonlinear Optical Polymeric Materials: From Chromophore Design to Commercial Applications. Vol. 158, pp. 1–86.
Dautzenberg, H. see Holm, C.: Vol. 166, pp. 113–171.
Davidson, J. M. see Prokop, A.: Vol. 160, pp. 119–174.
Den Decker, M. G. see Northolt, M. G.: Vol. 178, (in press).
Desai, S. M. and *Singh, R. P.*: Surface Modification of Polyethylene. Vol. 169, pp. 231–293.
DeSimone, J. M. see Canelas, D. A.: Vol. 133, pp. 103–140.

DeSimone, J. M. see Kennedy, K. A.: Vol. 175, pp. 329–346.
DiMari, S. see Prokop, A.: Vol. 136, pp. 1–52.
Dimonie, M. V. see Hunkeler, D.: Vol. 112, pp. 115–134.
Dingenouts, N., Bolze, J., Pötschke, D. and *Ballauf, M.*: Analysis of Polymer Latexes by Small-Angle X-Ray Scattering. Vol. 144, pp. 1–48.
Dodd, L. R. and *Theodorou, D. N.*: Atomistic Monte Carlo Simulation and Continuum Mean Field Theory of the Structure and Equation of State Properties of Alkane and Polymer Melts. Vol. 116, pp. 249–282.
Doelker, E.: Cellulose Derivatives. Vol. 107, pp. 199–266.
Dolden, J. G.: Calculation of a Mesogenic Index with Emphasis Upon LC-Polyimides. Vol. 141, pp. 189–245.
Domb, A. J., Amselem, S., Shah, J. and *Maniar, M.*: Polyanhydrides: Synthesis and Characterization. Vol. 107, pp. 93–142.
Domb, A. J. see Kumar, M. N. V. R.: Vol. 160, pp. 45–118.
Doruker, P. see Baschnagel, J.: Vol. 152, pp. 41–156.
Dubois, P. see Mecerreyes, D.: Vol. 147, pp. 1–60.
Dubrovskii, S. A. see Kazanskii, K. S.: Vol. 104, pp. 97–134.
Dudowicz, J. see Freed, K. F.: Vol. 183, pp. 63–126.
Dunkin, I. R. see Steinke, J.: Vol. 123, pp. 81–126.
Dunson, D. L. see McGrath, J. E.: Vol. 140, pp. 61–106.
Dziezok, P. see Rühe, J.: Vol. 165, pp. 79–150.

Eastmond, G. C.: Poly(e-caprolactone) Blends. Vol. 149, pp. 59–223.
Economy, J. and *Goranov, K.*: Thermotropic Liquid Crystalline Polymers for High Performance Applications. Vol. 117, pp. 221–256.
Ediger, M. D. and *Adolf, D. B.*: Brownian Dynamics Simulations of Local Polymer Dynamics. Vol. 116, pp. 73–110.
Edlund, U. and *Albertsson, A.-C.*: Degradable Polymer Microspheres for Controlled Drug Delivery. Vol. 157, pp. 53–98.
Edwards, S. F. see Aharoni, S. M.: Vol. 118, pp. 1–231.
Eisenbach, C. D. see Bohrisch, J.: Vol. 165, pp. 1–41.
Endo, T. see Yagci, Y.: Vol. 127, pp. 59–86.
Engelhardt, H. and *Grosche, O.*: Capillary Electrophoresis in Polymer Analysis. Vol. 150, pp. 189–217.
Engelhardt, H. and *Martin, H.*: Characterization of Synthetic Polyelectrolytes by Capillary Electrophoretic Methods. Vol. 165, pp. 211–247.
Eriksson, P. see Jacobson, K.: Vol. 169, pp. 151–176.
Erman, B. see Bahar, I.: Vol. 116, pp. 145–206.
Eschner, M. see Spange, S.: Vol. 165, pp. 43–78.
Estel, K. see Spange, S.: Vol. 165, pp. 43–78.
Ewen, B. and *Richter, D.*: Neutron Spin Echo Investigations on the Segmental Dynamics of Polymers in Melts, Networks and Solutions. Vol. 134, pp. 1–130.
Ezquerra, T. A. see Baltá-Calleja, F. J.: Vol. 108, pp. 1–48.

Fatkullin, N. see Kimmich, R.: Vol. 170, pp. 1–113.
Faust, R. see Charleux, B.: Vol. 142, pp. 1–70.
Faust, R. see Kwon, Y.: Vol. 167, pp. 107–135.
Fekete, E. see Pukánszky, B.: Vol. 139, pp. 109–154.
Fendler, J. H.: Membrane-Mimetic Approach to Advanced Materials. Vol. 113, pp. 1–209.
Fetters, L. J. see Xu, Z.: Vol. 120, pp. 1–50.

Fontenot, K. see *Schork, F. J.:* Vol. 175, pp. 129–255.
Förster, S., Abetz, V. and *Müller, A. H. E.:* Polyelectrolyte Block Copolymer Micelles. Vol. 166, pp. 173–210.
Förster, S. and *Schmidt, M.:* Polyelectrolytes in Solution. Vol. 120, pp. 51–134.
Freed, K. F. and *Dudowicz, J.:* Influence of Monomer Molecular Structure on the Miscibility of Polymer Blends. Vol. 183, pp. 63–126.
Freire, J. J.: Conformational Properties of Branched Polymers: Theory and Simulations. Vol. 143, pp. 35–112.
Frenkel, S. Y. see *Bronnikov, S. V.:* Vol. 125, pp. 103–146.
Frick, B. see *Baltá-Calleja, F. J.:* Vol. 108, pp. 1–48.
Fridman, M. L.: see *Terent'eva, J. P.:* Vol. 101, pp. 29–64.
Fuchs, G. see *Trimmel, G.:* Vol. 176, pp. 43–87.
Fukui, K. see *Otaigbe, J. U.:* Vol. 154, pp. 1–86.
Funke, W.: Microgels-Intramolecularly Crosslinked Macromolecules with a Globular Structure. Vol. 136, pp. 137–232.
Furusho, Y. see *Takata, T.:* Vol. 171, pp. 1–75.
Furuya, H. see *Abe, A.:* Vol. 181, pp. 121–152.

Galina, H.: Mean-Field Kinetic Modeling of Polymerization: The Smoluchowski Coagulation Equation. Vol. 137, pp. 135–172.
Gan, L. M. see *Chow, P. Y.:* Vol. 175, pp. 257–298.
Ganesh, K. see *Kishore, K.:* Vol. 121, pp. 81–122.
Gaw, K. O. and *Kakimoto, M.:* Polyimide-Epoxy Composites. Vol. 140, pp. 107–136.
Geckeler, K. E. see *Rivas, B.:* Vol. 102, pp. 171–188.
Geckeler, K. E.: Soluble Polymer Supports for Liquid-Phase Synthesis. Vol. 121, pp. 31–80.
Gedde, U. W. and *Mattozzi, A.:* Polyethylene Morphology. Vol. 169, pp. 29–73.
Gehrke, S. H.: Synthesis, Equilibrium Swelling, Kinetics Permeability and Applications of Environmentally Responsive Gels. Vol. 110, pp. 81–144.
Geil, P. H., Yang, J., Williams, R. A., Petersen, K. L., Long, T.-C. and *Xu, P.:* Effect of Molecular Weight and Melt Time and Temperature on the Morphology of Poly(tetrafluorethylene). Vol. 180, pp. 89–159.
de Gennes, P.-G.: Flexible Polymers in Nanopores. Vol. 138, pp. 91–106.
Georgiou, S.: Laser Cleaning Methodologies of Polymer Substrates. Vol. 168, pp. 1–49.
Geuss, M. see *Munz, M.:* Vol. 164, pp. 87–210.
Giannelis, E. P., Krishnamoorti, R. and *Manias, E.:* Polymer-Silicate Nanocomposites: Model Systems for Confined Polymers and Polymer Brushes. Vol. 138, pp. 107–148.
Godovsky, D. Y.: Device Applications of Polymer-Nanocomposites. Vol. 153, pp. 163–205.
Godovsky, D. Y.: Electron Behavior and Magnetic Properties Polymer-Nanocomposites. Vol. 119, pp. 79–122.
González Arche, A. see *Baltá-Calleja, F. J.:* Vol. 108, pp. 1–48.
Goranov, K. see *Economy, J.:* Vol. 117, pp. 221–256.
Gramain, P. see *Améduri, B.:* Vol. 127, pp. 87–142.
Grest, G. S.: Normal and Shear Forces Between Polymer Brushes. Vol. 138, pp. 149–184.
Grigorescu, G. and *Kulicke, W.-M.:* Prediction of Viscoelastic Properties and Shear Stability of Polymers in Solution. Vol. 152, p. 1–40.
Gröhn, F. see *Rühe, J.:* Vol. 165, pp. 79–150.
Grosberg, A. and *Nechaev, S.:* Polymer Topology. Vol. 106, pp. 1–30.
Grosche, O. see *Engelhardt, H.:* Vol. 150, pp. 189–217.
Grubbs, R., Risse, W. and *Novac, B.:* The Development of Well-defined Catalysts for Ring-Opening Olefin Metathesis. Vol. 102, pp. 47–72.

Gubler, U. and *Bosshard, C.*: Molecular Design for Third-Order Nonlinear Optics. Vol. 158, pp. 123–190.
Guida-Pietrasanta, F. and *Boutevin, B.*: Polysilalkylene or Silarylene Siloxanes Said Hybrid Silicones. Vol. 179, pp. 1–27.
van Gunsteren, W. F. see Gusev, A. A.: Vol. 116, pp. 207–248.
Gupta, B. and *Anjum, N.*: Plasma and Radiation-Induced Graft Modification of Polymers for Biomedical Applications. Vol. 162, pp. 37–63.
Gurtovenko, A. A. and *Blumen, A.*: Generalized Gaussian Structures: Models for Polymer Systems with Complex Topologies. Vol. 182, pp. 131–242.
Gusev, A. A., Müller-Plathe, F., van Gunsteren, W. F. and *Suter, U. W.*: Dynamics of Small Molecules in Bulk Polymers. Vol. 116, pp. 207–248.
Gusev, A. A. see Baschnagel, J.: Vol. 152, pp. 41–156.
Guillot, J. see Hunkeler, D.: Vol. 112, pp. 115–134.
Guyot, A. and *Tauer, K.*: Reactive Surfactants in Emulsion Polymerization. Vol. 111, pp. 43–66.

Hadjichristidis, N., Pispas, S., Pitsikalis, M., Iatrou, H. and *Vlahos, C.*: Asymmetric Star Polymers Synthesis and Properties. Vol. 142, pp. 71–128.
Hadjichristidis, N. see Xu, Z.: Vol. 120, pp. 1–50.
Hadjichristidis, N. see Pitsikalis, M.: Vol. 135, pp. 1–138.
Hahn, O. see Baschnagel, J.: Vol. 152, pp. 41–156.
Hakkarainen, M.: Aliphatic Polyesters: Abiotic and Biotic Degradation and Degradation Products. Vol. 157, pp. 1–26.
Hakkarainen, M. and *Albertsson, A.-C.*: Environmental Degradation of Polyethylene. Vol. 169, pp. 177–199.
Hall, H. K. see Penelle, J.: Vol. 102, pp. 73–104.
Hamley, I. W.: Crystallization in Block Copolymers. Vol. 148, pp. 113–138.
Hammouda, B.: SANS from Homogeneous Polymer Mixtures: A Unified Overview. Vol. 106, pp. 87–134.
Han, M. J. and *Chang, J. Y.*: Polynucleotide Analogues. Vol. 153, pp. 1–36.
Harada, A.: Design and Construction of Supramolecular Architectures Consisting of Cyclodextrins and Polymers. Vol. 133, pp. 141–192.
Haralson, M. A. see Prokop, A.: Vol. 136, pp. 1–52.
Hasegawa, N. see Usuki, A.: Vol. 179, pp. 135–195.
Hassan, C. M. and *Peppas, N. A.*: Structure and Applications of Poly(vinyl alcohol) Hydrogels Produced by Conventional Crosslinking or by Freezing/Thawing Methods. Vol. 153, pp. 37–65.
Hawker, C. J.: Dentritic and Hyperbranched Macromolecules Precisely Controlled Macromolecular Architectures. Vol. 147, pp. 113–160.
Hawker, C. J. see Hedrick, J. L.: Vol. 141, pp. 1–44.
He, G. S. see Lin, T.-C.: Vol. 161, pp. 157–193.
Hedrick, J. L., Carter, K. R., Labadie, J. W., Miller, R. D., Volksen, W., Hawker, C. J., Yoon, D. Y., Russell, T. P., McGrath, J. E. and *Briber, R. M.*: Nanoporous Polyimides. Vol. 141, pp. 1–44.
Hedrick, J. L., Labadie, J. W., Volksen, W. and *Hilborn, J. G.*: Nanoscopically Engineered Polyimides. Vol. 147, pp. 61–112.
Hedrick, J. L. see Hergenrother, P. M.: Vol. 117, pp. 67–110.
Hedrick, J. L. see Kiefer, J.: Vol. 147, pp. 161–247.
Hedrick, J. L. see McGrath, J. E.: Vol. 140, pp. 61–106.
Heine, D. R., Grest, G. S. and *Curro, J. G.*: Structure of Polymer Melts and Blends: Comparison of Integral Equation theory and Computer Sumulation. Vol. 173, pp. 209–249.

Heinrich, G. and *Klüppel, M.*: Recent Advances in the Theory of Filler Networking in Elastomers. Vol. 160, pp. 1–44.
Heller, J.: Poly (Ortho Esters). Vol. 107, pp. 41–92.
Helm, C. A. see Möhwald, H.: Vol. 165, pp. 151–175.
Hemielec, A. A. see Hunkeler, D.: Vol. 112, pp. 115–134.
Hergenrother, P. M., Connell, J. W., Labadie, J. W. and *Hedrick, J. L.*: Poly(arylene ether)s Containing Heterocyclic Units. Vol. 117, pp. 67–110.
Hernández-Barajas, J. see Wandrey, C.: Vol. 145, pp. 123–182.
Hervet, H. see Léger, L.: Vol. 138, pp. 185–226.
Hiejima, T. see Abe, A.: Vol. 181, pp. 121–152.
Hilborn, J. G. see Hedrick, J. L.: Vol. 147, pp. 61–112.
Hilborn, J. G. see Kiefer, J.: Vol. 147, pp. 161–247.
Hillborg, H. see Vancso, G. J.: Vol. 182, pp. 55–129.
Hiramatsu, N. see Matsushige, M.: Vol. 125, pp. 147–186.
Hirasa, O. see Suzuki, M.: Vol. 110, pp. 241–262.
Hirotsu, S.: Coexistence of Phases and the Nature of First-Order Transition in Poly-N-isopropylacrylamide Gels. Vol. 110, pp. 1–26.
Höcker, H. see Klee, D.: Vol. 149, pp. 1–57.
Holm, C., Hofmann, T., Joanny, J. F., Kremer, K., Netz, R. R., Reineker, P., Seidel, C., Vilgis, T. A. and *Winkler, R. G.*: Polyelectrolyte Theory. Vol. 166, pp. 67–111.
Holm, C., Rehahn, M., Oppermann, W. and *Ballauff, M.*: Stiff-Chain Polyelectrolytes. Vol. 166, pp. 1–27.
Hornsby, P.: Rheology, Compounding and Processing of Filled Thermoplastics. Vol. 139, pp. 155–216.
Houbenov, N. see Rühe, J.: Vol. 165, pp. 79–150.
Huber, K. see Volk, N.: Vol. 166, pp. 29–65.
Hugenberg, N. see Rühe, J.: Vol. 165, pp. 79–150.
Hui, C.-Y. see Creton, C.: Vol. 156, pp. 53–135.
Hult, A., Johansson, M. and *Malmström, E.*: Hyperbranched Polymers. Vol. 143, pp. 1–34.
Hünenberger, P. H.: Thermostat Algorithms for Molecular-Dynamics Simulations. Vol. 173, pp. 105–147.
Hunkeler, D., Candau, F., Pichot, C., Hemielec, A. E., Xie, T. Y., Barton, J., Vaskova, V., Guillot, J., Dimonie, M. V. and *Reichert, K. H.*: Heterophase Polymerization: A Physical and Kinetic Comparision and Categorization. Vol. 112, pp. 115–134.
Hunkeler, D. see Macko, T.: Vol. 163, pp. 61–136.
Hunkeler, D. see Prokop, A.: Vol. 136, pp. 1–52; 53–74.
Hunkeler, D. see Wandrey, C.: Vol. 145, pp. 123–182.

Iatrou, H. see Hadjichristidis, N.: Vol. 142, pp. 71–128.
Ichikawa, T. see Yoshida, H.: Vol. 105, pp. 3–36.
Ihara, E. see Yasuda, H.: Vol. 133, pp. 53–102.
Ikada, Y. see Uyama, Y.: Vol. 137, pp. 1–40.
Ikehara, T. see Jinnuai, H.: Vol. 170, pp. 115–167.
Ilavsky, M.: Effect on Phase Transition on Swelling and Mechanical Behavior of Synthetic Hydrogels. Vol. 109, pp. 173–206.
Imai, Y.: Rapid Synthesis of Polyimides from Nylon-Salt Monomers. Vol. 140, pp. 1–23.
Inomata, H. see Saito, S.: Vol. 106, pp. 207–232.
Inoue, S. see Sugimoto, H.: Vol. 146, pp. 39–120.
Irie, M.: Stimuli-Responsive Poly(N-isopropylacrylamide), Photo- and Chemical-Induced Phase Transitions. Vol. 110, pp. 49–66.

Ise, N. see Matsuoka, H.: Vol. 114, pp. 187–232.
Ishikawa, T.: Advances in Inorganic Fibers. Vol. 178, (in press).
Ito, H.: Chemical Amplification Resists for Microlithography. Vol. 172, pp. 37–245.
Ito, K. and *Kawaguchi, S.*: Poly(macronomers), Homo- and Copolymerization. Vol. 142, pp. 129–178.
Ito, K. see Kawaguchi, S.: Vol. 175, pp. 299–328.
Ito, S. and *Aoki, H.*: Nano-Imaging of Polymers by Optical Microscopy. Vol. 182, pp. 243–281.
Ito, Y. see Suginome, M.: Vol. 171, pp. 77–136.
Ivanov, A. E. see Zubov, V. P.: Vol. 104, pp. 135–176.

Jacob, S. and *Kennedy, J.*: Synthesis, Characterization and Properties of OCTA-ARM Polyisobutylene-Based Star Polymers. Vol. 146, pp. 1–38.
Jacobson, K., Eriksson, P., Reitberger, T. and *Stenberg, B.*: Chemiluminescence as a Tool for Polyolefin. Vol. 169, pp. 151–176.
Jaeger, W. see Bohrisch, J.: Vol. 165, pp. 1–41.
Jaffe, M., Chen, P., Choe, E.-W., Chung, T.-S. and *Makhija, S.*: High Performance Polymer Blends. Vol. 117, pp. 297–328.
Jancar, J.: Structure-Property Relationships in Thermoplastic Matrices. Vol. 139, pp. 1–66.
Jen, A. K.-Y. see Kajzar, F.: Vol. 161, pp. 1–85.
Jerome, R. see Mecerreyes, D.: Vol. 147, pp. 1–60.
de Jeu, W. H. see Li, L.: Vol. 181, pp. 75–120.
Jiang, M., Li, M., Xiang, M. and *Zhou, H.*: Interpolymer Complexation and Miscibility and Enhancement by Hydrogen Bonding. Vol. 146, pp. 121–194.
Jin, J. see Shim, H.-K.: Vol. 158, pp. 191–241.
Jinnai, H., Nishikawa, Y., Ikehara, T. and *Nishi, T.*: Emerging Technologies for the 3D Analysis of Polymer Structures. Vol. 170, pp. 115–167.
Jo, W. H. and *Yang, J. S.*: Molecular Simulation Approaches for Multiphase Polymer Systems. Vol. 156, pp. 1–52.
Joanny, J.-F. see Holm, C.: Vol. 166, pp. 67–111.
Joanny, J.-F. see Thünemann, A. F.: Vol. 166, pp. 113–171.
Johannsmann, D. see Rühe, J.: Vol. 165, pp. 79–150.
Johansson, M. see Hult, A.: Vol. 143, pp. 1–34.
Joos-Müller, B. see Funke, W.: Vol. 136, pp. 137–232.
Jou, D., Casas-Vazquez, J. and *Criado-Sancho, M.*: Thermodynamics of Polymer Solutions under Flow: Phase Separation and Polymer Degradation. Vol. 120, pp. 207–266.

Kaetsu, I.: Radiation Synthesis of Polymeric Materials for Biomedical and Biochemical Applications. Vol. 105, pp. 81–98.
Kaji, K. see Kanaya, T.: Vol. 154, pp. 87–141.
Kajzar, F., Lee, K.-S. and *Jen, A. K.-Y.*: Polymeric Materials and their Orientation Techniques for Second-Order Nonlinear Optics. Vol. 161, pp. 1–85.
Kakimoto, M. see Gaw, K. O.: Vol. 140, pp. 107–136.
Kaminski, W. and *Arndt, M.*: Metallocenes for Polymer Catalysis. Vol. 127, pp. 143–187.
Kammer, H. W., Kressler, H. and *Kummerloewe, C.*: Phase Behavior of Polymer Blends – Effects of Thermodynamics and Rheology. Vol. 106, pp. 31–86.
Kanaya, T. and *Kaji, K.*: Dynamcis in the Glassy State and Near the Glass Transition of Amorphous Polymers as Studied by Neutron Scattering. Vol. 154, pp. 87–141.
Kandyrin, L. B. and *Kuleznev, V. N.*: The Dependence of Viscosity on the Composition of Concentrated Dispersions and the Free Volume Concept of Disperse Systems. Vol. 103, pp. 103–148.

Kaneko, M. see Ramaraj, R.: Vol. 123, pp. 215–242.
Kang, E. T., Neoh, K. G. and *Tan, K. L.*: X-Ray Photoelectron Spectroscopic Studies of Electroactive Polymers. Vol. 106, pp. 135–190.
Karlsson, S. see Söderqvist Lindblad, M.: Vol. 157, pp. 139–161.
Karlsson, S.: Recycled Polyolefins. Material Properties and Means for Quality Determination. Vol. 169, pp. 201–229.
Kato, K. see Uyama, Y.: Vol. 137, pp. 1–40.
Kato, M. see Usuki, A.: Vol. 179, pp. 135–195.
Kautek, W. see Krüger, J.: Vol. 168, pp. 247–290.
Kawaguchi, S. see Ito, K.: Vol. 142, pp. 129–178.
Kawaguchi, S. and *Ito, K.*: Dispersion Polymerization. Vol. 175, pp. 299–328.
Kawata, S. see Sun, H.-B.: Vol. 170, pp. 169–273.
Kazanskii, K. S. and *Dubrovskii, S. A.*: Chemistry and Physics of Agricultural Hydrogels. Vol. 104, pp. 97–134.
Kennedy, J. P. see Jacob, S.: Vol. 146, pp. 1–38.
Kennedy, J. P. see Majoros, I.: Vol. 112, pp. 1–113.
Kennedy, K. A., Roberts, G. W. and *DeSimone, J. M.*: Heterogeneous Polymerization of Fluoroolefins in Supercritical Carbon Dioxide. Vol. 175, pp. 329–346.
Khokhlov, A., Starodybtzev, S. and *Vasilevskaya, V.*: Conformational Transitions of Polymer Gels: Theory and Experiment. Vol. 109, pp. 121–172.
Kiefer, J., Hedrick, J. L. and *Hiborn, J. G.*: Macroporous Thermosets by Chemically Induced Phase Separation. Vol. 147, pp. 161–247.
Kihara, N. see Takata, T.: Vol. 171, pp. 1–75.
Kilian, H. G. and *Pieper, T.*: Packing of Chain Segments. A Method for Describing X-Ray Patterns of Crystalline, Liquid Crystalline and Non-Crystalline Polymers. Vol. 108, pp. 49–90.
Kim, J. see Quirk, R. P.: Vol. 153, pp. 67–162.
Kim, K.-S. see Lin, T.-C.: Vol. 161, pp. 157–193.
Kimmich, R. and *Fatkullin, N.*: Polymer Chain Dynamics and NMR. Vol. 170, pp. 1–113.
Kippelen, B. and *Peyghambarian, N.*: Photorefractive Polymers and their Applications. Vol. 161, pp. 87–156.
Kirchhoff, R. A. and *Bruza, K. J.*: Polymers from Benzocyclobutenes. Vol. 117, pp. 1–66.
Kishore, K. and *Ganesh, K.*: Polymers Containing Disulfide, Tetrasulfide, Diselenide and Ditelluride Linkages in the Main Chain. Vol. 121, pp. 81–122.
Kitamaru, R.: Phase Structure of Polyethylene and Other Crystalline Polymers by Solid-State 13C/MNR. Vol. 137, pp. 41–102.
Klapper, M. see Rusanov, A. L.: Vol. 179, pp. 83–134.
Klee, D. and *Höcker, H.*: Polymers for Biomedical Applications: Improvement of the Interface Compatibility. Vol. 149, pp. 1–57.
Klemm, E., Pautzsch, T. and *Blankenburg, L.*: Organometallic PAEs. Vol. 177, pp. 53–90.
Klier, J. see Scranton, A. B.: Vol. 122, pp. 1–54.
v. Klitzing, R. and *Tieke, B.*: Polyelectrolyte Membranes. Vol. 165, pp. 177–210.
Klüppel, M.: The Role of Disorder in Filler Reinforcement of Elastomers on Various Length Scales. Vol. 164, pp. 1–86.
Klüppel, M. see Heinrich, G.: Vol. 160, pp. 1–44.
Knuuttila, H., Lehtinen, A. and *Nummila-Pakarinen, A.*: Advanced Polyethylene Technologies – Controlled Material Properties. Vol. 169, pp. 13–27.
Kobayashi, S., Shoda, S. and *Uyama, H.*: Enzymatic Polymerization and Oligomerization. Vol. 121, pp. 1–30.
Kobayashi, T. see Abe, A.: Vol. 181, pp. 121–152.

Köhler, W. and *Schäfer, R.*: Polymer Analysis by Thermal-Diffusion Forced Rayleigh Scattering. Vol. 151, pp. 1–59.
Koenig, J. L. see Bhargava, R.: Vol. 163, pp. 137–191.
Koenig, J. L. see Andreis, M.: Vol. 124, pp. 191–238.
Koike, T.: Viscoelastic Behavior of Epoxy Resins Before Crosslinking. Vol. 148, pp. 139–188.
Kokko, E. see Löfgren, B.: Vol. 169, pp. 1–12.
Kokufuta, E.: Novel Applications for Stimulus-Sensitive Polymer Gels in the Preparation of Functional Immobilized Biocatalysts. Vol. 110, pp. 157–178.
Konno, M. see Saito, S.: Vol. 109, pp. 207–232.
Konradi, R. see Rühe, J.: Vol. 165, pp. 79–150.
Kopecek, J. see Putnam, D.: Vol. 122, pp. 55–124.
Koßmehl, G. see Schopf, G.: Vol. 129, pp. 1–145.
Kostoglodov, P. V. see Rusanov, A. L.: Vol. 179, pp. 83–134.
Kozlov, E. see Prokop, A.: Vol. 160, pp. 119–174.
Kramer, E. J. see Creton, C.: Vol. 156, pp. 53–135.
Kremer, K. see Baschnagel, J.: Vol. 152, pp. 41–156.
Kremer, K. see Holm, C.: Vol. 166, pp. 67–111.
Kressler, J. see Kammer, H. W.: Vol. 106, pp. 31–86.
Kricheldorf, H. R.: Liquid-Cristalline Polyimides. Vol. 141, pp. 83–188.
Krishnamoorti, R. see Giannelis, E. P.: Vol. 138, pp. 107–148.
Krüger, J. and *Kautek, W.*: Ultrashort Pulse Laser Interaction with Dielectrics and Polymers, Vol. 168, pp. 247–290.
Kuchanov, S. I.: Modern Aspects of Quantitative Theory of Free-Radical Copolymerization. Vol. 103, pp. 1–102.
Kuchanov, S. I.: Principles of Quantitive Description of Chemical Structure of Synthetic Polymers. Vol. 152, pp. 157–202.
Kudaibergennow, S. E.: Recent Advances in Studying of Synthetic Polyampholytes in Solutions. Vol. 144, pp. 115–198.
Kuleznev, V. N. see Kandyrin, L. B.: Vol. 103, pp. 103–148.
Kulichkhin, S. G. see Malkin, A. Y.: Vol. 101, pp. 217–258.
Kulicke, W.-M. see Grigorescu, G.: Vol. 152, pp. 1–40.
Kumar, M. N. V. R., Kumar, N., Domb, A. J. and *Arora, M.*: Pharmaceutical Polymeric Controlled Drug Delivery Systems. Vol. 160, pp. 45–118.
Kumar, N. see Kumar, M. N. V. R.: Vol. 160, pp. 45–118.
Kummerloewe, C. see Kammer, H. W.: Vol. 106, pp. 31–86.
Kuznetsova, N. P. see Samsonov, G. V.: Vol. 104, pp. 1–50.
Kwon, Y. and *Faust, R.*: Synthesis of Polyisobutylene-Based Block Copolymers with Precisely Controlled Architecture by Living Cationic Polymerization. Vol. 167, pp. 107–135.

Labadie, J. W. see Hergenrother, P. M.: Vol. 117, pp. 67–110.
Labadie, J. W. see Hedrick, J. L.: Vol. 141, pp. 1–44.
Labadie, J. W. see Hedrick, J. L.: Vol. 147, pp. 61–112.
Lamparski, H. G. see O'Brien, D. F.: Vol. 126, pp. 53–84.
Laschewsky, A.: Molecular Concepts, Self-Organisation and Properties of Polysoaps. Vol. 124, pp. 1–86.
Laso, M. see Leontidis, E.: Vol. 116, pp. 283–318.
Lazár, M. and *Rychl, R.*: Oxidation of Hydrocarbon Polymers. Vol. 102, pp. 189–222.
Lechowicz, J. see Galina, H.: Vol. 137, pp. 135–172.
Léger, L., Raphaël, E. and *Hervet, H.*: Surface-Anchored Polymer Chains: Their Role in Adhesion and Friction. Vol. 138, pp. 185–226.

Lenz, R. W.: Biodegradable Polymers. Vol. 107, pp. 1–40.
Leontidis, E., de Pablo, J. J., Laso, M. and *Suter, U. W.*: A Critical Evaluation of Novel Algorithms for the Off-Lattice Monte Carlo Simulation of Condensed Polymer Phases. Vol. 116, pp. 283–318.
Lee, B. see Quirk, R. P.: Vol. 153, pp. 67–162.
Lee, K.-S. see Kajzar, F.: Vol. 161, pp. 1–85.
Lee, Y. see Quirk, R. P.: Vol. 153, pp. 67–162.
Lehtinen, A. see Knuuttila, H.: Vol. 169, pp. 13–27.
Leónard, D. see Mathieu, H. J.: Vol. 162, pp. 1–35.
Lesec, J. see Viovy, J.-L.: Vol. 114, pp. 1–42.
Li, L. and *de Jeu, W. H.*: Flow-induced mesophases in crystallizable polymers. Vol. 181, pp. 75–120.
Li, M. see Jiang, M.: Vol. 146, pp. 121–194.
Liang, G. L. see Sumpter, B. G.: Vol. 116, pp. 27–72.
Lienert, K.-W.: Poly(ester-imide)s for Industrial Use. Vol. 141, pp. 45–82.
Likhatchev, D. see Rusanov, A. L.: Vol. 179, pp. 83–134.
Lin, J. and *Sherrington, D. C.*: Recent Developments in the Synthesis, Thermostability and Liquid Crystal Properties of Aromatic Polyamides. Vol. 111, pp. 177–220.
Lin, T.-C., Chung, S.-J., Kim, K.-S., Wang, X., He, G. S., Swiatkiewicz, J., Pudavar, H. E. and *Prasad, P. N.*: Organics and Polymers with High Two-Photon Activities and their Applications. Vol. 161, pp. 157–193.
Lippert, T.: Laser Application of Polymers. Vol. 168, pp. 51–246.
Liu, Y. see Söderqvist Lindblad, M.: Vol. 157, pp. 139–161.
Long, T.-C. see Geil, P. H.: Vol. 180, pp. 89–159.
López Cabarcos, E. see Baltá-Calleja, F. J.: Vol. 108, pp. 1–48.
Lotz, B.: Analysis and Observation of Polymer Crystal Structures at the Individual Stem Level. Vol. 180, pp. 17–44.
Löfgren, B., Kokko, E. and *Seppälä, J.*: Specific Structures Enabled by Metallocene Catalysis in Polyethenes. Vol. 169, pp. 1–12.
Löwen, H. see Thünemann, A. F.: Vol. 166, pp. 113–171.
Luo, Y. see Schork, F. J.: Vol. 175, pp. 129–255.

Macko, T. and *Hunkeler, D.*: Liquid Chromatography under Critical and Limiting Conditions: A Survey of Experimental Systems for Synthetic Polymers. Vol. 163, pp. 61–136.
Majoros, I., Nagy, A. and *Kennedy, J. P.*: Conventional and Living Carbocationic Polymerizations United. I. A Comprehensive Model and New Diagnostic Method to Probe the Mechanism of Homopolymerizations. Vol. 112, pp. 1–113.
Makhija, S. see Jaffe, M.: Vol. 117, pp. 297–328.
Malmström, E. see Hult, A.: Vol. 143, pp. 1–34.
Malkin, A. Y. and *Kulichkhin, S. G.*: Rheokinetics of Curing. Vol. 101, pp. 217–258.
Maniar, M. see Domb, A. J.: Vol. 107, pp. 93–142.
Manias, E. see Giannelis, E. P.: Vol. 138, pp. 107–148.
Martin, H. see Engelhardt, H.: Vol. 165, pp. 211–247.
Marty, J. D. and *Mauzac, M.*: Molecular Imprinting: State of the Art and Perspectives. Vol. 172, pp. 1–35.
Mashima, K., Nakayama, Y. and *Nakamura, A.*: Recent Trends in Polymerization of a-Olefins Catalyzed by Organometallic Complexes of Early Transition Metals. Vol. 133, pp. 1–52.
Mathew, D. see Reghunadhan Nair, C. P.: Vol. 155, pp. 1–99.
Mathieu, H. J., Chevolot, Y., Ruiz-Taylor, L. and *Leónard, D.*: Engineering and Characterization of Polymer Surfaces for Biomedical Applications. Vol. 162, pp. 1–35.

Matsumoto, A.: Free-Radical Crosslinking Polymerization and Copolymerization of Multivinyl Compounds. Vol. 123, pp. 41–80.
Matsumoto, A. see Otsu, T.: Vol. 136, pp. 75.-138.
Matsuoka, H. and *Ise, N.*: Small-Angle and Ultra-Small Angle Scattering Study of the Ordered Structure in Polyelectrolyte Solutions and Colloidal Dispersions. Vol. 114, pp. 187–232.
Matsushige, K., Hiramatsu, N. and *Okabe, H.*: Ultrasonic Spectroscopy for Polymeric Materials. Vol. 125, pp. 147–186.
Mattice, W. L. see Rehahn, M.: Vol. 131/132, pp. 1–475.
Mattice, W. L. see Baschnagel, J.: Vol. 152, pp. 41–156.
Mattozzi, A. see Gedde, U. W.: Vol. 169, pp. 29–73.
Mauzac, M. see Marty, J. D.: Vol. 172, pp. 1–35.
Mays, W. see Xu, Z.: Vol. 120, pp. 1–50.
Mays, J. W. see Pitsikalis, M.: Vol. 135, pp. 1–138.
McGrath, J. E. see Hedrick, J. L.: Vol. 141, pp. 1–44.
McGrath, J. E., Dunson, D. L. and *Hedrick, J. L.*: Synthesis and Characterization of Segmented Polyimide-Polyorganosiloxane Copolymers. Vol. 140, pp. 61–106.
McLeish, T. C. B. and *Milner, S. T.*: Entangled Dynamics and Melt Flow of Branched Polymers. Vol. 143, pp. 195–256.
Mecerreyes, D., Dubois, P. and *Jerome, R.*: Novel Macromolecular Architectures Based on Aliphatic Polyesters: Relevance of the Coordination-Insertion Ring-Opening Polymerization. Vol. 147, pp. 1–60.
Mecham, S. J. see McGrath, J. E.: Vol. 140, pp. 61–106.
Menzel, H. see Möhwald, H.: Vol. 165, pp. 151–175.
Meyer, T. see Spange, S.: Vol. 165, pp. 43–78.
Mikos, A. G. see Thomson, R. C.: Vol. 122, pp. 245–274.
Milner, S. T. see McLeish, T. C. B.: Vol. 143, pp. 195–256.
Mison, P. and *Sillion, B.*: Thermosetting Oligomers Containing Maleimides and Nadiimides End-Groups. Vol. 140, pp. 137–180.
Miyasaka, K.: PVA-Iodine Complexes: Formation, Structure and Properties. Vol. 108, pp. 91–130.
Miller, R. D. see Hedrick, J. L.: Vol. 141, pp. 1–44.
Minko, S. see Rühe, J.: Vol. 165, pp. 79–150.
Möhwald, H., Menzel, H., Helm, C. A. and *Stamm, M.*: Lipid and Polyampholyte Monolayers to Study Polyelectrolyte Interactions and Structure at Interfaces. Vol. 165, pp. 151–175.
Monkenbusch, M. see Richter, D.: Vol. 174, (in press).
Monnerie, L. see Bahar, I.: Vol. 116, pp. 145–206.
Moore, J. S. see Ray, C. R.: Vol. 177, pp. 99–149.
Mori, H. see Bohrisch, J.: Vol. 165, pp. 1–41.
Morishima, Y.: Photoinduced Electron Transfer in Amphiphilic Polyelectrolyte Systems. Vol. 104, pp. 51–96.
Morton, M. see Quirk, R. P.: Vol. 153, pp. 67–162.
Motornov, M. see Rühe, J.: Vol. 165, pp. 79–150.
Mours, M. see Winter, H. H.: Vol. 134, pp. 165–234.
Müllen, K. see Scherf, U.: Vol. 123, pp. 1–40.
Müller, A. H. E. see Bohrisch, J.: Vol. 165, pp. 1–41.
Müller, A. H. E. see Förster, S.: Vol. 166, pp. 173–210.
Müller, M. see Thünemann, A. F.: Vol. 166, pp. 113–171.
Müller-Plathe, F. see Gusev, A. A.: Vol. 116, pp. 207–248.
Müller-Plathe, F. see Baschnagel, J.: Vol. 152, p. 41–156.
Mukerherjee, A. see Biswas, M.: Vol. 115, pp. 89–124.

Munz, M., Cappella, B., Sturm, H., Geuss, M. and *Schulz, E.*: Materials Contrasts and Nanolithography Techniques in Scanning Force Microscopy (SFM) and their Application to Polymers and Polymer Composites. Vol. 164, pp. 87–210.
Murat, M. see Baschnagel, J.: Vol. 152, p. 41–156.
Mylnikov, V.: Photoconducting Polymers. Vol. 115, pp. 1–88.

Nagy, A. see Majoros, I.: Vol. 112, pp. 1–11.
Naka, K. see Uemura, T.: Vol. 167, pp. 81–106.
Nakamura, A. see Mashima, K.: Vol. 133, pp. 1–52.
Nakayama, Y. see Mashima, K.: Vol. 133, pp. 1–52.
Narasinham, B. and *Peppas, N. A.*: The Physics of Polymer Dissolution: Modeling Approaches and Experimental Behavior. Vol. 128, pp. 157–208.
Nechaev, S. see Grosberg, A.: Vol. 106, pp. 1–30.
Neoh, K. G. see Kang, E. T.: Vol. 106, pp. 135–190.
Netz, R. R. see Holm, C.: Vol. 166, pp. 67–111.
Netz, R. R. see Rühe, J.: Vol. 165, pp. 79–150.
Newman, S. M. see Anseth, K. S.: Vol. 122, pp. 177–218.
Nijenhuis, K. te: Thermoreversible Networks. Vol. 130, pp. 1–252.
Ninan, K. N. see Reghunadhan Nair, C. P.: Vol. 155, pp. 1–99.
Nishi, T. see Jinnai, H.: Vol. 170, pp. 115–167.
Nishikawa, Y. see Jinnai, H.: Vol. 170, pp. 115–167.
Noid, D. W. see Otaigbe, J. U.: Vol. 154, pp. 1–86.
Noid, D. W. see Sumpter, B. G.: Vol. 116, pp. 27–72.
Nomura, M., Tobita, H. and *Suzuki, K.*: Emulsion Polymerization: Kinetic and Mechanistic Aspects. Vol. 175, pp. 1–128.
Northolt, M. G., Picken, S. J., Den Decker, M. G., Baltussen, J. J. M. and *Schlatmann, R.*: The Tensile Strength of Polymer Fibres. Vol. 178, (in press).
Novac, B. see Grubbs, R.: Vol. 102, pp. 47–72.
Novikov, V. V. see Privalko, V. P.: Vol. 119, pp. 31–78.
Nummila-Pakarinen, A. see Knuuttila, H.: Vol. 169, pp. 13–27.

O'Brien, D. F., Armitage, B. A., Bennett, D. E. and *Lamparski, H. G.*: Polymerization and Domain Formation in Lipid Assemblies. Vol. 126, pp. 53–84.
Ogasawara, M.: Application of Pulse Radiolysis to the Study of Polymers and Polymerizations. Vol. 105, pp. 37–80.
Okabe, H. see Matsushige, K.: Vol. 125, pp. 147–186.
Okada, M.: Ring-Opening Polymerization of Bicyclic and Spiro Compounds. Reactivities and Polymerization Mechanisms. Vol. 102, pp. 1–46.
Okano, T.: Molecular Design of Temperature-Responsive Polymers as Intelligent Materials. Vol. 110, pp. 179–198.
Okay, O. see Funke, W.: Vol. 136, pp. 137–232.
Onuki, A.: Theory of Phase Transition in Polymer Gels. Vol. 109, pp. 63–120.
Oppermann, W. see Holm, C.: Vol. 166, pp. 1–27.
Oppermann, W. see Volk, N.: Vol. 166, pp. 29–65.
Osad'ko, I. S.: Selective Spectroscopy of Chromophore Doped Polymers and Glasses. Vol. 114, pp. 123–186.
Osakada, K. and *Takeuchi, D.*: Coordination Polymerization of Dienes, Allenes, and Methylenecycloalkanes. Vol. 171, pp. 137–194.
Otaigbe, J. U., Barnes, M. D., Fukui, K., Sumpter, B. G. and *Noid, D. W.*: Generation, Characterization, and Modeling of Polymer Micro- and Nano-Particles. Vol. 154, pp. 1–86.

Otsu, T. and *Matsumoto, A.*: Controlled Synthesis of Polymers Using the Iniferter Technique: Developments in Living Radical Polymerization. Vol. 136, pp. 75–138.

de Pablo, J. J. see Leontidis, E.: Vol. 116, pp. 283–318.
Padias, A. B. see Penelle, J.: Vol. 102, pp. 73–104.
Pascault, J.-P. see Williams, R. J. J.: Vol. 128, pp. 95–156.
Pasch, H.: Analysis of Complex Polymers by Interaction Chromatography. Vol. 128, pp. 1–46.
Pasch, H.: Hyphenated Techniques in Liquid Chromatography of Polymers. Vol. 150, pp. 1–66.
Paul, W. see Baschnagel, J.: Vol. 152, pp. 41–156.
Pautzsch, T. see Klemm, E.: Vol. 177, pp. 53–90.
Penczek, P., Czub, P. and *Pielichowski, J.*: Unsaturated Polyester Resins: Chemistry and Technology. Vol. 184, pp. 1–95.
Penczek, P. see Batog, A. E.: Vol. 144, pp. 49–114.
Penczek, P. see Bogdal, D.: Vol. 163, pp. 193–263.
Penelle, J., Hall, H. K., Padias, A. B. and *Tanaka, H.*: Captodative Olefins in Polymer Chemistry. Vol. 102, pp. 73–104.
Peppas, N. A. see Bell, C. L.: Vol. 122, pp. 125–176.
Peppas, N. A. see Hassan, C. M.: Vol. 153, pp. 37–65.
Peppas, N. A. see Narasimhan, B.: Vol. 128, pp. 157–208.
Petersen, K. L. see Geil, P. H.: Vol. 180, pp. 89–159.
Pet'ko, I. P. see Batog, A. E.: Vol. 144, pp. 49–114.
Pheyghambarian, N. see Kippelen, B.: Vol. 161, pp. 87–156.
Pichot, C. see Hunkeler, D.: Vol. 112, pp. 115–134.
Picken, S. J. see Northolt, M. G.: Vol. 178, (in press)
Pielichowski, J. see Bogdal, D.: Vol. 163, pp. 193–263.
Pielichowski, J. see Penczek, P.: Vol. 184, pp. 1–95.
Pieper, T. see Kilian, H. G.: Vol. 108, pp. 49–90.
Pispas, S. see Pitsikalis, M.: Vol. 135, pp. 1–138.
Pispas, S. see Hadjichristidis, N.: Vol. 142, pp. 71–128.
Pitsikalis, M., Pispas, S., Mays, J. W. and *Hadjichristidis, N.*: Nonlinear Block Copolymer Architectures. Vol. 135, pp. 1–138.
Pitsikalis, M. see Hadjichristidis, N.: Vol. 142, pp. 71–128.
Pleul, D. see Spange, S.: Vol. 165, pp. 43–78.
Plummer, C. J. G.: Microdeformation and Fracture in Bulk Polyolefins. Vol. 169, pp. 75–119.
Pötschke, D. see Dingenouts, N.: Vol. 144, pp. 1–48.
Pokrovskii, V. N.: The Mesoscopic Theory of the Slow Relaxation of Linear Macromolecules. Vol. 154, pp. 143–219.
Pospíšil, J.: Functionalized Oligomers and Polymers as Stabilizers for Conventional Polymers. Vol. 101, pp. 65–168.
Pospíšil, J.: Aromatic and Heterocyclic Amines in Polymer Stabilization. Vol. 124, pp. 87–190.
Powers, A. C. see Prokop, A.: Vol. 136, pp. 53–74.
Prasad, P. N. see Lin, T.-C.: Vol. 161, pp. 157–193.
Priddy, D. B.: Recent Advances in Styrene Polymerization. Vol. 111, pp. 67–114.
Priddy, D. B.: Thermal Discoloration Chemistry of Styrene-co-Acrylonitrile. Vol. 121, pp. 123–154.
Privalko, V. P. and *Novikov, V. V.*: Model Treatments of the Heat Conductivity of Heterogeneous Polymers. Vol. 119, pp. 31–78.
Prociak, A. see Bogdal, D.: Vol. 163, pp. 193–263.

Prokop, A., Hunkeler, D., DiMari, S., Haralson, M. A. and *Wang, T. G.*: Water Soluble Polymers for Immunoisolation I: Complex Coacervation and Cytotoxicity. Vol. 136, pp. 1–52.
Prokop, A., Hunkeler, D., Powers, A. C., Whitesell, R. R. and *Wang, T. G.*: Water Soluble Polymers for Immunoisolation II: Evaluation of Multicomponent Microencapsulation Systems. Vol. 136, pp. 53–74.
Prokop, A., Kozlov, E., Carlesso, G. and *Davidsen, J. M.*: Hydrogel-Based Colloidal Polymeric System for Protein and Drug Delivery: Physical and Chemical Characterization, Permeability Control and Applications. Vol. 160, pp. 119–174.
Pruitt, L. A.: The Effects of Radiation on the Structural and Mechanical Properties of Medical Polymers. Vol. 162, pp. 65–95.
Pudavar, H. E. see Lin, T.-C.: Vol. 161, pp. 157–193.
Pukánszky, B. and *Fekete, E.*: Adhesion and Surface Modification. Vol. 139, pp. 109–154.
Putnam, D. and *Kopecek, J.*: Polymer Conjugates with Anticancer Acitivity. Vol. 122, pp. 55–124.
Putra, E. G. R. see Ungar, G.: Vol. 180, pp. 45–87.

Quirk, R. P., Yoo, T., Lee, Y., M., Kim, J. and *Lee, B.*: Applications of 1,1-Diphenylethylene Chemistry in Anionic Synthesis of Polymers with Controlled Structures. Vol. 153, pp. 67–162.

Ramaraj, R. and *Kaneko, M.*: Metal Complex in Polymer Membrane as a Model for Photosynthetic Oxygen Evolving Center. Vol. 123, pp. 215–242.
Rangarajan, B. see Scranton, A. B.: Vol. 122, pp. 1–54.
Ranucci, E. see Söderqvist Lindblad, M.: Vol. 157, pp. 139–161.
Raphaël, E. see Léger, L.: Vol. 138, pp. 185–226.
Rastogi, S. and *Terry, A. E.*: Morphological implications of the interphase bridging crystalline and amorphous regions in semi-crystalline polymers. Vol. 180, pp. 161–194.
Ray, C. R. and *Moore, J. S.*: Supramolecular Organization of Foldable Phenylene Ethynylene Oligomers. Vol. 177, pp. 99–149.
Reddinger, J. L. and *Reynolds, J. R.*: Molecular Engineering of p-Conjugated Polymers. Vol. 145, pp. 57–122.
Reghunadhan Nair, C. P., Mathew, D. and *Ninan, K. N.*: Cyanate Ester Resins, Recent Developments. Vol. 155, pp. 1–99.
Reichert, K. H. see Hunkeler, D.: Vol. 112, pp. 115–134.
Rehahn, M., Mattice, W. L. and *Suter, U. W.*: Rotational Isomeric State Models in Macromolecular Systems. Vol. 131/132, pp. 1–475.
Rehahn, M. see Bohrisch, J.: Vol. 165, pp. 1–41.
Rehahn, M. see Holm, C.: Vol. 166, pp. 1–27.
Reineker, P. see Holm, C.: Vol. 166, pp. 67–111.
Reitberger, T. see Jacobson, K.: Vol. 169, pp. 151–176.
Reynolds, J. R. see Reddinger, J. L.: Vol. 145, pp. 57–122.
Richter, D. see Ewen, B.: Vol. 134, pp. 1–130.
Richter, D., Monkenbusch, M. and *Colmenero, J.*: Neutron Spin Echo in Polymer Systems. Vol. 174, (in press).
Riegler, S. see Trimmel, G.: Vol. 176, pp. 43–87.
Risse, W. see Grubbs, R.: Vol. 102, pp. 47–72.
Rivas, B. L. and *Geckeler, K. E.*: Synthesis and Metal Complexation of Poly(ethyleneimine) and Derivatives. Vol. 102, pp. 171–188.
Roberts, G. W. see Kennedy, K. A.: Vol. 175, pp. 329–346.

Robin, J. J.: The Use of Ozone in the Synthesis of New Polymers and the Modification of Polymers. Vol. 167, pp. 35–79.
Robin, J. J. see *Boutevin, B.*: Vol. 102, pp. 105–132.
Rodríguez-Pérez, M. A.: Crosslinked Polyolefin Foams: Production, Structure, Properties, and Applications. Vol. 184, pp. 97–126.
Roe, R.-J.: MD Simulation Study of Glass Transition and Short Time Dynamics in Polymer Liquids. Vol. 116, pp. 111–114.
Roovers, J. and *Comanita, B.*: Dendrimers and Dendrimer-Polymer Hybrids. Vol. 142, pp. 179–228.
Rothon, R. N.: Mineral Fillers in Thermoplastics: Filler Manufacture and Characterisation. Vol. 139, pp. 67–108.
de Rosa, C. see *Auriemma, F.*: Vol. 181, pp. 1–74.
Rozenberg, B. A. see *Williams, R. J. J.*: Vol. 128, pp. 95–156.
Rühe, J., Ballauff, M., Biesalski, M., Dziezok, P., Gröhn, F., Johannsmann, D., Houbenov, N., Hugenberg, N., Konradi, R., Minko, S., Motornov, M., Netz, R. R., Schmidt, M., Seidel, C., Stamm, M., Stephan, T., Usov, D. and *Zhang, H.*: Polyelectrolyte Brushes. Vol. 165, pp. 79–150.
Ruckenstein, E.: Concentrated Emulsion Polymerization. Vol. 127, pp. 1–58.
Ruiz-Taylor, L. see *Mathieu, H. J.*: Vol. 162, pp. 1–35.
Rusanov, A. L.: Novel Bis (Naphtalic Anhydrides) and Their Polyheteroarylenes with Improved Processability. Vol. 111, pp. 115–176.
Rusanov, A. L., Likhatchev, D., Kostoglodov, P. V., Müllen, K. and *Klapper, M.*: Proton-Exchanging Electrolyte Membranes Based on Aromatic Condensation Polymers. Vol. 179, pp. 83–134.
Russel, T. P. see *Hedrick, J. L.*: Vol. 141, pp. 1–44.
Russum, J. P. see *Schork, F. J.*: Vol. 175, pp. 129–255.
Rychly, J. see *Lazár, M.*: Vol. 102, pp. 189–222.
Ryner, M. see *Stridsberg, K. M.*: Vol. 157, pp. 27–51.
Ryzhov, V. A. see *Bershtein, V. A.*: Vol. 114, pp. 43–122.

Sabsai, O. Y. see *Barshtein, G. R.*: Vol. 101, pp. 1–28.
Saburov, V. V. see *Zubov, V. P.*: Vol. 104, pp. 135–176.
Saito, S., Konno, M. and *Inomata, H.*: Volume Phase Transition of N-Alkylacrylamide Gels. Vol. 109, pp. 207–232.
Samsonov, G. V. and *Kuznetsova, N. P.*: Crosslinked Polyelectrolytes in Biology. Vol. 104, pp. 1–50.
Santa Cruz, C. see *Baltá-Calleja, F. J.*: Vol. 108, pp. 1–48.
Santos, S. see *Baschnagel, J.*: Vol. 152, p. 41–156.
Sato, T. and *Teramoto, A.*: Concentrated Solutions of Liquid-Christalline Polymers. Vol. 126, pp. 85–162.
Schaller, C. see *Bohrisch, J.*: Vol. 165, pp. 1–41.
Schäfer, R. see *Köhler, W.*: Vol. 151, pp. 1–59.
Scherf, U. and *Müllen, K.*: The Synthesis of Ladder Polymers. Vol. 123, pp. 1–40.
Schlatmann, R. see *Northolt, M. G.*: Vol. 178, (in press).
Schmidt, M. see *Förster, S.*: Vol. 120, pp. 51–134.
Schmidt, M. see *Rühe, J.*: Vol. 165, pp. 79–150.
Schmidt, M. see *Volk, N.*: Vol. 166, pp. 29–65.
Scholz, M.: Effects of Ion Radiation on Cells and Tissues. Vol. 162, pp. 97–158.
Schönherr, H. see *Vancso, G. J.*: Vol. 182, pp. 55–129.

Schopf, G. and *Koßmehl, G.*: Polythiophenes – Electrically Conductive Polymers. Vol. 129, pp. 1–145.

Schork, F. J., Luo, Y., Smulders, W., Russum, J. P., Butté, A. and *Fontenot, K.*: Miniemulsion Polymerization. Vol. 175, pp. 127–255.

Schulz, E. see Munz, M.: Vol. 164, pp. 97–210.

Schwahn, D.: Critical to Mean Field Crossover in Polymer Blends. Vol. 183, pp. 1–61.

Seppälä, J. see Löfgren, B.: Vol. 169, pp. 1–12.

Sturm, H. see Munz, M.: Vol. 164, pp. 87–210.

Schweizer, K. S.: Prism Theory of the Structure, Thermodynamics, and Phase Transitions of Polymer Liquids and Alloys. Vol. 116, pp. 319–378.

Scranton, A. B., Rangarajan, B. and *Klier, J.*: Biomedical Applications of Polyelectrolytes. Vol. 122, pp. 1–54.

Sefton, M. V. and *Stevenson, W. T. K.*: Microencapsulation of Live Animal Cells Using Polycrylates. Vol. 107, pp. 143–198.

Seidel, C. see Holm, C.: Vol. 166, pp. 67–111.

Seidel, C. see Rühe, J.: Vol. 165, pp. 79–150.

Shamanin, V. V.: Bases of the Axiomatic Theory of Addition Polymerization. Vol. 112, pp. 135–180.

Shcherbina, M. A. see Ungar, G.: Vol. 180, pp. 45–87.

Sheiko, S. S.: Imaging of Polymers Using Scanning Force Microscopy: From Superstructures to Individual Molecules. Vol. 151, pp. 61–174.

Sherrington, D. C. see Cameron, N. R.: Vol. 126, pp. 163–214.

Sherrington, D. C. see Lin, J.: Vol. 111, pp. 177–220.

Sherrington, D. C. see Steinke, J.: Vol. 123, pp. 81–126.

Shibayama, M. see Tanaka, T.: Vol. 109, pp. 1–62.

Shiga, T.: Deformation and Viscoelastic Behavior of Polymer Gels in Electric Fields. Vol. 134, pp. 131–164.

Shim, H.-K. and *Jin, J.*: Light-Emitting Characteristics of Conjugated Polymers. Vol. 158, pp. 191–241.

Shoda, S. see Kobayashi, S.: Vol. 121, pp. 1–30.

Siegel, R. A.: Hydrophobic Weak Polyelectrolyte Gels: Studies of Swelling Equilibria and Kinetics. Vol. 109, pp. 233–268.

de Silva, D. S. M. see Ungar, G.: Vol. 180, pp. 45–87.

Silvestre, F. see Calmon-Decriaud, A.: Vol. 207, pp. 207–226.

Sillion, B. see Mison, P.: Vol. 140, pp. 137–180.

Simon, F. see Spange, S.: Vol. 165, pp. 43–78.

Simon, G. P. see Becker, O.: Vol. 179, pp. 29–82.

Simonutti, R. see Sozzani, P.: Vol. 181, pp. 153–177.

Singh, R. P. see Sivaram, S.: Vol. 101, pp. 169–216.

Singh, R. P. see Desai, S. M.: Vol. 169, pp. 231–293.

Sinha Ray, S. see Biswas, M.: Vol. 155, pp. 167–221.

Sivaram, S. and *Singh, R. P.*: Degradation and Stabilization of Ethylene-Propylene Copolymers and Their Blends: A Critical Review. Vol. 101, pp. 169–216.

Slugovc, C. see Trimmel, G.: Vol. 176, pp. 43–87.

Smulders, W. see Schork, F. J.: Vol. 175, pp. 129–255.

Soares, J. B. P. see Anantawaraskul, S.: Vol. 182, pp. 1–54.

Sozzani, P., Bracco, S., Comotti, A. and *Simonutti, R.*: Motional Phase Disorder of Polymer Chains as Crystallized to Hexagonal Lattices. Vol. 181, pp. 153–177.

Söderqvist Lindblad, M., Liu, Y., Albertsson, A.-C., Ranucci, E. and *Karlsson, S.*: Polymer from Renewable Resources. Vol. 157, pp. 139–161.

Spange, S., Meyer, T., Voigt, I., Eschner, M., Estel, K., Pleul, D. and *Simon, F.*: Poly(Vinylformamide-co-Vinylamine)/Inorganic Oxid Hybrid Materials. Vol. 165, pp. 43–78.
Stamm, M. see Möhwald, H.: Vol. 165, pp. 151–175.
Stamm, M. see Rühe, J.: Vol. 165, pp. 79–150.
Starodybtzev, S. see Khokhlov, A.: Vol. 109, pp. 121–172.
Stegeman, G. I. see Canva, M.: Vol. 158, pp. 87–121.
Steinke, J., Sherrington, D. C. and *Dunkin, I. R.*: Imprinting of Synthetic Polymers Using Molecular Templates. Vol. 123, pp. 81–126.
Stelzer, F. see Trimmel, G.: Vol. 176, pp. 43–87.
Stenberg, B. see Jacobson, K.: Vol. 169, pp. 151–176.
Stenzenberger, H. D.: Addition Polyimides. Vol. 117, pp. 165–220.
Stephan, T. see Rühe, J.: Vol. 165, pp. 79–150.
Stevenson, W. T. K. see Sefton, M. V.: Vol. 107, pp. 143–198.
Stridsberg, K. M., Ryner, M. and *Albertsson, A.-C.*: Controlled Ring-Opening Polymerization: Polymers with Designed Macromoleculars Architecture. Vol. 157, pp. 27–51.
Sturm, H. see Munz, M.: Vol. 164, pp. 87–210.
Suematsu, K.: Recent Progress of Gel Theory: Ring, Excluded Volume, and Dimension. Vol. 156, pp. 136–214.
Sugimoto, H. and *Inoue, S.*: Polymerization by Metalloporphyrin and Related Complexes. Vol. 146, pp. 39–120.
Suginome, M. and *Ito, Y.*: Transition Metal-Mediated Polymerization of Isocyanides. Vol. 171, pp. 77–136.
Sumpter, B. G., Noid, D. W., Liang, G. L. and *Wunderlich, B.*: Atomistic Dynamics of Macromolecular Crystals. Vol. 116, pp. 27–72.
Sumpter, B. G. see Otaigbe, J. U.: Vol. 154, pp. 1–86.
Sun, H.-B. and *Kawata, S.*: Two-Photon Photopolymerization and 3D Lithographic Microfabrication. Vol. 170, pp. 169–273.
Suter, U. W. see Gusev, A. A.: Vol. 116, pp. 207–248.
Suter, U. W. see Leontidis, E.: Vol. 116, pp. 283–318.
Suter, U. W. see Rehahn, M.: Vol. 131/132, pp. 1–475.
Suter, U. W. see Baschnagel, J.: Vol. 152, pp. 41–156.
Suzuki, A.: Phase Transition in Gels of Sub-Millimeter Size Induced by Interaction with Stimuli. Vol. 110, pp. 199–240.
Suzuki, A. and *Hirasa, O.*: An Approach to Artifical Muscle by Polymer Gels due to Micro-Phase Separation. Vol. 110, pp. 241–262.
Suzuki, K. see Nomura, M.: Vol. 175, pp. 1–128.
Swiatkiewicz, J. see Lin, T.-C.: Vol. 161, pp. 157–193.

Tagawa, S.: Radiation Effects on Ion Beams on Polymers. Vol. 105, pp. 99–116.
Taguet, A., Ameduri, B. and *Boutevin, B.*: Crosslinking of Vinylidene Fluoride-Containing Fluoropolymers. Vol. 184, pp. 127–211.
Takata, T., Kihara, N. and *Furusho, Y.*: Polyrotaxanes and Polycatenanes: Recent Advances in Syntheses and Applications of Polymers Comprising of Interlocked Structures. Vol. 171, pp. 1–75.
Takeuchi, D. see Osakada, K.: Vol. 171, pp. 137–194.
Tan, K. L. see Kang, E. T.: Vol. 106, pp. 135–190.
Tanaka, H. and *Shibayama, M.*: Phase Transition and Related Phenomena of Polymer Gels. Vol. 109, pp. 1–62.
Tanaka, T. see Penelle, J.: Vol. 102, pp. 73–104.
Tauer, K. see Guyot, A.: Vol. 111, pp. 43–66.

Teramoto, A. see *Sato, T.*: Vol. 126, pp. 85–162.
Terent'eva, J. P. and *Fridman, M. L.*: Compositions Based on Aminoresins. Vol. 101, pp. 29–64.
Terry, A. E. see *Rastogi, S.*: Vol. 180, pp. 161–194.
Theodorou, D. N. see *Dodd, L. R.*: Vol. 116, pp. 249–282.
Thomson, R. C., Wake, M. C., Yaszemski, M. J. and *Mikos, A. G.*: Biodegradable Polymer Scaffolds to Regenerate Organs. Vol. 122, pp. 245–274.
Thünemann, A. F., Müller, M., Dautzenberg, H., Joanny, J.-F. and *Löwen, H.*: Polyelectrolyte complexes. Vol. 166, pp. 113–171.
Tieke, B. see *v. Klitzing, R.*: Vol. 165, pp. 177–210.
Tobita, H. see *Nomura, M.*: Vol. 175, pp. 1–128.
Tokita, M.: Friction Between Polymer Networks of Gels and Solvent. Vol. 110, pp. 27–48.
Traser, S. see *Bohrisch, J.*: Vol. 165, pp. 1–41.
Tries, V. see *Baschnagel, J.*: Vol. 152, p. 41–156.
Trimmel, G., Riegler, S., Fuchs, G., Slugovc, C. and *Stelzer, F.*: Liquid Crystalline Polymers by Metathesis Polymerization. Vol. 176, pp. 43–87.
Tsuruta, T.: Contemporary Topics in Polymeric Materials for Biomedical Applications. Vol. 126, pp. 1–52.

Uemura, T., Naka, K. and *Chujo, Y.*: Functional Macromolecules with Electron-Donating Dithiafulvene Unit. Vol. 167, pp. 81–106.
Ungar, G., Putra, E. G. R., de Silva, D. S. M., Shcherbina, M. A. and *Waddon, A. J.*: The Effect of Self-Poisoning on Crystal Morphology and Growth Rates. Vol. 180, pp. 45–87.
Usov, D. see *Rühe, J.*: Vol. 165, pp. 79–150.
Usuki, A., Hasegawa, N. and *Kato, M.*: Polymer-Clay Nanocomposites. Vol. 179, pp. 135–195.
Uyama, H. see *Kobayashi, S.*: Vol. 121, pp. 1–30.
Uyama, Y.: Surface Modification of Polymers by Grafting. Vol. 137, pp. 1–40.

Vancso, G. J., Hillborg, H. and *Schönherr, H.*: Chemical Composition of Polymer Surfaces Imaged by Atomic Force Microscopy and Complementary Approaches. Vol. 182, pp. 55–129.
Varma, I. K. see *Albertsson, A.-C.*: Vol. 157, pp. 99–138.
Vasilevskaya, V. see *Khokhlov, A.*: Vol. 109, pp. 121–172.
Vaskova, V. see *Hunkeler, D.*: Vol. 112, pp. 115–134.
Verdugo, P.: Polymer Gel Phase Transition in Condensation-Decondensation of Secretory Products. Vol. 110, pp. 145–156.
Vettegren, V. I. see *Bronnikov, S. V.*: Vol. 125, pp. 103–146.
Vilgis, T. A. see *Holm, C.*: Vol. 166, pp. 67–111.
Viovy, J.-L. and *Lesec, J.*: Separation of Macromolecules in Gels: Permeation Chromatography and Electrophoresis. Vol. 114, pp. 1–42.
Vlahos, C. see *Hadjichristidis, N.*: Vol. 142, pp. 71–128.
Voigt, I. see *Spange, S.*: Vol. 165, pp. 43–78.
Volk, N., Vollmer, D., Schmidt, M., Oppermann, W. and *Huber, K.*: Conformation and Phase Diagrams of Flexible Polyelectrolytes. Vol. 166, pp. 29–65.
Volksen, W.: Condensation Polyimides: Synthesis, Solution Behavior, and Imidization Characteristics. Vol. 117, pp. 111–164.
Volksen, W. see *Hedrick, J. L.*: Vol. 141, pp. 1–44.
Volksen, W. see *Hedrick, J. L.*: Vol. 147, pp. 61–112.
Vollmer, D. see *Volk, N.*: Vol. 166, pp. 29–65.
Voskerician, G. and *Weder, C.*: Electronic Properties of PAEs. Vol. 177, pp. 209–248.

Waddon, A. J. see Ungar, G.: Vol. 180, pp. 45–87.
Wagener, K. B. see Baughman, T. W.: Vol. 176, pp. 1–42.
Wake, M. C. see Thomson, R. C.: Vol. 122, pp. 245–274.
Wandrey, C., Hernández-Barajas, J. and *Hunkeler, D.*: Diallyldimethylammonium Chloride and its Polymers. Vol. 145, pp. 123–182.
Wang, K. L. see Cussler, E. L.: Vol. 110, pp. 67–80.
Wang, S.-Q.: Molecular Transitions and Dynamics at Polymer/Wall Interfaces: Origins of Flow Instabilities and Wall Slip. Vol. 138, pp. 227–276.
Wang, S.-Q. see Bhargava, R.: Vol. 163, pp. 137–191.
Wang, T. G. see Prokop, A.: Vol. 136, pp. 1–52; 53–74.
Wang, X. see Lin, T.-C.: Vol. 161, pp. 157–193.
Webster, O. W.: Group Transfer Polymerization: Mechanism and Comparison with Other Methods of Controlled Polymerization of Acrylic Monomers. Vol. 167, pp. 1–34.
Weder, C. see Voskericиan, G.: Vol. 177, pp. 209–248.
Whitesell, R. R. see Prokop, A.: Vol. 136, pp. 53–74.
Williams, R. A. see Geil, P. H.: Vol. 180, pp. 89–159.
Williams, R. J. J., Rozenberg, B. A. and *Pascault, J.-P.*: Reaction Induced Phase Separation in Modified Thermosetting Polymers. Vol. 128, pp. 95–156.
Winkler, R. G. see Holm, C.: Vol. 166, pp. 67–111.
Winter, H. H. and *Mours, M.*: Rheology of Polymers Near Liquid-Solid Transitions. Vol. 134, pp. 165–234.
Wittmeyer, P. see Bohrisch, J.: Vol. 165, pp. 1–41.
Wood-Adams, P. M. see Anantawaraskul, S.: Vol. 182, pp. 1–54.
Wu, C.: Laser Light Scattering Characterization of Special Intractable Macromolecules in Solution. Vol. 137, pp. 103–134.
Wunderlich, B. see Sumpter, B. G.: Vol. 116, pp. 27–72.

Xiang, M. see Jiang, M.: Vol. 146, pp. 121–194.
Xie, T. Y. see Hunkeler, D.: Vol. 112, pp. 115–134.
Xu, P. see Geil, P. H.: Vol. 180, pp. 89–159.
Xu, Z., Hadjichristidis, N., Fetters, L. J. and *Mays, J. W.*: Structure/Chain-Flexibility Relationships of Polymers. Vol. 120, pp. 1–50.

Yagci, Y. and *Endo, T.*: N-Benzyl and N-Alkoxy Pyridium Salts as Thermal and Photochemical Initiators for Cationic Polymerization. Vol. 127, pp. 59–86.
Yamaguchi, I. see Yamamoto, T.: Vol. 177, pp. 181–208.
Yamamoto, T., Yamaguchi, I. and *Yasuda, T.*: PAEs with Heteroaromatic Rings. Vol. 177, pp. 181–208.
Yamaoka, H.: Polymer Materials for Fusion Reactors. Vol. 105, pp. 117–144.
Yannas, I. V.: Tissue Regeneration Templates Based on Collagen-Glycosaminoglycan Copolymers. Vol. 122, pp. 219–244.
Yang, J. see Geil, P. H.: Vol. 180, pp. 89–159.
Yang, J. S. see Jo, W. H.: Vol. 156, pp. 1–52.
Yasuda, H. and *Ihara, E.*: Rare Earth Metal-Initiated Living Polymerizations of Polar and Nonpolar Monomers. Vol. 133, pp. 53–102.
Yasuda, T. see Yamamoto, T.: Vol. 177, pp. 181–208.
Yaszemski, M. J. see Thomson, R. C.: Vol. 122, pp. 245–274.
Yoo, T. see Quirk, R. P.: Vol. 153, pp. 67–162.

Yoon, D. Y. see *Hedrick, J. L.*: Vol. 141, pp. 1–44.
Yoshida, H. and *Ichikawa, T.*: Electron Spin Studies of Free Radicals in Irradiated Polymers. Vol. 105, pp. 3–36.

Zhang, H. see *Rühe, J.*: Vol. 165, pp. 79–150.
Zhang, Y.: Synchrotron Radiation Direct Photo Etching of Polymers. Vol. 168, pp. 291–340.
Zheng, J. and *Swager, T. M.*: Poly(arylene ethynylene)s in Chemosensing and Biosensing. Vol. 177, pp. 151–177.
Zhou, H. see *Jiang, M.*: Vol. 146, pp. 121–194.
Zhou, Z. see *Abe, A.*: Vol. 181, pp. 121–152.
Zubov, V. P., Ivanov, A. E. and *Saburov, V. V.*: Polymer-Coated Adsorbents for the Separation of Biopolymers and Particles. Vol. 104, pp. 135–176.

Subject Index

Accelerator 48, 63
Acid anhydrides 5
Acid resistance, VDF-based
  fluoroelastomers 197
Acid value 15
Acrylic acid 28
Activation energy 50, 75
Adipic acid 26
Aminodiols 65
Autoacceleration 71
Azodicarbonamide 101

Barrier properties 42
Benzylamine 166
Bis-cinnamylidene hexamethylene
  diamine 151
Bismaleimide 20
Bis-peroxycarbamates 156
Bisphenol A 34
Bisphenols, srosslinking 127, 170
Block copolyetherester 35
Block copolymers 53
Boltzman integral model,
  force-deformation 114
Butylamine 145

Calcium carbonate 87
CARB LEVII 203
Carcinogen 42
Castor oil 39
Cell wall thickness, PO 108
Chemical resistance 10, 26, 46, 47
Collapse 112
Compression set resistance 163
Compressive strength 11
Copolymerization 13, 72
CPOF 103
– applications 121
– dynamical mechanical behavior 117

– stress-starin 111
Cracking 19, 46, 60, 68,74
Creep, PO foams 116
Crosslinking 61, 97
– density 60, 152
Cyclopentadiene 16, 21

Decomposition 50
Dehydrofluorination 136
DETA 164
Diallylmelamine 179
Diamines, crosslinking 146
Dibutylamine 148
Dicarboxylic esters 5
Diethyl fumarate 52
Diethylcyclohexylamine 148
Diethylene glycol 7
Diethylene triamine 159
Diode array detection (DAD) 58
Divinylbenzene 179
DMAC 192
Dynamic light scattering 51
Dynamic mechanical analysis 59, 69

Elastomer 21, 32
Electron beam radiation,
  fluoroelastomers,VDF-based 187
Epoxy resin 6, 9, 30, 39, 50, 66, 70–75
Epoxyfumarate resins 8-10
ESI resin 104
Ethylene glycol 13
Ethylenediamine 154
Ethylene-styrene 104, 120
EVA 99, 111
Expanded graphite 44

Ferrocene 63
Filler 86
Fire retardant 8, 82

FKM gum   167
Flammability   43
Flex fuels   203
Flexural strength   10, 16, 17, 42
Fluoroelastomers   130
– diamines   151
– VDF-based, crosslinking   136
– irradiation   190
Fluoropolymers, VDF   127
Fly ash   86
Foam, heat transfer   118
– injection molded   120
– open-cell   106
Foam bulk modulus   113
Foam microstructure   113
Foaming, crosslinking   105
Foaming agents   97, 105
Force-deformation curves   114
Fracture energy   32
Free radical copolymerization   62

Gas diffusion, PO   116
Gel time   23, 58, 64
Glass transition temperature   20, 48, 69, 77
Grafting, fluoroelastomers, VDF-based   186

HBTBP   156
HDPE   99, 120
Hexamethyldiamine   128
Hexamethylene diamine (HMDA)   151
HFP   166
Hybrid resin   74
Hydrocarbon resins   21
Hydrogen maleate   6

Impact strength   37, 39, 42
Impregnating   41
Injection molded foams   120
Intramolecular crosslinking   61
Irradiation, fluoroelastomers, VDF-based   190
Irritation   67

Kelvin model   113
Kinetic analysis   64

LDPE   99
– cell shapes   109
– cushion curves   115
– stress-strain   117
Light scattering   61
LLDPE   99

Magic angle spinning (MAS)   57
Maleate   9
Maleic anhydride   6–16, 19, 22,24, 27–30, 36, 37, 47–49, 56, 57, 66, 80
Maleopimaric acid   1, 12, 31
Mechanical properties   40
Mechanical strength   38
Methanol   203
Methylethylketone peroxide   73
Microgels   68
Microstructure   52
Molecular weight   19, 52, 61, 71, 85
Monoamines   145
Morphology   36, 77

Natural fibers   84
Nuclear magnetic resonance (NMR)   56

ODR, fluoroelastomers, bisphenol-cured   173
Optical microscopy   79
O-rings   203

Particle board   89
PE, metallocene   104
– foams, thermal expansion coefficient   119
Pentaerythritol diacetate   17, 18
Perfluoropolyethers   55
Peroxide-cure, coagents   179
Peroxides   30
– crosslinking, VDF   128, 185
Peroxides/TAIC, crosslinking   184
Peroxycarbamates   156
PET   1
– glycolyzate   81
Phase separation   36, 37, 70
Photoinitiator   85
Photopolymerization   66, 67
Phthalic anhydride   22, 27, 66
Piperidine   9, 148, 159
PO, running shoes   122
Polarizing microscopy   49
Poly(oxyethylene) diol   35
Poly(TFE-*co*-P)   182
Poly(VDF-*co*-CTFE)   145

Subject Index

Poly(VDF-*co*-HFP),
  dehydrofluorination 136
Polycondensation 5, 12, 17, 19, 29, 41, 53
Polydispersity 17
Polyester resins, unsaturated 1
Polyesterification 23, 74
Polyethylene foams 97
– open-cell 106
Polyolefin foams, cell wall/edges 108
– crosslinked 97
– crosslinked closed-shell 102
– structure 107
Polypropylene 99
– foams 104
Polyurethanes 5, 38, 40, 51, 72
Polyvinylidene fluoride (PVDF) 128, 131
Profilometry 59
PVDF 128, 131
PVDF film, irradiation 190

Radiation crosslinking,
  fluoroelastomers, VDF-based 186
Rapeseed oil 25, 27
Recycling 88, 120
Reference model, force-deformation 114
Refractive index 6
Resin transfer molding 70

Sealings 203
SEC, FKM gum 167
Sedimentation 82
Sewage systems 84
Sheet molding compounds 5, 88
Solubility parameters 47, 87
Storage modulus, CPOF 117
Storage stability 65
Strain rates, PO 111
Stress-starin curves,
  compressive, CPOF 111
Succinic acid 11

TAC 179
TAIC 128, 179, 180
Tensile strength 18, 34
Tetrahydrophthalic anhydride 6
Tetramethylethyldiamine 159
Thermal conductivity 118
Thermal degradation 43, 46
Thermal resistance 14, 31
Thermal stability 68
Thermomechanical properties 31
Thermoplastic additives 79
Thermoplastics 89
Thiol ene systems, VDF-based
  fluoroelastomers 196
Torque, ODR 173
Triallylcyanurate (TAC) 179
Triallylisocyanurate (TAIC) 128, 179, 180
Triethylamine 148

Ultrasonic technique 60
Unsaturated polyesters 16
UPRs (unsaturated polyester resins) 5
UPs (unsaturated polyesters) 5

VDF 127
Vinyl ester reins 1
Vinyl monomers 62
Viscosity 43
Volume shrinkage 78, 80

Waste utilization 88

Young's modulus 112, 114

Zinc stannate 44

RETURN TO: **CHEMISTRY LIBRARY**
100 Hildebrand Hall • 510-642-3753

LOAN PERIOD 1

2 HOUR

DUE AS STAMPED BELOW.

MAY 18		

FORM NO. DD 10
3M 5-04

UNIVERSITY OF CALIFORNIA, BERKELEY
Berkeley, California 94720–6000